THE PATTERN BOOK
FRACTALS, ART, and NATURE

THE PATTERN BOOK
FRACTALS, ART, and NATURE

Editor

Clifford A Pickover

IBM Thomas J. Watson Research Center

Published by

World Scientific Publishing Co. Pte. Ltd.
P O Box 128, Farrer Road, Singapore 912805
USA office: Suite 1B, 1060 Main Street, River Edge, NJ 07661
UK office: 57 Shelton Street, Covent Garden, London WC2H 9HE

Library of Congress Cataloging-in-Publication Data
The Pattern Book : fractals, art, and nature / editor, Clifford A. Pickover.
p. cm.
Includes index.
ISBN 981021426X
1. Computer graphics. 1. Pickover, Clifford A.
T385.P376
745.4--dc20 94-47114
CIP

British Library Cataloguing-in-Publication Data
A catalogue record for this book is available from the British Library.

First published 1995
First reprint 1997

Printed in Singapore by Uto-Print

Clifford A. Pickover

The Pattern Book:

Fractals, Art, and Nature

Introduction

> *"Art and science will eventually be seen to be as closely connected as arms to the body. Both are vital elements of order and its discovery. The word 'art' derives from the Indo-European base 'ar', meaning to join or fit together. In this sense, science, in the attempt to learn how and why things fit, becomes art. And when art is seen as the ability to do, make, apply or portray in a way that withstands the test of time, its connection with science becomes more clear."*
>
> Sven Carlson, *Science News* (1987)

This book will allow you to travel through time and space. To facilitate your journey, I have scoured the four corners of the earth in a quest for unusual people and their fascinating patterns. From Mozambique, to Asia, to many European countries, the contributors to *The Pattern Book* include world-famous cancer researchers, little-known artists, and eclectic computer programmers. Some of the patterns are ultramodern, while others are centuries old. Many of the patterns are drawn from the universe of mathematics. To start you on the journey, I will first provide some relevant background material on computers, pattern, science, and art.

The line between science and art is a fuzzy one; the two are fraternal philosophies formalized by ancient Greeks like Pythagoreas and Ictinus. Today, computer graphics is one method through which scientists and artists reunite these philosophies by providing scientific ways to represent natural and artistic objects. In fact many of this book's patterns were generated on small computers using simple algorithms. Other (equally interesting) patterns were generated by human hands, and these patterns often illustrate ornaments of both modern and ancient civilizations. Sometimes these patterns consist of symmetrical and repeating designs, for example, Moorish, Persian, and other motifs in tiled floors and cloths.

This book serves as an introductory catalog to some of the many facets of geometrical patterns, and you are urged to explore the ideas in greater depth than can be presented in this compendium. Perhaps I should attempt to define "pattern" before proceeding. You can find many definitions when consulting a dictionary, for example, "an artistic or mechanical design" or "a natural or chance configuration". The patterns in this book have such a great diversity that colleagues have debated whether the shapes should really be called "patterns" at all. However, I take the broad view, and include visually interesting shapes and themes from all areas of human, natural, and mathematical realms. Although the emphasis is on computer-generated patterns, the book is informal, and the intended audience spans several fields. This book might be used by students, graphic artists, illustrators, and craftspeople in search of visually intriguing designs, or anyone fascinated by optically provocative art. In addition, the book may be used by scientists, artists, laypeople, programmers and students. In the same spirit as Gardner's book, *Mathematical Circus*, or Pappas' book, *The Joy of Mathematics, The Pattern Book* combines old and new ideas — with emphasis on the fun that the true pattern lover finds in doing, rather than in reading about the doing! The book is organized into three main parts: **Representing Nature** (for those patterns which describe or show real physical phenomena, e.g., visualizations of protein motion, sea lilies, etc.), **Mathematics and Symmetry** (for those patterns which describe or show mathematical behavior, e.g., fractals), and **Human Art** (for those patterns which are artistic works of humans and made without the aid of a computer, e.g., Moslem tiling patterns). I provide a comprehensive glossary to help ease readers into technical or unfamiliar waters.

When deciding how to arrange material within the three parts of *The Pattern Book*, many divisions came to mind — computer and non-computer generated forms, science and art, nature and mathematics. However, the line between all of these categories becomes indistinct or artificial, and I have therefore randomly arranged the patterns within each part of the book to retain the playful spirit of the book and to give the reader unexpected pleasures. Some patterns could easily be placed in either of the three main sections of the book.

The reader is forewarned that some of the presented material in this book's catalog of shapes involves sophisticated concepts (e.g., "The Reversible Greensberg-Hastings Cellular Automaton" by Drs. P. Tamayo and

H. Hartman) while other patterns (e.g., "Satanic Flowers" by Dr. Harold J. McWhinnie) require little mathematical knowledge in order to appreciate or construct the shapes. Readers are free to pick and choose from the smorgasbord of patterns. Many of the pattern descriptions are brief and give the reader just a flavor of an application or method. Additional information is often in the referenced publications. In order to encourage reader involvement, computational hints and recipes for producing many of the computer-drawn figures are provided. For many readers, seeing pseudocode will clarify the concepts in a way which mere words cannot.

Currently, I know of no book which presents such a large range of patterns and instructions for generating the patterns. There are, however, numerous books available that publish patterns in specific categories. Most are inexpensive paperbacks published by Dover Publications, and many are reprints of nineteenth century books. I think you will enjoy these. Some are listed in the reference section.

Before concluding this preface, I should point out that today scientists and artists seem to have a growing fascination with symmetry and repetition in design. On the topic of art, there are the modern isometric designs of John Locke and the geometrical ornaments of Russian artist, Chernikow (where simple forms create complex interweavings),

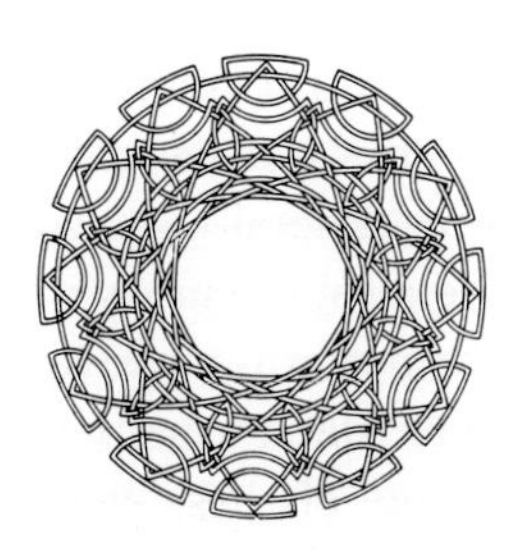

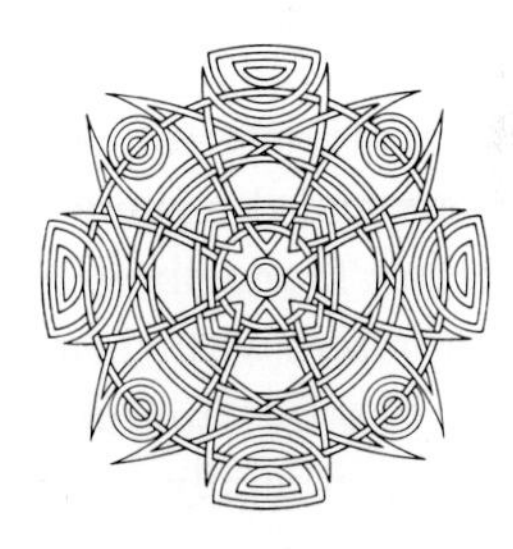

and a variety of popular art deco designs. Also "controlled accident" has found its place in many areas of the modern arts (O'Brien, 1968). For example, Dadaist and Surrealist painters such as Miro, Masson, and Arp capitalized on the elements of chance, and the works they created provide challenges for the mind as well as the eye. In the area of science, researchers are intrigued by the way nature often expresses itself in terms of repeating symmetries — and the cross section of plants, phase transitions, standing waves on metal plates, muscle striations, snow crystals, and dendritic ice are just a few examples.

From the branching of rivers and blood vessels, to the highly convoluted surface of brains and bark, the physical world contains intricate patterns formed from simple shapes through the repeated application of dynamic procedures. Questions about the fundamental rules underlying the variety of nature have led to the search to identify, measure, and define these patterns in precise scientific terms.

One final observation on patterns in nature. Our physical world around us often seems chaotic, exhibiting a limitless and complex array of patterns. However, you should note that our world is also actually highly structured. From an evolutionary standpoint, biological themes, structures, and "solutions" are repeated when possible, and inanimate forms such as mountains and snowflakes are constrained by physical laws to a finite class of patterns. The apparently intricate fabric of nature and the universe is produced from a limited variety of threads which are, in turn, organized into a multitude of combinations. You will see some of these threads throughout this book.

The World of Fractals and Chaos

Many of the patterns in this book come from the exciting mathematical fields of fractal geometry and chaos. This section is intended as a brief introduction to these fields.

These days computer-generated fractal patterns are everywhere. From squiggly designs on computer art posters, to illustrations in the most serious of physics journals, interest continues to grow among scientists and, rather surprisingly, artists and designers. The word "fractal" was coined in 1975 by IBM scientist Benoit Mandelbrot to describe a set of curves rarely seen before the advent of computers with their ability to perform massive numbers of calculations quickly. Fractals are bumpy objects which usually show a wealth of detail as they are continually magnified. Some of these shapes exist only in abstract geometric space, but others can be used to model complex natural shapes such as coastlines and mountains.

Chaos and fractal geometry go hand-in-hand. Both fields deal with intricately shaped objects, and chaotic processes often produce fractal patterns. To ancient humans, chaos represented the unknown, the spirit world — menacing, nightmarish visions that reflected man's fear of the irrational and the need to give shape and form to his apprehensions. Today, chaos theory is a growing field which involves the study of a range of phenomena exhibiting a sensitive dependence on initial conditions. This means that some *natural* systems, such as the weather, are so sensitive to even small local fluctuations that we will never be able to accurately predict what they will do in the future. For certain *mathematical* systems, if you change a parameter ever-so-slightly, the results can be very different. Although chaos seems totally "random", it often obeys strict mathematical rules derived from equations that can be formulated and studied. One important research tool to aid the study of chaos is computer graphics. From chaotic toys with randomly blinking lights to wisps and eddies of cigarette smoke, chaotic behavior is irregular and disorderly. Other examples include certain neurological and cardiac activity, the stock market, and some electrical networks of computers. Chaos theory has also often been applied to a wide range of visual art.

So extensive is the interest in fractals and chaos that keeping up with the literature on the subject is rapidly becoming a full-time task. In 1989, the world's scientific journals published about 1,200 articles with the words "chaos" or "fractal(s)" in the title. The figure here shows the number of papers with

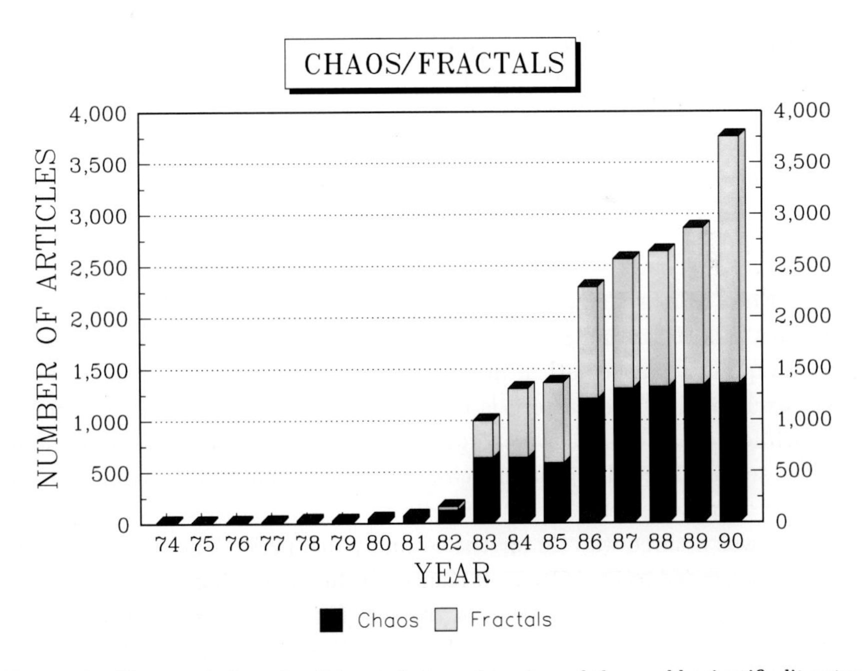

Figure 1. Chaos and fractal article explosion. A review of the world scientific literature between 1973 and 1990 shows the number of chaos and fractal articles rising dramatically between the years 1982 and 1990. (Figure from *Computers and the Imagination* by C. Pickover, ©1991 St. Martin's Press. All rights reserved.)

titles containing the words "chaos" or "fractal(s)" for the years 1975–1990, the 1990 values estimated from data for January–June 1990.

The Best of the Best

After the patterns in this book were compiled, I invited six distinguished judges to help select the "best" patterns in the book. Naturally selecting from such a diverse collection was not an easy or well-defined task. The judges selected patterns that they considered beautiful, novel, and/or scientifically interesting. I thank the following individuals on the "board of judges" for singling out their favorite patterns for special recognition in the book.

1. Professor Akhlesh Lakhtakia, Pennsylvania State University. Editor-in-Chief, *Speculations in Science and Technology*; Milestone Editor, *Selected Papers on Natural Optical Activity*; Co-author, *Time-harmonic Electromagnetic Fields in Chiral Media*; member, The Electromagnetics Academy.
2. Dr. Roger Malina, Editor of *Leonardo*, a journal of the International Society for the Arts, Sciences, and Technology.
3. Ivars Peterson, Author of *The Mathematical Tourist* and *Islands of Truth*, and mathematics and physics editor at *Science News.*
4. Chris Severud, President of Bourbaki Software, Inc.
5. Dawn Friedman, Chemistry Department, Harvard University. Naturalist, science-writer, theoretical chemist, futurist.
6. Phil LoPiccolo, Editor, *Computer Graphics World.*

The **First Place Prize** was awarded to Steven Schiller of Adobe Systems, California. His "Gaussian Fraction" pattern was judged the best because of its aesthetic quality, mathematical interest, and novelty. One of the more poetic judges exclaimed: "The pattern is a perfect daydream generator. It reminds me of Dirac's ocean of negative-energy electrons, with stray points fountaining from the surface like solar prominences, and unseen quantum events producing shimmers deep undersea."

There was a three-way tie for the second place prize. The **Second Place Prize** was awarded to Earl Glynn of Kansas for his "Spiraling Tree/Biomorphic Cells" pattern, Dr. Ian Entwistle of the UK for his pattern, "Serpents and Dragons: A mapping of $f(z) \rightarrow \sinh(z) + c$ in the complex plane", and to the late Ernst Haeckel for the "Sea-Lilies" pattern. Of the "Serpents and Dragons", one judge noted: "Among dozens of stunning patterns based on Julia sets, this was simply the most beautiful. Beauty and order seem to swim upwards out of a chaotic sea like vigorous fish, growing and uncurling like fern fronds reaching for the light." Of the sea-lilies, another judge noted that "these beautifully intricate forms owe as much to the artist's eyes as they do to the natural shapes of the organisms themselves; this is typical of the precision and artistry found in much 19th-century illustration".

The remaining prizes go to Stefan Muller, Theo Plesser, and Benno Hess of Germany for their patterns on "Rotating Spiral Waves in the Belousov-Zhabotinskii Reaction", A. K. Dewdney of Canada for his "Informal Tesselation of Cats", John MacManus of Canada for his "Jungle Canopy", and Henrik Bohr and Soren Brunak of Denmark for their "Patterns of Protein Conformations". Of the rotating spiral wave patterns, one judge noted: "Ever since I first heard of and then actually demonstrated for myself the existence of oscillating chemical reactions, I have found both the chemistry and the mathematics of these patterns irresistible. The spirals have wonderfully complicated symmetries." Another judge remarked:

> "Naturally, as a chemist, I am delighted to see a beautiful chemical pattern, its spiral forms enhanced by an ingenious and skillful piece of laboratory-cum-video work. But I chose this pattern for another reason as well. The aim of this book, other than to delight, must be to elucidate. What happens in the murky swirling fluid inside a reaction vessel is the ancient mystery of chemistry, darker and more forbidding than the linked equations of thermodynamics or the fortress walls of quantum matrices. Many students of chemistry are reduced to following recipes and pouring ingredients, hoping the magic will work as promised. The discovery of structure, pattern, beauty, inside that murky flask, the casting of light into the darkness, must encourage us all."

Of Dewdney's cats, one judge noted:

> "I'm sure I won't be the first or the last to pick this one. I admire Dr. Dewdney's unspoken point: if you are going to tile a plane with a species, by all means choose one that is naturally graceful, flexible, and incapable of uncomfortable angularities."

For Further Reading

> *"Some people can read a musical score and in their minds hear the music ... Others can see, in their mind's eye, great beauty and structure in certain mathematical functions ... Lesser folk, like me, need to hear music played and see numbers rendered to appreciate their structures."*
>
> Peter B. Schroeder, *Byte Magazine* (1986)

As you will see in many patterns from this book, mathematical formulas can sometimes be used to simulate natural forms. For example, computer graphics provides a way to represent biological objects. For an excellent book on techniques for simulating nature, see Rivlin (1986). Researchers have explored the use of rules based on the laws of nature, such as logarithmic spirals for sea shells (Kawaguchi, 1982) or tree branching patterns determined from the study of living specimens (Aono, 1984). Other papers describe the generation of plant leaf vein patterns (Kolata, 1987) and woodgrains (Yessios, 1979). Bloomenthal (1985) describes methods for simulating tree bark, leaves, and limbs. Other sophisticated approaches to botanical structure generation exist, for example, beautiful "particle systems" consisting of trajectories of particles influenced by the pull of gravity (Reeves, 1985). See also (Viennot *et al.*, 1989; Prusinkiewicz *et al.*, 1988). For references on symmetry in historical ornaments, see Audsley (1968) and Rozsa (1986). Audsley's book includes illustrations of ancient Egyptian patterns from the painted ceiling of various tombs, interlaced Celtic designs typical of those used to illuminate manuscripts, and various Japanese ornaments. For a fascinating collection of Persian designs and motifs, see Dowlatshahi (1979). Symmetrical ornaments, such as those presented in *The Pattern Book*, have persisted from ancient to modern times. The different kinds of symmetries have been most fully explored in Arabic and Moorish design. The later Islamic artists were forbidden by religion to represent the human form, so they naturally turned to elaborate geometric themes. To explore the full range of symmetry in historic ornaments, you may wish to study the work of Gombrich who discusses the psychology of decorative art and presents several additional examples of five-fold symmetry.

The following reference list includes books and papers describing patterns in a range of scientific and artistic fields.

Allen, J. (1988) *Designer's Guide to Japanese Patterns.* Chronicle Books: San Francisco.

Audsley, W. (1968) *Designs and Patterns from Historic Ornament.* Dover: New York.

Barnard, J. (1973) *The Decorative Tradition.* The Architectural Press: London.

Barratt, K. (1980) *Logic and Design in Art, Science, and Mathematics.* Design Press: New York. (The pictures alone will stimulate readers to experiment further.)

Blossfeldt, K. (1985) *Art Forms in the Plant World.* Dover: New York.

Doczi, C. (1988) *The Language of Ornament.* Portland House: New York.

Doczi, G. (1986) "Seen and unseen symmetries", *Computers and Mathematics with Applications* **12B**:39–62.

Dowlatshahi, A. (1979) *Persian Designs and Motifs.* Dover: New York.

Durant, S. (1986) *Ornament.* McDonald and Company: London.

Gardner, M. (1969) "Spirals", in *The Unexpected Hanging.* Simon and Schuster: New York.

Gardner, M. (1970) *Mathematical Circus.* Penguin Books: New York.

Glazier, R. (1980) *Historic Ornament.* Coles: Toronto.

Gombrich, E. (1979) *The Sense of Order: A Study in the Psychology of Decorative Art.* Cornell University Press: New York.

Grunbaum, B., Grunbaum, Z. and Shephard, G. (1986) "Symmetry in Moorish and other ornaments", *Computers and Mathematics with Applications* **12B**:641–653. Hargittai, I. (1989) *Symmetry 2: Unifying Human Understanding.* Oxford: New York.

Hargittai, I. and Pickover, C. (1992) *Spiral Symmetry.* World Scientific: New Jersey.

Harlow, W. (1976) *Art Forms from Plant Life.* Dover: New York.

Hayes, B. (1986) "On the bathtub algorithm for dot-matrix holograms", *Computer Language* **3**:21–25.

Justema, W. (1976) *Pattern: A Historical Perspective.* New York Graphic Society: Massachusetts.

Lockwood, E. and Macmillan, R. (1978) *Geometric Symmetry.* Cambridge University Press: New York.

Lewis, P. and Darley, G. (1986) *Historic Ornament: A Pictorial Archive.* Pantheon Books: New York.

Makovicky, E. (1986) "Symmetrology of art: coloured and generalized symmetries", *Computers and Mathematics with Applications* **12B**:949–980.

Mamedov, K. (1986) "Crystallographic patterns", *Computers and Mathematics with Applications* **12B**:511–529.

Mandelbrot, B. (1983) *The Fractal Geometry of Nature.* Freeman: San Francisco.

Moon, F. (1987) *Chaotic Vibrations.* John Wiley and Sons: New York.

Peterson, I. (1988) *The Mathematical Tourist.* Freeman: New York.

Pickover, C. (1990) *Computers, Pattern, Chaos, and Beauty.* St. Martin's Press: New York.

Pickover, C. (1991) *Computers and the Imagination.* St. Martin's Press: New York.

Pickover, C. (1992) *Mazes for the Mind.* St. Martin's Press: New York.

Pickover, C. (1994) *Chaos in Wonderland: Visual Adventures in a Fractal World.* St. Martin's Press: New York.

Pickover, C. (1995) *Keys to Infinity.* Wiley: New York.

Postle, D. (1976) *Fabric of the Universe.* Crown: New York.

Racinet, A. (1988) *The Encyclopedia of Ornament.* Portland House: New York.

Reitman, E. (1989) *Exploring the Geometry of Nature.* Windcrest Books: Pennsylvania.

Reichardt, J. (1969) *Cybernetic Serendipity: The Computer and the Arts.* Prager: New York.

Rucker, R. (1982) *Infinity and the Mind.* Bantam: New York.

Shaw, A. (1984) *The Dripping Faucet as a Model Chaotic System.* Aerial Press: California.

Stevens, C. (1989) *Fractal Programming in C.* M and T Books: California. (This book is a dream come true for computer programmers interested in fractals.)

Steinhaus, H. (1983) *Mathematical Snapshots*, 3rd ed. Oxford University Press: New York. (Topics include tesselations, soap-bubbles, maps, screws, spiders, honeycombs, and platonic solids.)

"Symmetries and Asymmetries" (1985) *Mosaic* **16** (An entire issue on the subject of fractals, symmetry and chaos. *Mosaic* is published six times a year as a source of information for scientific and educational communities served by the National Science Foundation, Washington DC 20550).

Tufte, E. (1983) *The Visual Display of Quantitative Information.* Graphics Press: Connecticut.

Niman, J. Norman, J. and Stahl, S. (1978) "The teaching of mathematics through art" (a report on the conference March 20–21, 1978) Metropolitan

Museum of Art and Hunter College of the City University of New York. pp. 1–55.

O'Brien, J. (1968) *How to Design by Accident.* Dover: New York.

Pappas, T. (1990) *The Joy of Mathematics.* Wide World Publishing: California.

Peachey, D. (1985) "Solid texturing of complex surfaces", *Computer Graphics (ACM SIGGRAPH)* **19**(3): 279–286.

Perlin, K. (1985) "An image synthesizer", *Computer Graphics (ACM SIGGRAPH)* **19**(3): 287–296.

Postle, D. (1976) *The Fabric of the Universe.* Crown Publishers Inc.: New York.

Rozsa, E. (1986) "Symmetry in Muslim arts", *Computers and Mathematics with Applications* **12B**:725–750.

Viennot, X., Eyrolles, G., Janey, N. and Arques, D. (1989) "Combinatorial analysis of ramified patterns and computer imagery of trees", *Computer Graphics (ACM-SIGGRAPH)* **23**(3): 31–40.

Zvilna, J. (1986) "Colored symmetries in space-time", *Computers and Mathematics with Applications* **12B**:895–911.

Acknowledgments

Some of the figures in the book come from the Dover Pictorial Archive of modern and ancient art. This series constitutes a collection of the world's greatest designers from Ancient Egypt to Art Deco designs. For more information, write to Dover Publications, 31 East 2nd Street, Mineola, New York 11501.

The opening quotation by Sven G. Carlson on art and science appeared in his letter to *Science News* (Vol. 132, 1987, p. 382).

Contents

Proceeding pages (pp. xxx–xxxvi)
Some of the winning designs: Gaussian Fractions; Haeckel's Sea-Lilies; Serpents and Dragons; Patterns of Protein Conformations; Spiraling Tree/Biomorphic Cells; Jungle Canopy (*landscape*); An Informal Tesselation of Cats.

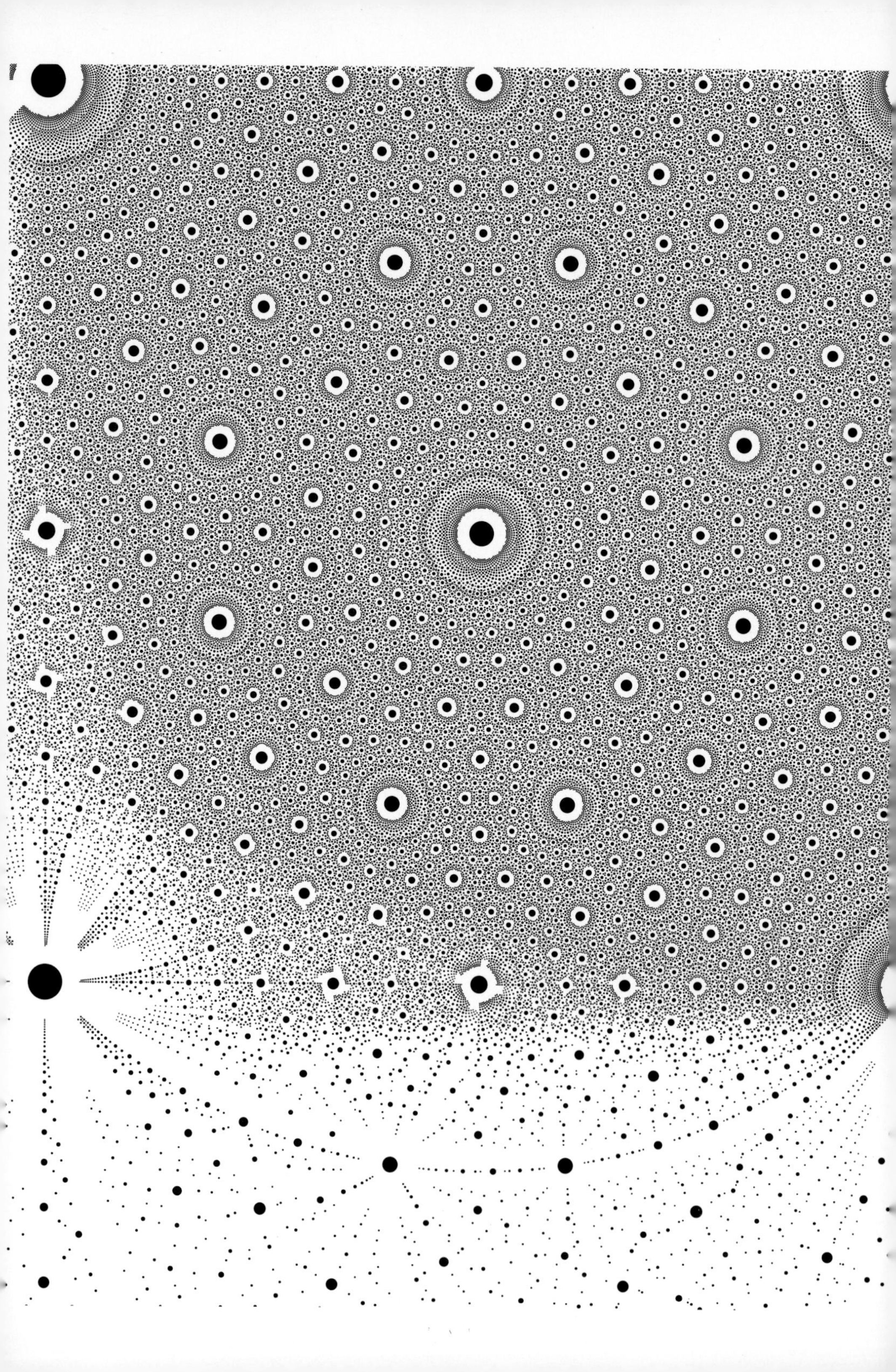

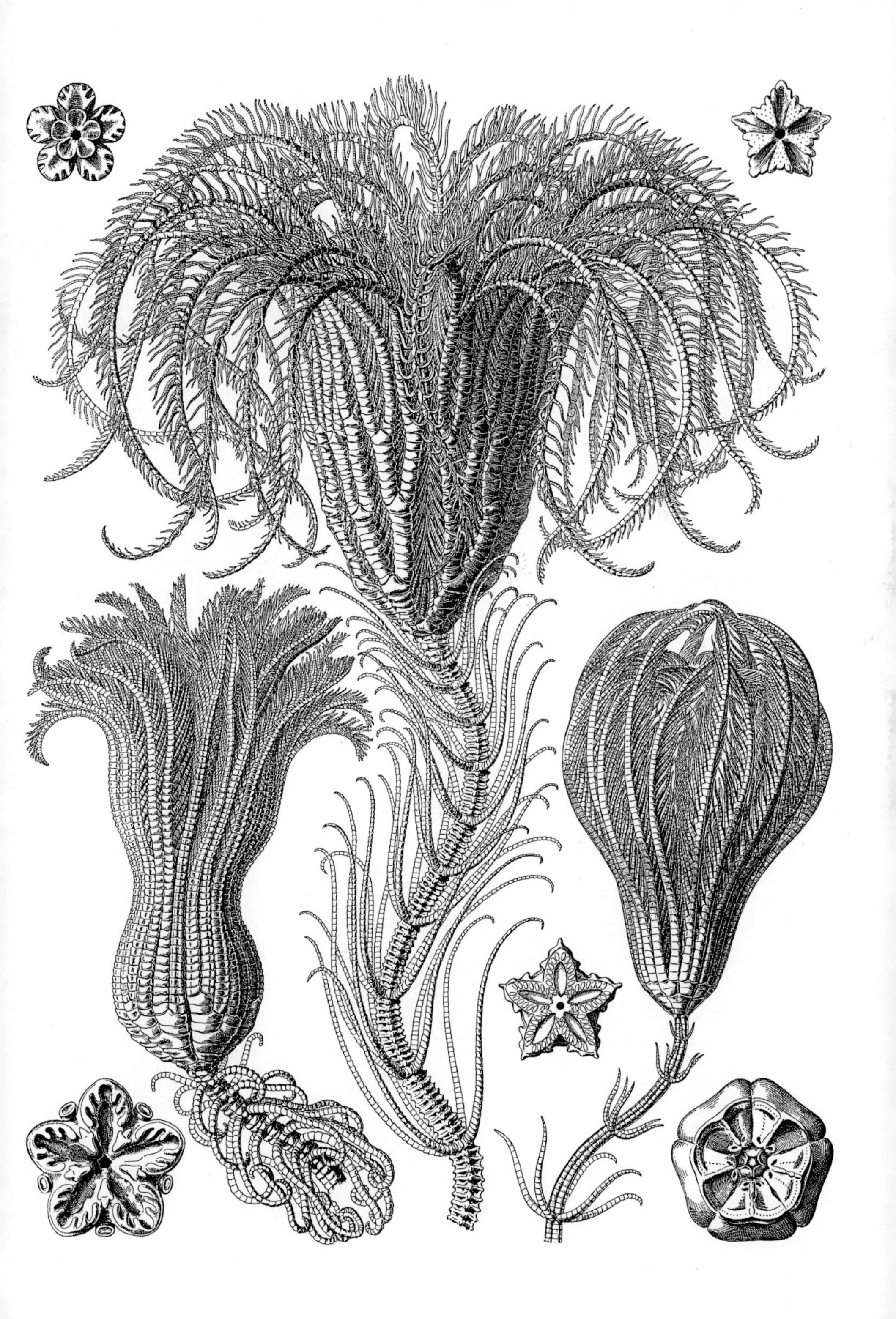

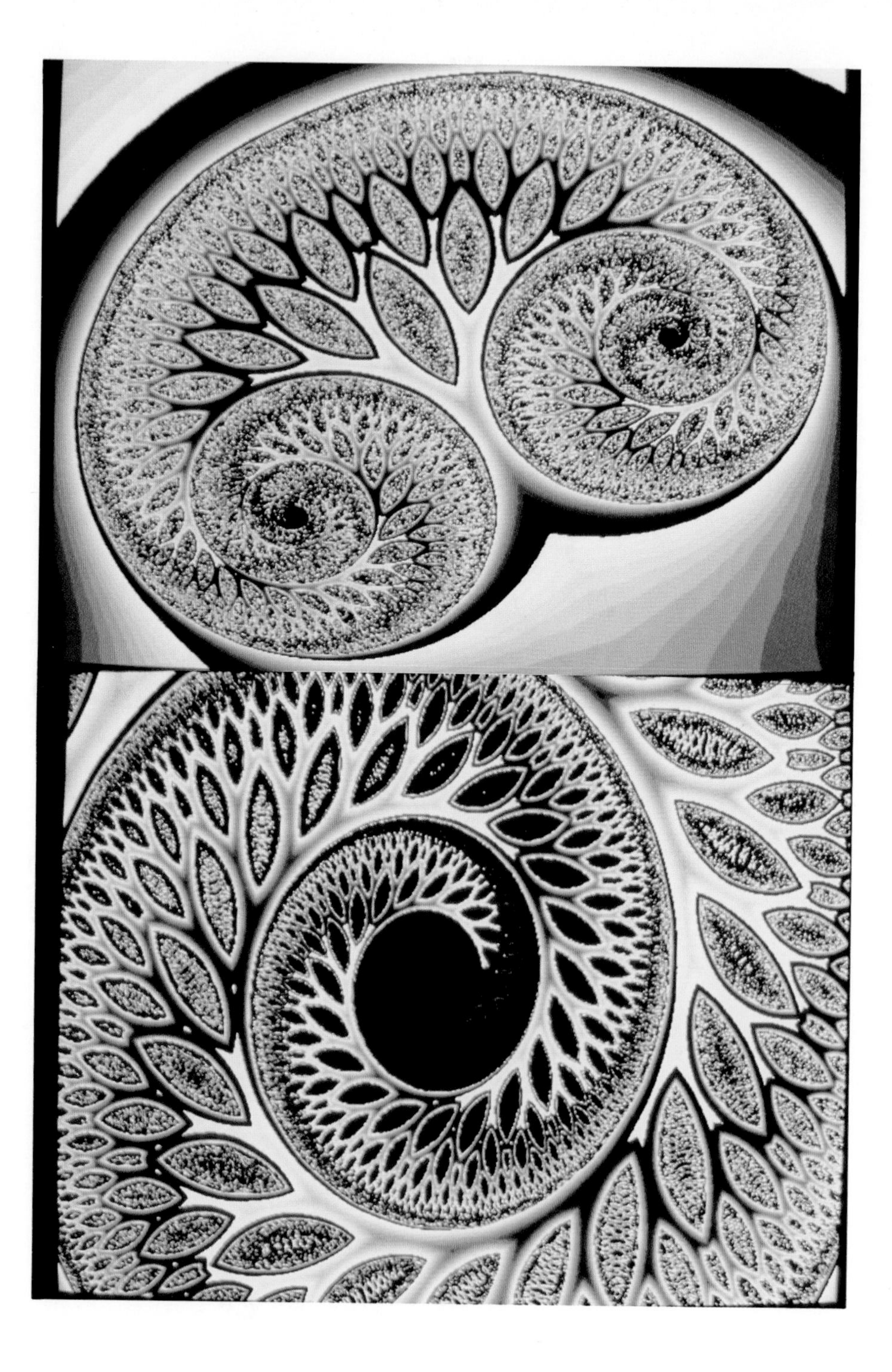

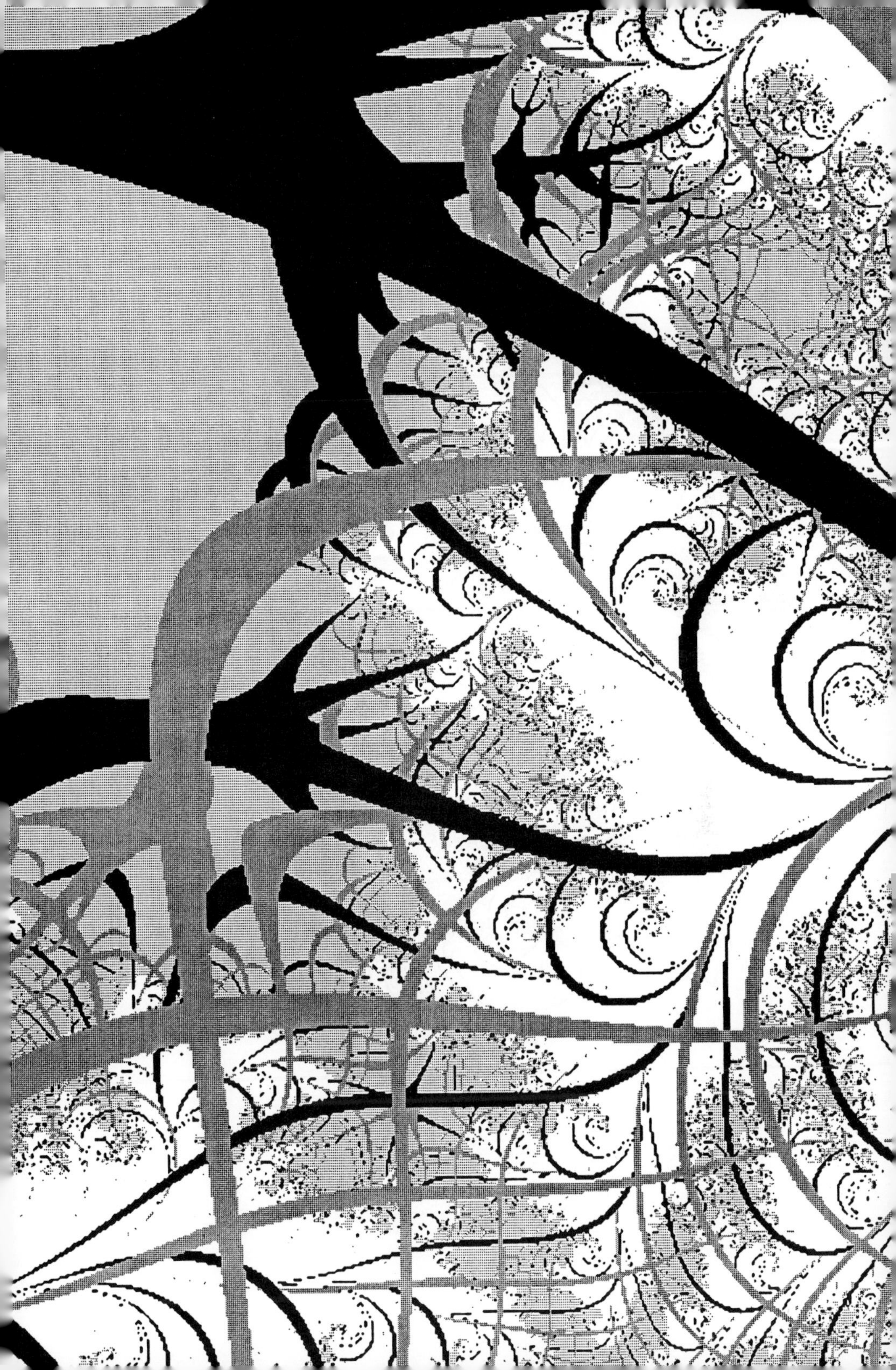

PART I
REPRESENTING NATURE

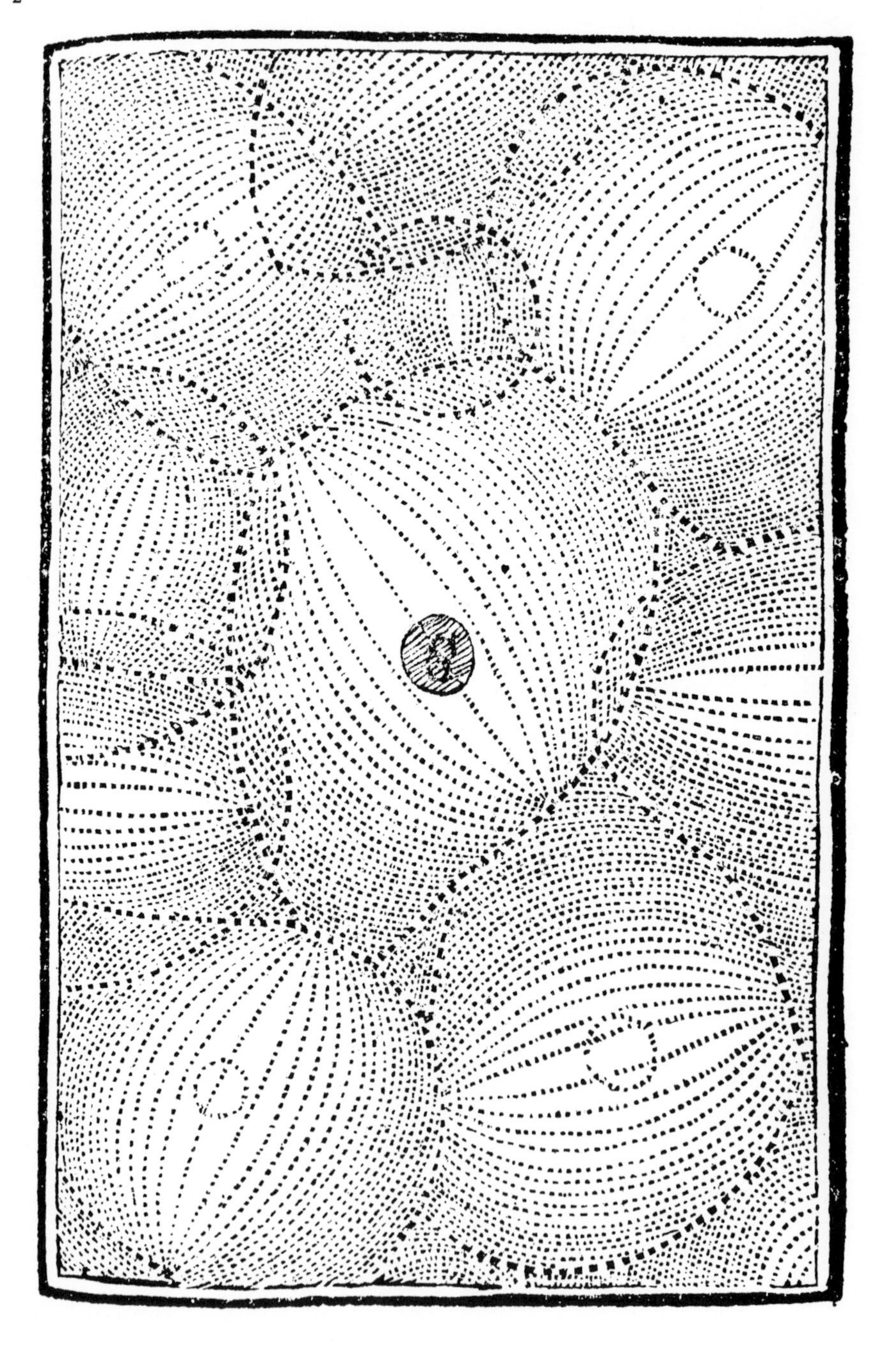

Kevin L. Cope

Evolution of the Solar and Planetary Vortices

This pattern represents the emergence of heavenly bodies from primitive celestial vortices as imagined by Gabriel Daniel, a late seventeenth-century commentator on Descartes. More than a few controversialists of the early-modern era made their careers by reviewing, revising, and interpreting Descartes' remarkable vortex theory. These overly ambitious scribblers applied a plethora of methods to the French physicist-philosopher, trying everything from science to satire. Descartes had argued that the universe was a plenum, an infinitely large container completely filled with extended matter and uncorrupted by empty spaces. To explain the differences in the densities of objects in a universe allegedly packed as tightly as possible, Descartes conjectured that his cosmos might be pock-marked with "vortices". Energetic, unstable, and even dangerous, these whirlpools might draw some collections of matter into tighter quarters than others. Turbulence reigns supreme in a Cartesian universe. According to vortecticians, matter begins as a plasma-like whirlpool, condenses into a furious sun, calms into a jiggling "terrela" (a luminous, gaseous, whirling, or incomplete planet like Jupiter or Saturn), and, at last, settles into a planetary lump. The pattern shown here portrays the dynamic, even terrifying interference of the great, originative vortices that were eventually to become the components of our solar system. Gabriel Daniel was himself a rather unstable fellow. After writing several hundred pages of satire against Descartes' theories, he apparently developed an affection for the great geometer, consoling himself that Descartes had never really died, but had only been carried up into some higher vortex by means of his own whirling tobacco smoke [1].

Reference

1. G. Daniel, *A Voyage to the World of Cartesius* (Thomas Bennett, 1694) (translated by Taylor, T.).

Stefan C. Müller, Theo Plesser and Benno Hess

Rotating Spiral Waves in the Belousov-Zhabotinskii Reaction

Describe here is a macroscopic chemical pattern occurring on a 1 mm scale in a chemical excitable solution which is maintained under conditions far from thermodynamic equilibrium. The Belousov-Zhabotinskii reaction has become the most prominent chemical model example of an excitable medium displaying spatio-temporal self-organization. In this reaction due to the nonlinear interaction of complex reaction kinetics with molecular diffusion, spiral-shaped waves of chemical activity are readily observed [1, 2]. For quantitative pattern analysis a computerized, video-based two-dimensional spectrophotometer has been designed which yields digitized high-resolution data of the spatio-temporal evolution of the patterns [3, 4]. Picture A of the figure shows a complex wave pattern in a thin solution layer of this reaction, as observed in transmitted light of appropriately selected wavelength and recorded by a video camera.

In order to characterize the properties of the core regions of the displayed spirals, a digital overlay technique is applied to a series of stored images. The result of this overlay is depicted in picture B of the figure, showing a black spot close to each of the spiral tips of picture A. These spots indicate the singular features of the spiral cores in that, at these particular sites, the chemical state of the medium remains constant in time, whereas at all points outside the core region the medium undergoes an oscillatory transition between two states of maximum and minimum excitation. By applying three-dimensional perspective techniques [5] the structure of the cores can be visualized in detail (picture C of the figure), thus emphasizing their singular properties.

References

1. R. J. Field and M. Burger, eds., *Oscillations and Travelling Waves in Chemical Systems* (John Wiley, 1986).
2. J. Ross, S. C. Müller and C. Vidal, "Chemical waves", *Science* **240** (1988) 460–465.
3. S. C. Müller, Th. Plesser and B. Hess, "Two-dimensional spectrophotometry of spiral wave propagation in the Belousov-Zhabotinskii reaction. I. Experiments and digital data representation", *Physica D* **24** (1987) 71–86.

4. S. C. Müller, Th. Plesser and B. Hess, "Two-dimensional spectrophotometry and pseudo-color representation of chemical reaction patterns", *Naturwissenschaften* **73** (1986) 165–179.
5. S. C. Müller, Th. Plesser and B. Hess, "Three-dimensional representation of chemical gradients", *Biophys. Chem.* **26** (1987) 357–365.

Figure 1. (A) Snapshot of a spiral wave pattern in the Belousov-Zhabotinskii reaction. (B) Digital overlay of six consecutive images covering one revolution of the spirals of image (A). (C) Three-dimensional perspective representation of the composite image (B) seen from a view point located below the upper edge of (B).

Henrik Bohr and Søren Brunak

Patterns of Protein Conformations

The figure visualizes an ensemble of three-dimensional conformations of the protein, *Avian Pancreatic Polypeptide* (APP), obtained by a method of optimization which is based on an analogy of the Traveling Salesman Problem.

The dynamical behavior of complex systems like proteins is in general extremely hard to simulate on a computer. It has been suggested that the reason might be that many complex systems actually function as computers, performing computations which cannot be completed in fewer logical steps than the systems are using themselves. In other words, the computations are irreducible and the computational results cannot be obtained by the use of short cuts, because they do not exist [1].

In the case of the dynamics of proteins, where parts of the proteins move relatively to each other, the question is: What is the protein computing while it moves, and what is the final result of the computation?

The result of the protein folding process is clearly a topology. The process looks for good neighbors for the various parts, and the system ends up in a conformational state with high stability in which parts with an affinity for each other (or a dislike of the solvent) are brought together, and where mutually repulsive parts seek positions with the highest possible distance between them.

Conformational substates of a protein are partly determined by the positions of the side-chains [2]. The multiple configurations of these side-chains and their stability is the subject of this study [3]. A change in the position of one side-chain can have global effects on the positions of other side-chains. We have considered a model where only the nearest neighbor interactions are included. This assures that the dynamics of the side-chain configurations are collective. We are then left with the problem of choosing a suitable neighbor topology for the three-dimensional configuration of side-chains, being the result of a global optimization of interaction energy throughout the protein.

In the classical optimization problem of the traveling salesman, a similar situation arises: how to choose — for each city — two neighboring cities which minimizes the overall length of a round trip between the cities. In this analogy, the (moving) protein side-chains take the place of the cities, and the cost function becomes the sum of the interaction between neighboring side-chains.

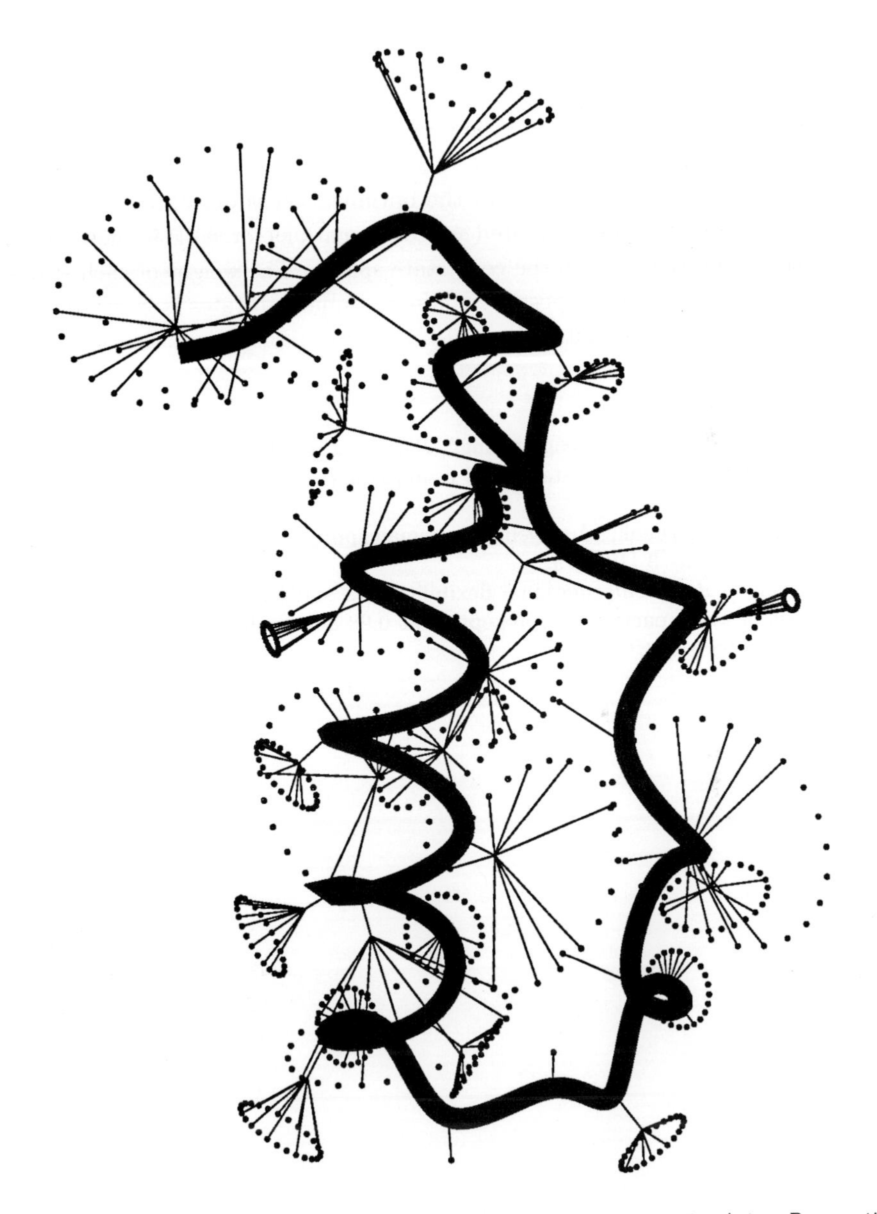

Figure 1. The ensemble of conformational substates for the protein *Avian Pancreatic Polypeptide* comprising 36 amino acids. The cones represent degrees of freedom for each side-chain.

The actual optimization procedure starts with the side-chains in random positions on the cones (representing the degrees of freedom for each side-chain) and a random choice for the nearest neighbor topology. The optimization proceeds as simulated annealing and the tour becomes stable at a certain temperature where all the side-chains have relaxed accordingly into a conformation of low global energy. The result of the optimization is presented in the figure, where a sample of one hundred subconformations for APP is shown with the distribution over sites in the cone state spaces. The weight of each site is not shown. This picture of possible conformational substates is in reasonable agreement with existing data [4].

References

1. S. Wolfram, "Complexity engineering", *Physica* **22D** (1986) 385.
2. A. Ansani *et al.*, "Protein states and protein quakes", *Proc. Natl. Sci.* **82** (1985) 5000.
3. H. Bohr and S. Brunak, "A travelling salesman approach to protein conformation", *Complex Systems* **3** (1989) 9.
4. I. Glover *et al.*, "Conformational flexibility in a small globular hormone: X-ray analysis of avian pancreatic polypeptide at 0.98 angstroms resolution", *Biopolymers* **22** (1983) 293.

Richard Dawkins

Blind Watchmaker Biomorphs

These patterns, called "biomorphs", are all generated by the same recursive tree algorithm, familiar from computer science textbooks and most easily understood with reference to the simple tree in the middle of the figure:

```
procedure Tree(x, y, length, dir: integer; dx, dy: array [0..7] of integer);
{Tree is called with the arrays dx and dy specifying the form of the tree, and
 the starting value of length. Thereafter, tree calls itself recursively
 with a progressively decreasing value of length};
var xnew,ynew: integer;
begin if dir < 0 then dir:=dir + 8; if dir >=8 then dir:=dir - 8;
xnew:=x + length * dx[dir];   ynew:=y + length * dy[dir];
MoveTo(x, y);   LineTo(xnew, ynew);
if length > 0 then {now follow the two recursive calls, drawing to left
and right respectively}
        begin
        tree(xnew, ynew, length - 1, dir - 1) {this initiates a series of inner calls}
        tree(xnew, ynew, length - 1, dir + 1)
        end
end {tree};
```

The only difference between the biomorphs (with some additions noted below) is in the quantitative parameters fed into the procedure as the arrays *dx* and *dy*. The program was written as a demonstration of the power of Darwinian evolution by artificial selection. The basic tree algorithm constitutes the "embryology" of the organisms. The quantitative paramters are thought of as "genes", passed from "parent" to "child" in a sexual reproduction. In every generation, a parent biomorph is displayed in the centre of the screen, surrounded by a litter of its own offspring which may, with some random probability, have mutated. A human then chooses which one to breed from. It glides to the center of the screen and "spawns" a new generation of mutant progeny. The process continues until, after a few dozen generations of this selective breeding, a radically different shape has evolved to the taste of the chooser.

Many of the biomorphs in the figure were generated by a slightly extended version of the program [1], described in Appendix [2] to the American edition of my book, *The Blind Watchmaker.* As well as the nine genes of the original version, the later version has additional genes controlling "segmentation" and symmetry in various planes [3].

> *"When I wrote the program, I never thought that it would evolve anything more than a variety of tree-like shapes. I had hoped for weeping willows, cedars of Lebanon, Lombardy poplars, seaweeds, perhaps deer antlers. Nothing in my biologist's intuition, nothing in my 20 years' experience of programming computers, and nothing in my wildest dreams, prepared me for what actually emerged on the screen. I can't remember exactly when in the sequence it first began to dawn on me that an evolved resemblance to something like an insect was possible. With a wild surmise, I began to breed, generation after generation, from whichever child looked most like an insect. My incredulity grew in parallel with the evolving resemblance ... I still cannot conceal from you my feeling of exultation as I first watched these exquisite creatures emerging before my eyes. I distinctly heard the triumphal opening chords of "Also sprach Zarathustra" (the "2001" theme) in my mind. I couldn't eat, and that night "my" insects swarmed behind my eyelids as I tried to sleep ... There are computer games on the market in which the player has the illusion that he is wandering about in an underground labyrinth, which has a definite if complex geography and in which he encounters dragons, minotaurs or other mythic adversaries. In these games the monsters are rather few in number. They are all designed by a human programmer, and so is the geography of the labyrinth. In the evolution game, whether the computer version or the real thing, the player (or observer) obtains the same feeling of wandering metaphorically through a labyrinth of branching passages, but the number of possible pathways is all but infinite, and the monsters that one encounters are undesigned and unpredictable. On my wanderings through the backwaters of Biomorph Land, I have encountered fairy shrimps, Aztec temples, Gothic church windows, aboriginal drawings of kangaroos, and, on one memorable but unrecapturable occasion, a passable caricature of the Wykeham Professor of Logic"* (from *The Blind Watchmaker*, pp. 59–60).

References

1. This extended program, called *The Blind Watchmaker* is available for the Apple Macintosh computer, from W W Norton and Co., 500 Fifth Avenue, New York 10110, USA. The unextended version is available for the IBM PC from the same address. In the UK, both versions can be obtained from Software Production Associates, P.O. Box 59, Leamington Spa CV31 3QA, UK.
2. R. Dawkins, *The Blind Watchmaker* (W W Norton 1986).
3. R. Dawkins, "The evolution of evolvability", *Artificial Life*, ed. C. Langton (Addison Wesley 1989) pp. 201–220.

Figure 1. Twenty-one "biomorphs", distant "cousins" of one another, generated by the program "Blind Watchmaker".

Phil Brodatz

Apple Tree Pattern

Wood patterns are infinitely varied, offering myriad opportunities for novel designs. Shown here is a photograph of a cross-section of an apple tree burl. Infrared film with A25 filter is used to eliminate the gray tones and to accent the harder wood lines.

Reference

1. P. Brodatz, *Wood and Wood Grains* (Dover, 1971).

WW73 Cross section of apple tree burl.

Infrared film with A25 filter.

Dawn Friedman

Spring Tiles the Planes of Sunlight: Decussate and Tricussate Phyllotaxy in New Growth

Plant forms are characterized by arrangements of leaves around a stem: alternating, whorled, or spiraling. The various arrangements, or phyllotaxies, form patterns with different degrees of visual symmetry. Though these patterns may not be explicitly considered by gardeners and artists, a plant derives much of its beauty and its visual "feel" — formal or free, ornate or simple — from its phyllotaxy. When the phyllotaxy along a single axis is duplicated over and over, spectacular patterns can be formed.

The plants in these figures are members of a ubiquitous garden group: mound-forming perennials. As perennials, each spring, they grow back from their roots to fill an ever-larger share of sunlit space. As mounding plants, rather than growing a main stem and reaching for the sun with that single "limb", they make dozens or hundreds of small stems radiating upward in all directions. Each stem is an independent phyllotactic axis, unfurling leaves in its set pattern until they overlap the leaves of its neighboring stems. The shoots grow taller, the plant larger, but the surface of the mound is always a hemisphere crammed with leafy phyllotaxies. Whether the sun shines from the east or west, there is a leaf to intercept the light. Every possible plane of incident sunlight is tiled with leaves.

Figure 1(a) shows a catmint (*Nepeta* species) with the characteristic phyllotaxy of the mint family: decussate, or whorls of two. Each stem produces pairs of leaves on opposite sides of the stem, and each new pair is staggered, rotated 90 degrees relative to the one before. In Fig. 1(b), another member of the mint family, *Veronica Latifolia*, tiles a mound with its decussate stems. The multiplication of right angles in this phyllotaxy, emphasized by long, arrow-like leaves, seems to contradict the rounded form of the plant as a whole.

Figure 2(a) shows the tricussate phyllotaxy of *Sedum Seiboldii.* The leaves may seem to spiral at first glance, but in fact they appear in whorls of three, staggered with respect to their predecessors by a 60-degree rotation. It is the propeller tilt of the whorl members that gives a spiral effect. Figure 2(b) shows the entire sedum plant, a compact mound of more than one hundred tricussate stems. These round whorls of three rounded leaves create a hemisphere of circles. The alert reader will discover a stem with four-leafed whorls near the center of the plant. Like the four-leafed clover, there is always an exception to the rule.

All four photographs are by the author, taken at Princeton University and at the Mary Flagler Cary Arboretum, Millbrook, New York.

(a)

(b)

Figure 1.

(a)

(b)

Figure 2.

Peter Desain and Henkjan Honing

Trajectories of a Neural Network Quantizer in Rhythm Space

Described here is a pattern showing different trajectories in a space of all possible rhythms of four notes. This space is traversed by a quantization system that seeks a metrical interpretation of a performed rhythm. The durations of the notes are adjusted to an equilibrium state in which many of the adjacent induced time intervals have a small integer ratio. The system is implemented in a connectionist, distributed way. A network of cells — each with very simple individual behavior — is used, in which the cells that represent adjacent intervals interact with each other.

The three degrees of freedom are mapped to two dimensions by normalizing the total length of the rhythm. Each point (x, y) represents a rhythm of three inter-onset invervals $x : y : 1 - x - y$ in a net of interacting cells. Plotting the trajectories of different rhythms exhibits the behavior of the network and the stable attractor points in this two-dimensional space. They are positioned on straight lines that represent rhythms with an integer ratio of two durations or their sums ($x = y$, $x + y = z$, $2x = y$, etc.). A graphical front end to the system was used to produce the figures.

Figure 1 shows the space in which the system is given an initial state in an interactive way (clicking with a mouse at a certain point in the rhythm space), making it possible to explore the space, evaluate the performance of the quantizer, search for maximally ambiguous rhythms, etc.

Figure 2 shows an automated run through a large systematic set of possible rhythms. One can see relatively large areas around the simple rhythms and relatively small areas around more complex rhythms.

The system differs from other methods of the quantization of musical time in that it combines the following three characteristics: it is *context sensitive*, has *no musical knowledge* and exhibits *graceful degradation*. It was first described in [1] together with a micro version of the connectionist quantizer coded in CommonLISP. Methods for studying its behavior appeared in [2]. Non-connectionist methods for quantization of musical rhythms are described in [3] using a system with tempo-tracking, in [4] using an expert system, and in [5] using a divisive search method with backtracking. Clark [6] gives an idea of all the processes that are involved in adding expressive information to a musical score in performance.

Note that the resemblance to actual microscopic pictures of neurons is purely accidental.

References

1. P. Desain and H. Honing, "Quantization of musical time: A connectionist approach", *Computer Music Journal* **13**, 3 (1989) 56–66.
2. P. Desain, H. Honing and K. de Rijk, "A connectionist quantizer", *Proc. 1989 Int. Computer Music Conf.*, San Francisco, California (Computer Music Association, 1989) pp. 80–85.
3. R. B. Dannenberg and B. Mont-Reynaud, "An on-line algorithm for real-time accompaniment", *Proc. 1987 Int. Computer Music Conf.*, San Francisco, California (Computer Music Association, 1987) pp. 241–248.
4. J. Chowning *et al.*, "Intelligent systems for the analysis of digitized acoustical signals", CCRMA Report No. STAN-M-15, Stanford, California (1984).
5. H. C. Longuet-Higgins, *Mental Processes* (MIT Press, 1987).
6. E. Clarke "Levels of structure in the organization of musical time", *Contemporary Music Review* **2** (1987) 212–238.

Figure 1. Exploring the quantization of rhythm space interactively.

Figure 2. Systematic behavior mapping of the Quantizer in rhythm space.

Jacques Boivin

An Internally Mechanistic Nucleon

Represented here is the first visualization of a nucleon structure developed according to the principles of the Heart Single Field Theory. In this illustration, each small circle represents a spherical space containing an electron-like component closely bound to its neighbors, their number totalling 1836 to form a proton. These components are "electron-like" only in terms of that aspect of their structure which is responsible for their mass; that aspect of their structure, normally responsible for "charge", has been altered into an intra-nucleonic weave. This model was meant as an example of a type of theoretical construct that could be sought within the context of a coherent all-encompassing theory on the nature of the universe. Uncertainty, simultaneity and dualities (such as "wave/particle") may not be considered as absolute features of elementary reality for the purpose of such models; rather, the behavior of precise fundamental structures must give rise to apparent uncertainty, instantaneity and duality.

Briefly stated, the Heart Single Field Theory holds that all events in the universe derive from the structural configurations of aggregates of Hearts, a Heart being the sole building block of the universe and existing in only one state, a basic self-consistent field. Great numbers of these identical elementary units join end to end to form strings that organize themselves along a natural progression of complexity. Dimensional configuration determines the properties of each particular entity; a string of Hearts with both ends unattached becomes a traveling helix (a "photon") while loops of Heartstrings form into a hierarchy of shapes, such as toroidal coils, which combine in various ways to generate all existent particles and fields.

This "neoclassical" vision conceives of reality as ultimately amenable to conscious understanding without resorting to counter-intuitive or compartmentalized abstractions. Quarks, gluons, gravitons and their ilk are postulated to exist in only one area of the universe: the cranial cookie canisters of twentieth-century physicists. Currently, the only version of the Heart Single Field Theory in print is the schematic early version from which this "nucleon structure" is taken [1]; a much more elaborate version has been promised for many years but a number of mundane preoccupations have been holding it up [2].

References

1. J. Boivin, *The Heart Single Field Theory: Some Speculations on the Essential Unity of the Universe* (self-published). Reprinted in *Speculations in Science and Technology* **3**, 2 (1980) 185–204.
2. J. Boivin, *Hearts Not Quarks* (Tentative title) (to appear).

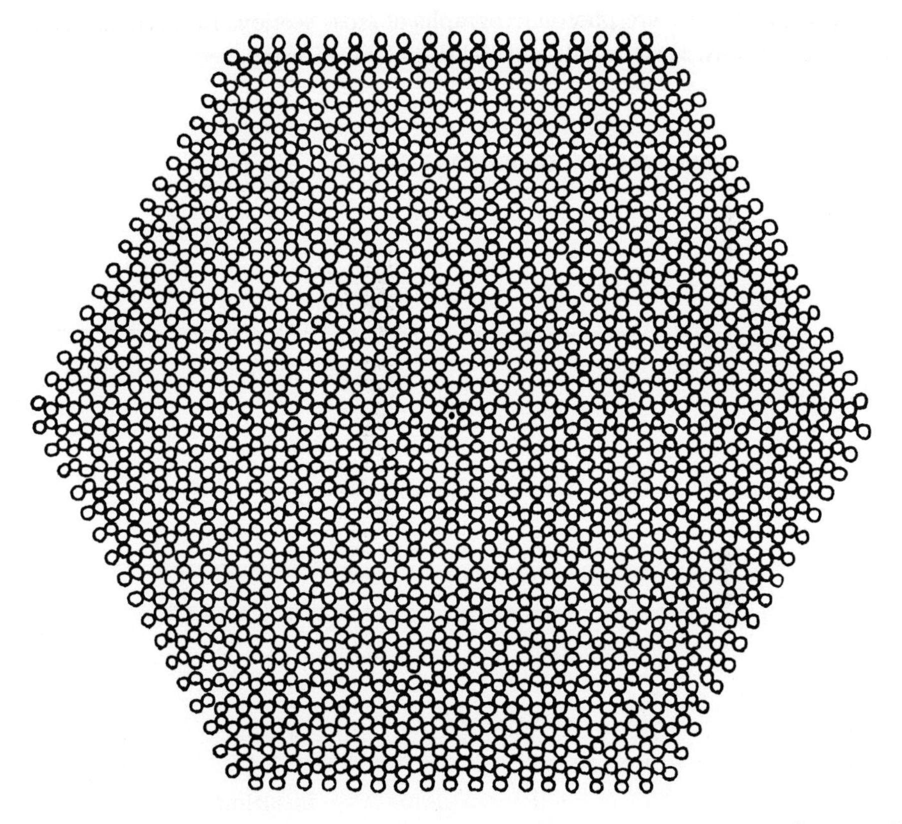

Figure 1. Nucleon structure. ○ = electron (total: 1836). • = center of nucleon (location of stabilizing particle in neutron).

Phil Brodatz

Wood Pattern

Wood patterns are infinitely varied, offering myriad opportunities for novel designs. Shown here are photomicrographs of cross sections for various woods including Urnday, Orey wood, and Paldo or Guinea wood.

Reference

1. P. Brodatz, *Wood and Wood Grains* (Dover, 1971).

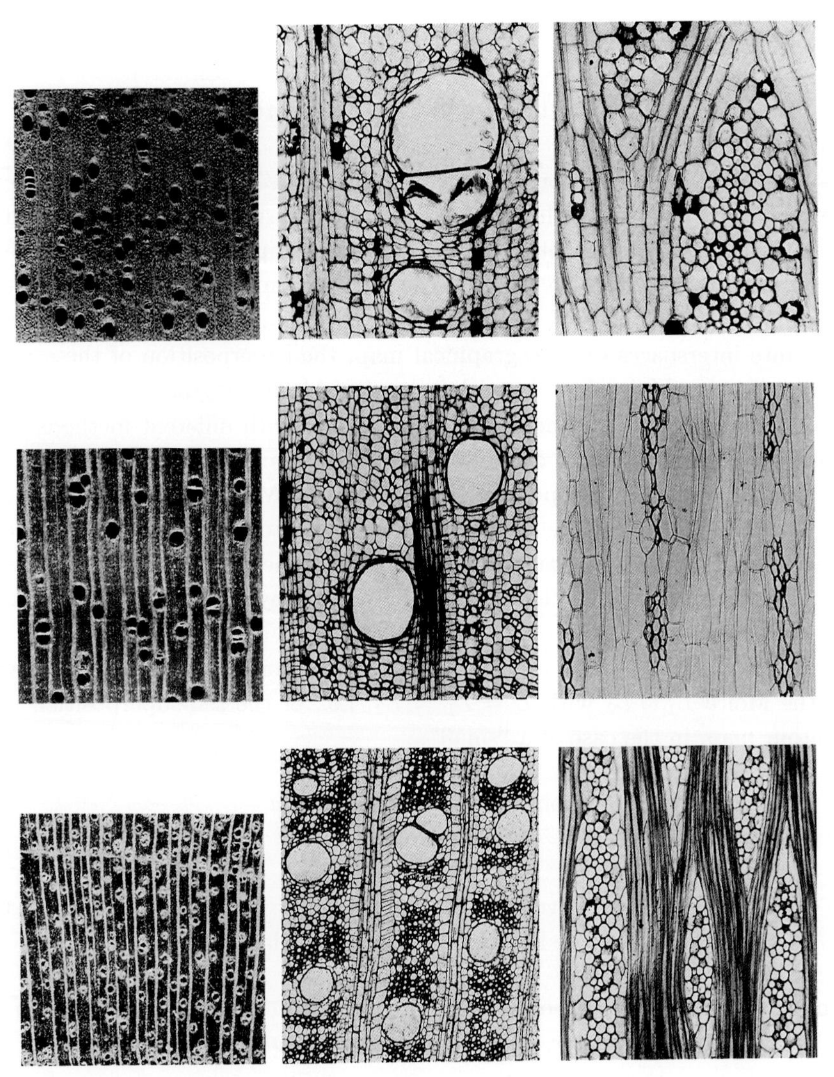

Top to bottom: Ceiba, Balsa, and Freijo.

Hans Giger

Moiré Pattern

The pattern described belongs to the Moiré patterns first investigated by Lord Raleigh in 1874.

The Moiré phenomenon is based on purely geometrical principles in as much as the image processing in the eye of the observer need not be taken into account: Two textures with black and white components are geometrically superposed by forming the union or the intersection of the black pointsets.

If, for instance, each of the textures L_1 and L_2 are the black contours and its white interspaces of a geographical map, the superposition of these maps as films form a Moiré $L_1 \cup L_2$ with its typical Moiré fringes.

Harthong and the author independently and with different methods have proven a generalized form of the following theorem: If the two contour maps L_1 and L_2 of two landscapes are superposed, the Moiré fringes of the resulting Moiré $L_1 \cup L_2$ can be interpreted as the contour map of the landscape determined by the differences of the heights of each pair of points with the same projection point on the map, the first point laying on L_1 and the second on L_2.

From this interpretation results the following corollary: The Moiré fringes of the Moiré $L_1 \cup L_2$ with $L = L_1 = L_2$, i.e., of the self-superposition of a contour map, in the case of a "small",

1) *translation* is the contour map of the *directional*,
2) *rotation* relative to a given turning point is the contour map of the *rotational*,
3) *radial stretching* relative to a given fixed point is the contour map of the *radial derivative* of the function describing the landscape.

The three effects are demonstrated with the map of the landscape given by cylindrical coordinates $z(r, \varphi) = r \cdot \varphi$, $0 < r$, $0 \leq \varphi < 2\pi$. (Figs. 1, 2, 3).

Moiré 1) direction $\varphi = 0$, resembles the sourcefield of an electrical charge;
Moiré 2) turning point $r = 0$, is the contour map of the surface of a circular cone;
Moiré 3) fixed point $r = 0$, reproduces the contour map of the given landscape.

References

1. J. Harthong, "Le Moiré", *Advances in Applied Mathematics* **2** (1981) 24–75.
2. H. Giger, "Moirés", *Comp. & Maths. with Appl.*, **12B,** 1 & 2 (1986) 329–361.

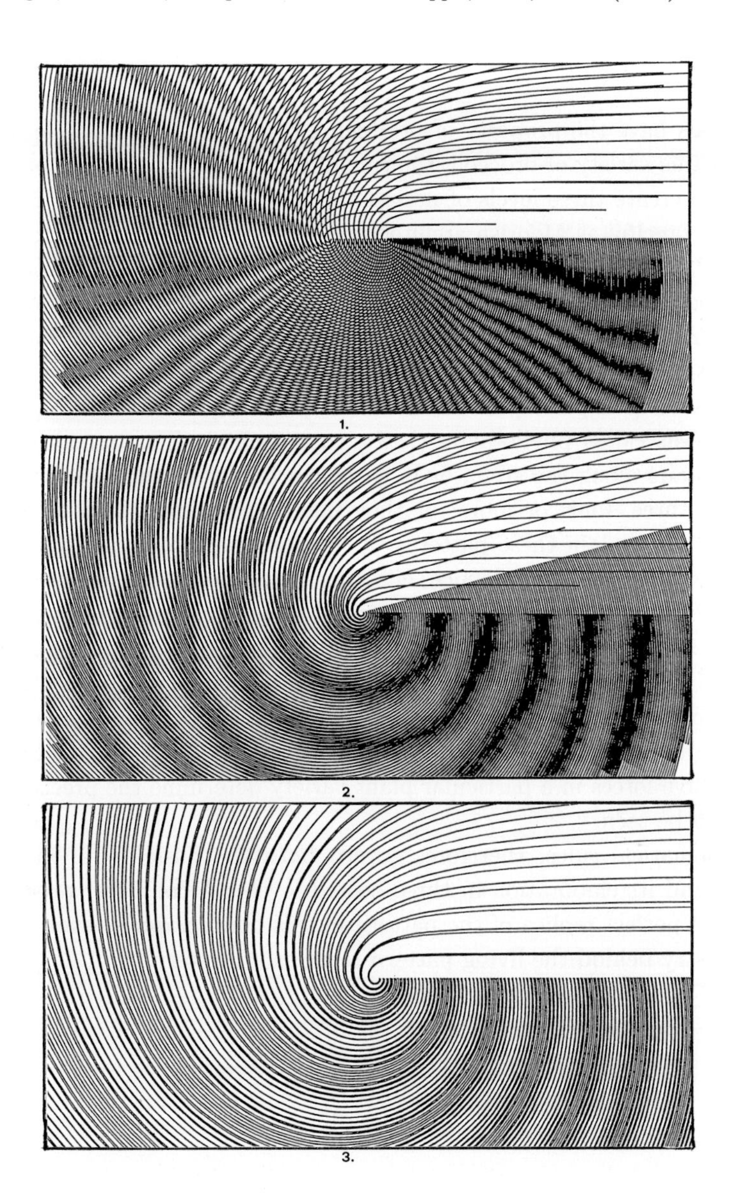

Dawn Friedman

Fractal Phyllotactic Spirals: Broccoli "Minaret"

Fractals have the property of producing complexity from simple, iterated rules. In this variety of broccoli, the rule of self-similarity operates on a spiral pattern to create three levels of nested spiraling florets. The effect is as elaborate and fanciful as Moorish architecture, giving the plant its unbroccoli-like name of Minaret. But the fractal generation of rich complexity from simple rules appears everywhere in nature. What is most remarkable about Broccoli Minaret is that the fundamental spiral pattern on which the fractal rules operate is itself the product of a few simple rules.

Each variety of plant produces its leaves, flowers, and shoots in characteristic patterns called phyllotaxies (leaf arrangements). Shoots may appear in opposite pairs along the stem, or in a zig-zag ladder pattern; they may form whorls of three, four, or more; or they may climb along the stem in a spiral helix. To explain these patterns, it has long been assumed that existing shoots exert a repulsive influence, ensuring that new shoots will be placed at a distance from old ones. The biochemical or biophysical mechanism of this influence is still unknown. But recently it has been shown that a single set of equations describing the behavior of the repulsive effect can generate each of the phyllotactic patterns seen in plants, from the alternating leaves of a leek to the tight spiral of broccoli florets seen here. The strength and range of the repulsive forces in a particular plant variety determine the precise pattern which will be seen.

Mathematical rules, therefore, govern both the basic spiral in Broccoli Minaret and its elaboration in three nested levels. A computer program can generate a perfect replica of the pattern in this photograph — even while the biochemistry behind the living pattern remains a mystery.

Broccoli Minaret is reproduced here by written permission of *Johnny's Selected Seeds.*
Photograph: *Johhny's Selected Seeds.*

Ulrich Melcher

Puppy Representation of DNA Nucleotide Sequences

Described here is a pattern showing a useful representation of the sequence of nucleotides in the DNA of a plant virus. The DNA is the genetic material of the virus and encodes all functions necessary for the virus to reproduce itself in host plants. A DNA strand is a linear polymer of four nucleotides: adenosine, guanosine, thymidine, and cytidine monophosphates. The information content of DNA lies in the order of the nucleotides. To depict the order, the upper case Roman character for the first letter of the English name of a nucleotide usually represents that nucleotide. Such representations are difficult to scan for interesting features, inefficient in the amount of space they occupy and in some type fonts lead to mistaken reading of "G" and "C". In this pattern [1], the top line of the sequence row has circles for nucleotides containing purine bases ("A" and "G"), while circles on the bottom line of the sequence row indicate pyrimidine nucleotides ("T" and "C"). The middle line of the row contains an additional circle for "G" and "C" residues. Most DNA molecules, including that of this virus, contain two anti-parallel DNA strands. The sequence of one strand is complementary to that of the other, according to the rules that "A" pairs with "T" and "C" with "G". The Puppy representation shows the sequence of both strands; reading the diagram upside down gives the complementary strand. This feature also means that small self-complementary sequences (often targets for enzyme recognition), such as the "stair-step" sequence TCGA, have recognizable symmetric appearances. Single circles represent nucleotides that form weaker base pairs in double-stranded DNA, while pairs of circles show the stronger base pair formers. Thus, circle density of a region of sequence is an indicator of the thermal stability of that region. The sequence shown in the figure is that of the Cabbage S isolate of cauliflower mosaic virus [2] and is 8,024 nucleotides in length. Visual inspection of the representation allows identification of many sequence features. A particularly striking feature is the comb-like region of repeated alternating "C" and "T" residues. Visual inspection led to the identification of a region that, in the RNA transcript of this DNA, is probably internally base-paired.

References

1. U. Melcher "A readable and space-efficient DNA sequence representation: Application to caulimoviral DNAs", *Comput. Appl. Biosci.* **4** (1988) 93–96.
2. A. Franck *et al.*, "Nucleotide sequence of cauliflower mosaic virus", *Cell* **21** (1980) 285–294.

Kazunori Miyata

Stone Wall Pattern

Described here is an algorithm for generating stone wall patterns. This algorithm requires only a few parameters as input data. The output data are a bump plane, which represents each stone's height data, and an attribute plane, which represents each stone's attributes. The method is an enhancement of C. I. Yessios' work [1].

The algorithm models a stone wall's joint pattern by a "node and link" model, shown in Fig. 1. Each node has a position and link data. The links are restricted to four directions: upper, lower, right, and left. Each enclosed space of the joint pattern is equivalent to the space occupied by a stone in the wall.

The generation procedure has the following six steps. Figure 2 shows the flow chart.

1. The basic joint pattern is generated by using the average size of a stone in the wall and the variance of its size (Fig. 3).
2. The basic joint pattern is deformed by relocating its nodes. After node relocation, each line segment is subdivided recursively, using the fractal [2] method (Fig. 4).
3. The space occupied by each stone is found by using the link information of the basic joint pattern. The stone space is a polygon formed by nodes and line segments.
4. The texture of individual stones is generated by subdividing the stone primitive recursively [3]. For this, the fractal method and the roughness value of the stone are used.
5. The stone texture is clipped by cut polygons, which are contracted polygons of the stone spaces.
6. The height data and the attributes of the clipped stones are placed in the bump plane and the attribute plane respectively, by the scan-line method.

Bump data are used for the shading process, and attribute data are used to change the color of each stone, its optical features, and so on. An example of a generated stone wall pattern is shown in Fig. 5. Highly realistic images of walls, pavements, and steps can be obtained by mapping the generated patterns (Fig. 6). More details are given in [4].

References

1. C. I. Yessios, "Computer drafting of stones, wood, plant and ground materials", *Computers and Graphics* **3**, 2 (1979) 190–198.
2. B. B. Mandelbrot, *Fractals Form, Chance, and Dimension* (Freemann, 1977).
3. A. Fournier, D. Fussell and L. Carpenter, "Computer rendering of stochastic models", *Commun. ACM* **25**, 6 (1982) 371–384.
4. K. Miyata, "A method of generating stone wall patterns", *Computer Graphics* **24**, 4 (1990) 387–394.

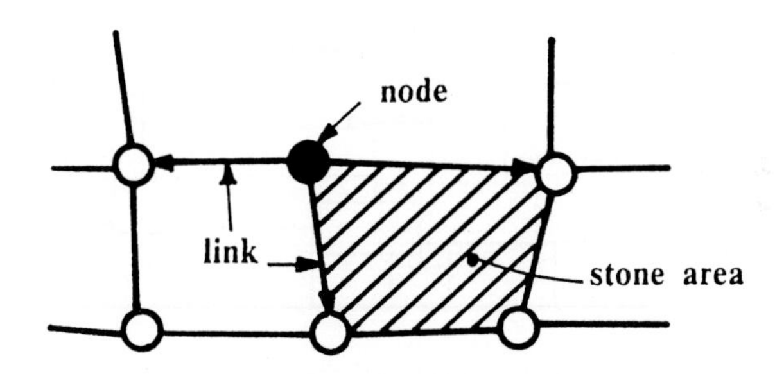

Figure 1. Node and link model.

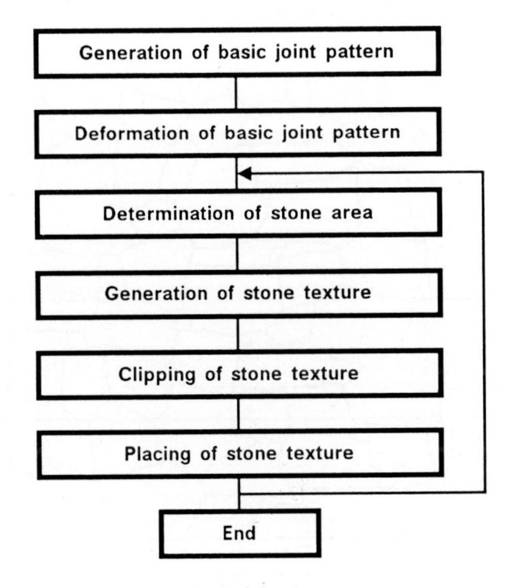

Figure 2. Procedure flow.

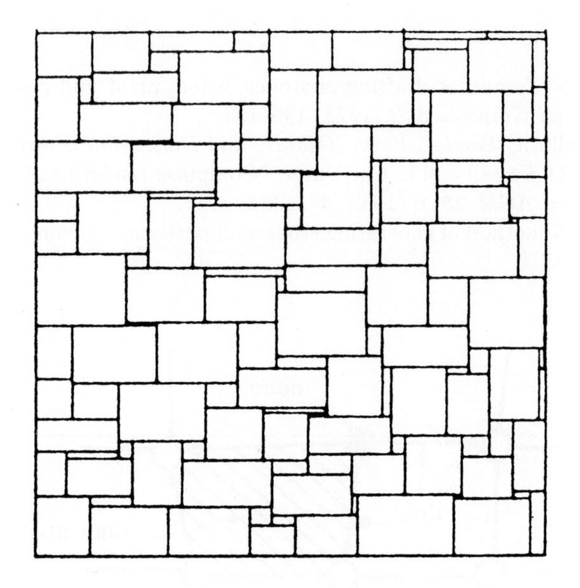

Figure 3. Example of a basic joint pattern.

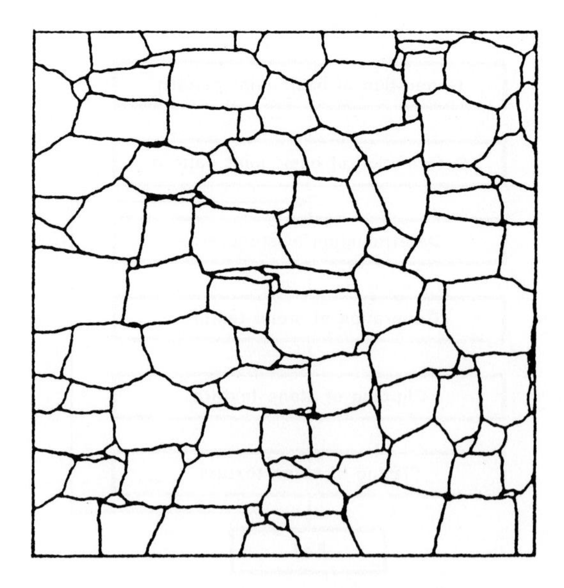

Figure 4. Example of a final joint pattern.

Figure 5. Example of a generated stone wall pattern.

Figure 6. Edo castle.

Alan Peevers

Novel Representations, Pattern, Sound

Various techniques for music visualization, music transcription, melody storage, and melody matching have been proposed in the past (see references). A few weeks ago, I was playing with an audio data acquisition board in conjunction with a music synthesizer — acquiring about 7 seconds each of various voices. After I adjusted the pitch so that the period was a multiple of 512 samples, I was able to display the time-varying waveform as a $512 \times 512 \times 8$ greyscale image on my NTSC frame buffer. Some of the patterns I looked at are really quite beautiful (see figure).

The image is taken from 256K 16-bit samples at a 50 kHz sampling rate (a bit over 5 seconds of sound). The source of the audio is a Yamaha (TM) DX-100 FM synthesizer programmed to modulate the timbre of the sound using a Low Frequency triangle wave giving an overall "wa-wa" effect. I adjusted the pitch so that an integral number of fundamental periods fit into 512 samples, which correspond to one scanline in the image. So, successive scanlines are just successive "slices" of the sound, giving an impression of the overall temporal evolution of the sound. I wrote an assembly language program on the PC that takes the 16-bit 2's complement samples, converts them to offset binary, scales them, and truncates the values to 8 bits, suitable for displaying on a greyscale display. There is only one axis, time. Successive scanlines (the y axis) represent successive "slices" of the sound. Each slice represents some multiple of the fundamental period.

References

1. R. Cogan, *New Images of Musical Sound* (Harvard University Press, 1984).
2. C. Roades, "Research in music and artificial intelligence", *ACM Comput. Surv.* **17** (1985) 163–190.
3. D. Starr, "Computer chorus", *Omni Magazine* (May 1984) 41.
4. C. Pickover, "Representation of melody patterns using topographic spectral distribution functions", *Computer Music Journal* **10**, 3 (1986) 72–78.
5. J. Pierce, *The Science of Musical Sound* (Scientific American Library, 1983).

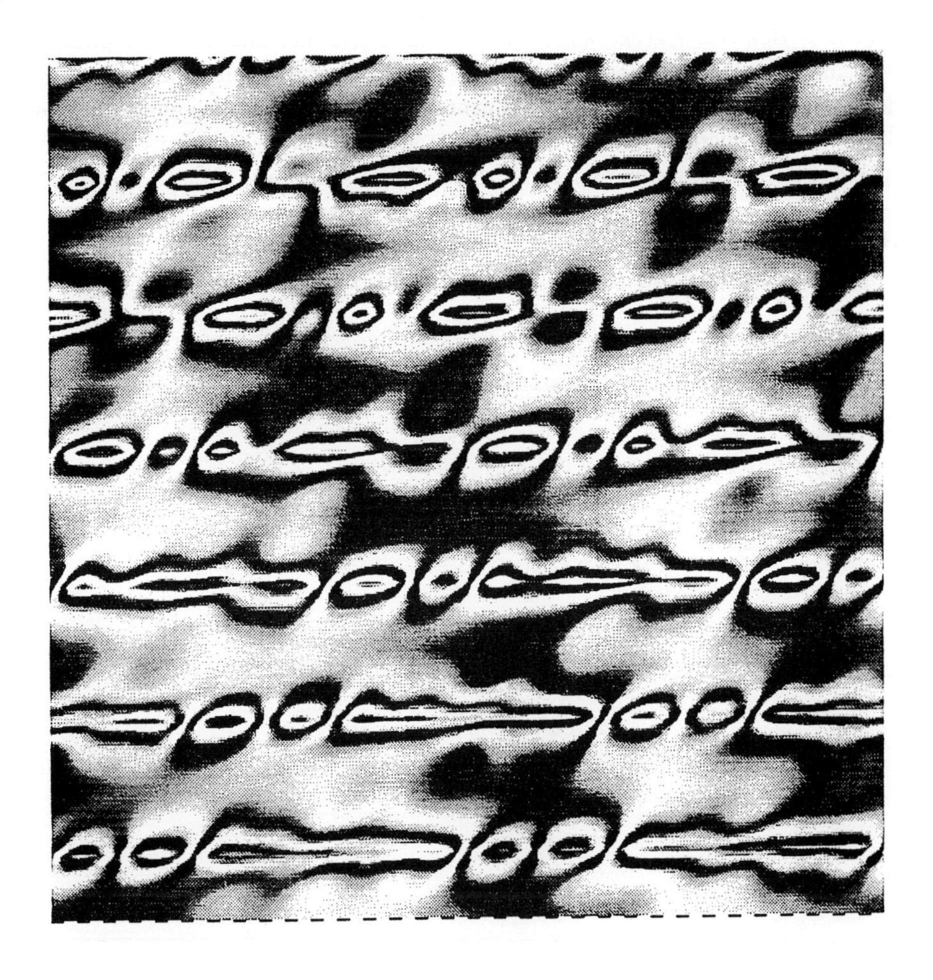

Clifford A. Pickover

Haeckel's Sea-Lilies

Ernst Heinrich Haeckel (1834–1919) was a German biologist and philosopher interested in the beauty of natural forms. Throughout his career he made detailed drawings of a range of organisms. He seemed particularly interested in deep-sea and microscopic life. Shown here are some of his drawings of various species of sea-lilies (animals related to starfishes and sea-urchins).

Reference

1. E. Haeckel, *Art Forms in Nature* (Dover, 1974).

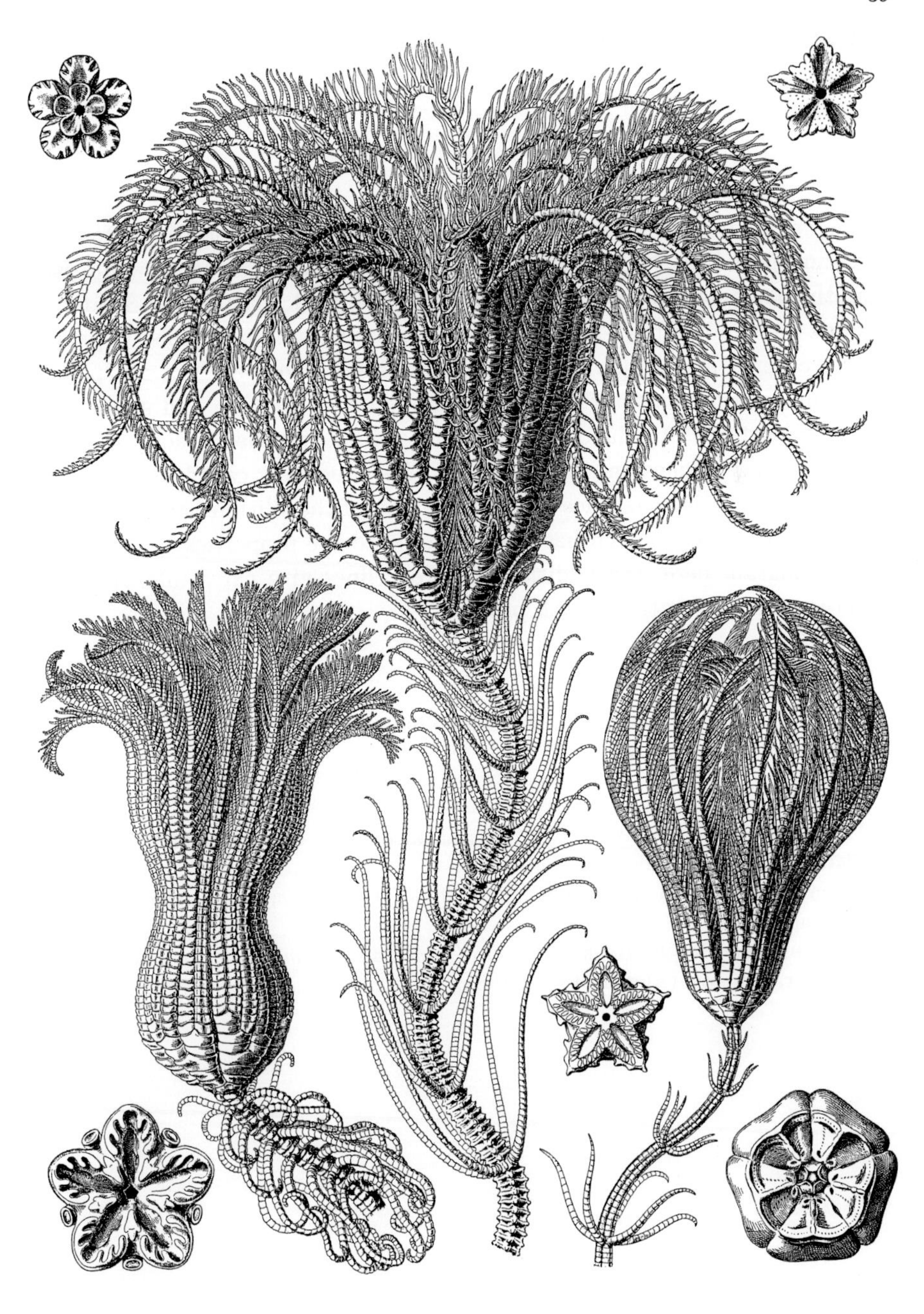

Daniel Platt

Diffusion Limited Aggregation

Described here is a pattern produced by chaotic behavior arising from aggregating particles. "Aggregation" is a term used to describe growth arising from the agglomeration of diffusing particles. In 1981, Witten and Sander developed a computer model for aggregation starting with a single seed particle at the center of a space [1]. Their computer program introduces a new particle which moves randomly until it approaches another particle and sticks to it. The first particles attach to the seed, but soon a branched, fractal structure evolves with a dimension of about 1.7. This process is called *diffusion limited aggregation* (DLA). Since the introduction of the Witten-Sander model in 1981, considerable research has been devoted to the properties of DLA [2–6]. In the diagram here, the additional contour lines give extra information to scientists concerning the growth process.

References

1. T. Witten and L. Sander, "Diffusion limited aggregation: A kinetic critical phenomenon", *Phys. Rev. Lett.* **47** (1981) 1400.
2. M. Batty "Fractals — geometry between dimensions", *New Scientist* (April 1985) 31–40.
3. E. Stanley and P. Meakin, "Multifractal phenomena in physics and chemistry", *Nature* **335** (September 1988) 404–409.
4. F. Family and D. Landau, eds., *Kinetics of Aggregation and Gelation* (North-Holland, 1984).
5. H. Stanley and N. Ostrowsky, eds., *On Growth and Form* (Martinus-Nijhof, 1985).
6. J. Feder, *Fractals* (Plenum Press, 1988).

Mels Sluyser and Erik L. L. Sonnhammer

RNA Structure Based on Prime Number Sequence

This display represents the secondary structure of an RNA sequence based on the first 2000 prime numbers.

The relationship between prime numbers has previously been investigated by a large number of methods [1]. Here, we describe a new approach in which we designate pairs of primes by the symbols A, G, C or U, depending on the property of that prime pair as described below. These symbols were chosen because they are commonly used to designate nucleotide bases in RNA. The computer program FOLD has been designed to investigate the properties of ribonucleic acid sequences [2, 3]; here we implement this program to analyze the series of primes.

To calculate the parity of prime numbers they were expressed in the binary system. If the sum of the coefficients (S_c) is odd, the parity (Pa) of the prime is assigned 1 (example: prime $7 = 1 \times 2^0 + 1 \times 2^1 + 1 \times 2^2$; $S_c(7) = 3$; $Pa(7) = 1$). If S_c is even, the parity of the prime is assigned 0 (example: prime $3 = 1 \times 2^0 + 1 \times 2^1$; $S_c(3) = 2$; $Pa(3) = 0$). Successive prime numbers (P) were coupled pairwise: [2, 3], [5, 7], [11, 13], [17, 19], [23, 29]. ... and expressed as parities. This yields $Pa(P)$: [1, 0], [0, 1], [1, 1], [0, 1], [0, 0] ... . Parity pairs [1, 1], [0, 0], [1, 0] and [0, 1] were assigned the symbols A, U, C and G, respectively. The first fifty letters of the sequence are thus:

CGAGUACCACAUCACACGCAAAGAAAAUGUCCCCUUCCGGAAAGGUACGGG

The sequence of nucleotides of an RNA molecule determines its secondary structure. Computer methods have been used to minimize free energies for the prediction of the secondary structure of biological RNA molecules. Zuker and Stiegler [2] implemented a dynamic programming algorithm for this purpose. We used Zuker's FOLD program (UWGCG) [3] with the free energy parameters calculated by Freier *et al.* [4] to investigate the prime parity sequence described above. The FOLD program calculates a secondary structure which exhibits a base-paring structure where an energy minimum exists. The figure shows the optimal folding of the RNA sequence representing the first 1000 prime pairs.

References

1. P. Ribenboim, *The Book of Prime Number Records*, 2nd ed. (Springer-Verlag, 1989).
2. M. Zuker and P. Stiegler, "Optimal computer folding of large RNA sequences using thermodynamics and auxiliary information", *Nucl. Acids Res.* **9** (1981) 133–148.
3. J. Devereux, P. Haeberli and O. Smithies, "A comprehensive set of sequence analysis programs for the VAX", *Nucl. Acids Res.* **12** (1984) 387–395.
4. S. M. Freier *et al.*, "Improved free-energy parameters for predictions of RNA duplex stability", *Proc. Natl. Acad. Sci. U.S.A.* **83** (1986) 9373–9377.

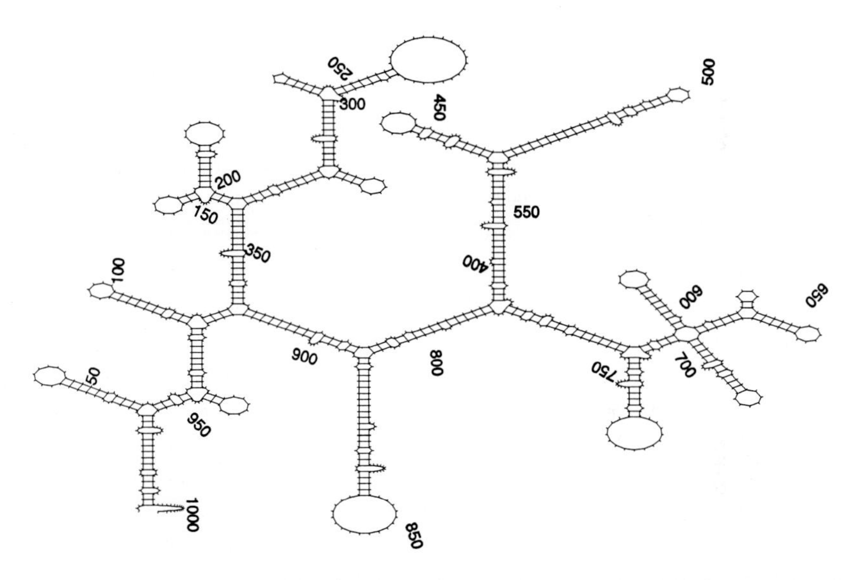

Duplex display of a pseudo RNA structure obtained from the first 1000 prime pairs. This SQUIGGLES display was obtained by analysis with the FOLD program. A biological RNA molecule with this sequence is predicted to have a free energy of −256.9 kcal/mole.

Thomas D. Schneider

Genetic Patterns as shown by Sequence Logos

The two patterns shown here represent two different views of a control switch in the genetic material of a virus called T7. Like many other viruses, T7 is made of proteins that coat and protect its genetic material DNA (deoxyribonucleic acid). The virus looks like a lunar-lander landing on the moon. When a T7 particle contacts the surface of the bacterium *Escherichia coli*, it sticks there. Once it has landed, it injects its DNA into the cell, much like a hypodermic needle does. The DNA contains instructions for taking over the bacterial cell. The cellular machinery unwittingly copies these instructions into RNA. The RNA is then used to make proteins that stop the normal cell mechanisms. Other instructions tell the cell to make copies of the viral DNA and to make the proteins of the viral coat. These are assembled to produce perhaps a hundred new T7 particles. The final instruction causes the cell to burst open, just as in the movie "Alien".

T7's strategy for taking over the cell is to replace the bacteria's RNA-making machine with a new one. This new "RNA polymerase" binds to certain spots on the viral DNA, and then makes the RNA molecules. Other machinery then translates these RNAs into the proteins that make up the virus.

The entire genetic material of T7 consists of exactly 39,936 chemical "letters" named A, C, G and T. There are 17 places on this string of letters to which the RNA polymerase binds. Their sequences are:

```
  ------------
  221111111111--------- ++++++
  1098765432109876543210123456
  ............................
1 ttattaatacaactcactataaggagag
2 aaatcaatacgactcactatagagggac
3 cggttaatacgactcactataggagaac
4 gaagtaatacgactcagtatagggacaa
5 ctggtaatacgactcactaaaggaggta
6 cgcttaatacgactcactaaaggagaca
```

```
 7 gaagtaatacgactcactattagggaag
 8 taattaattgaactcactaaagggagac
 9 gagacaatccgactcactaaagagagag
10 attctaatacgactcactaaaggagaca
11 aatactattcgactcactataggagata
12 aaattaatacgactcactatagggagat
13 aatttaatacgactcactatagggagac
14 aaattaatacgactcactatagggagac
15 aaattaatacgactcactatagggagaa
16 gaaataatacgactcactatagggagag
17 aaattaatacgactcactatagggagag
```

The coordinates of each position are written vertically above the sequences.

The "sequence logo" pattern in Fig. 1 was created from these sequences. The sequence logo consists of stacks of letters on top of one another. The height of the stack is the consistency of preservation of the pattern, measured in bits. The vertical bar is 2 bits high. The heights of individual letters are proportional to the number of times they appear in the sequences given above. The error bars show the expected variation of the stack heights. In Fig. 2, they are an underestimate of the variation.

The area of the sequence logo in Fig. 1, which measures the total conservation of these genetic patterns, is 35 bits. In contrast, the amount of conservation needed for the job of finding the promoters in the cellular DNA is only about 17 bits. So there is twice as much sequence pattern as there should be. To test this, many T7 promoter DNAs were made by chemical synthesis. The logo for the functional ones is shown in Fig. 2. The area of this logo is 18 ± 2 bits, which is about the amount needed. So the RNA polymerase does use only 17 to 18 bits of information, and the extra pattern in the upper logo must have some other unknown function.

References

1. T. D. Schneider, G. D. Stormo, L. Gold and A. Ehrenfeucht, "Information content of binding sites on nucleotide sequences", *J. Mol. Biol.* **188** (1986) 415–431.
2. T. D. Schneider, "Information and entropy of patterns in genetic switches", *Maximum-Entropy and Bayesian Methods in Science and Engineering*, Vol. 2, eds. G. J. Erickson and C. R. Smith (Kluwer Academic Publishers, 1988) pp. 147–154.

3. T. D. Schneider and G. D. Stormo, "Excess information at bacteriophage T7 genomic promoters detected by a random cloning technique", *Nucl. Acids Res.* **17** (1989) 659–674.
4. T. D. Schneider and R. M. Stephens, "Sequence logos: A new way to display consensus sequences", *Nucl. Acids Res.* **18** (1990) 6097–6100.

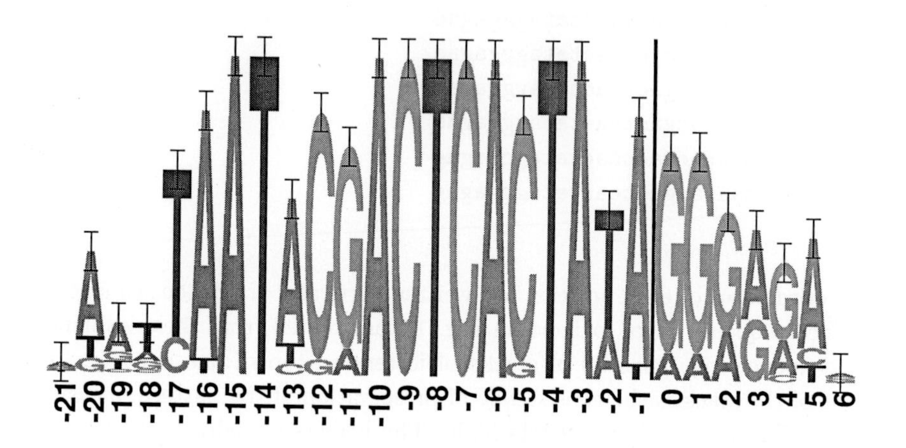

Figure 1. Pattern at T7 RNA polymerase binding sites.

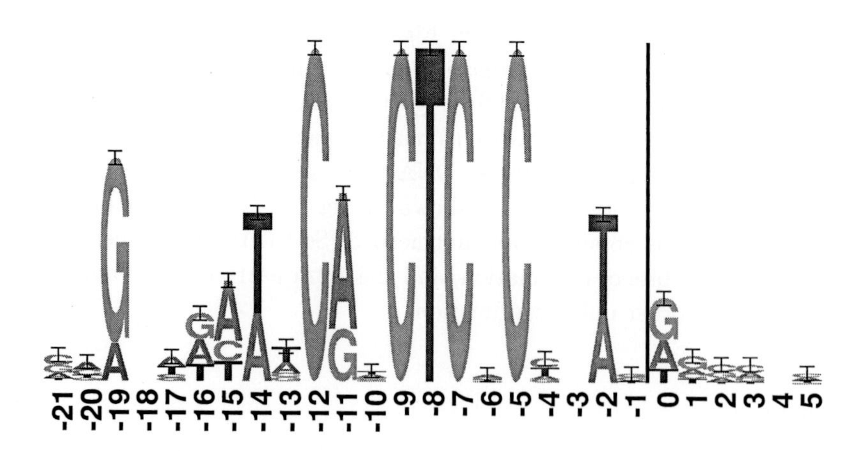

Figure 2. Pattern required by T7 RNA polymerase to function.

PART II
MATHEMATICS AND SYMMETRY

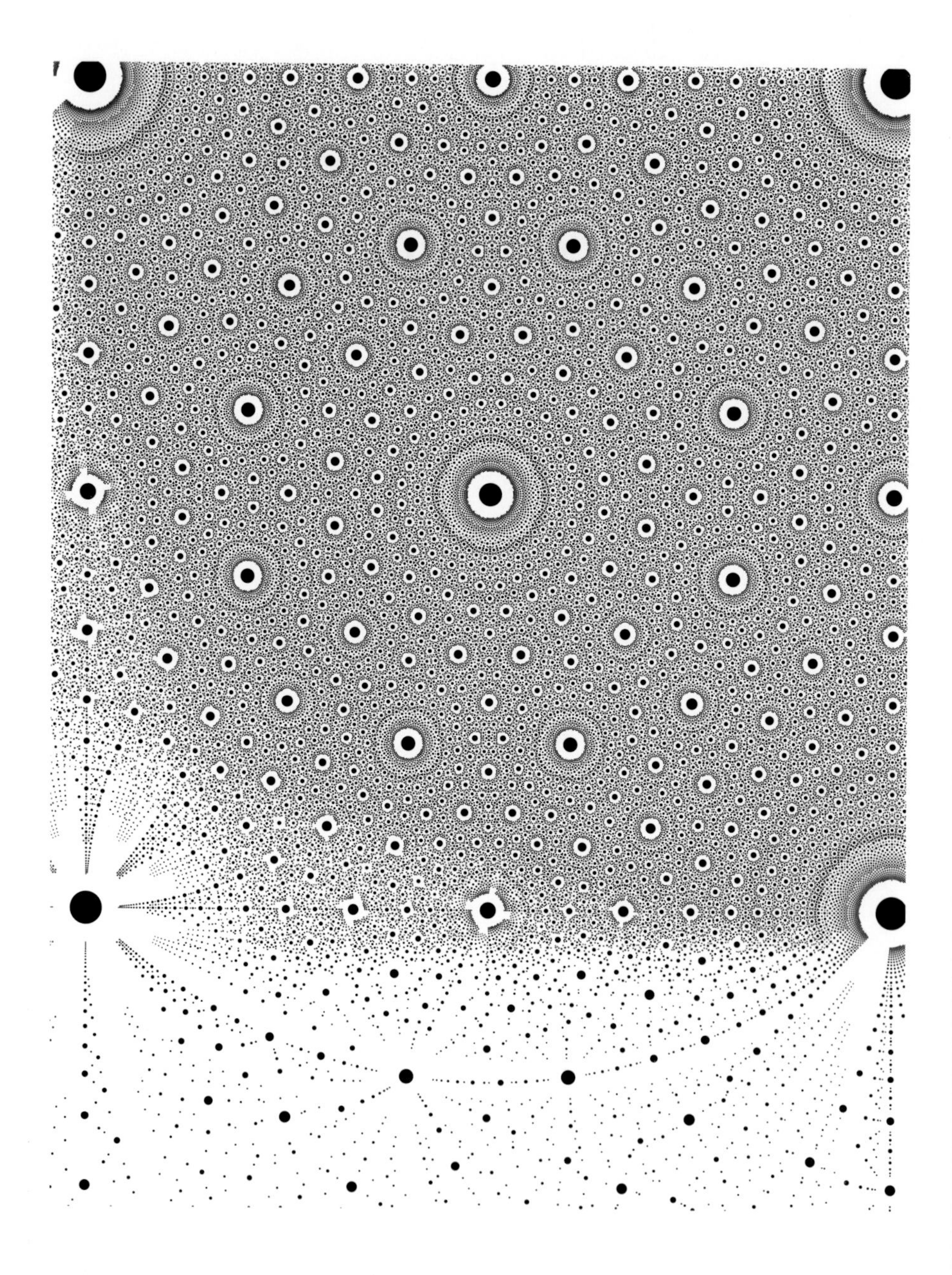

Stephen Schiller

Gaussian Fractions

This pattern shows the distribution of fractions of Gaussian integers whose denominators are limited to a certain size. A Gaussian integer is a complex number whose real and imaginary parts are both integers. A fraction of Gaussian integers, A/B, is simply A divided by B using the rules of complex arithmetic. The size of a Gaussian integer referred to above is its complex absolute value, given by $\sqrt{(x^2 + y^2)}$ for the complex number $x + iy$. In the pattern presented here the size limit was 25. The individual fractions were plotted by using the real part as an x coordinate and the imaginary part as a y coordinate. The page corresponds to the complex plane as follows: the origin of the complex plane is at the center of the large dot in the lower left corner of the page. The number 1 + i0 would be the center of the large dot in the lower right corner and the number 0 + i1 would be the large dot in the upper left.

This pattern arose from my desire to know how well fractions of Gaussian integers, or Gaussian fractions, can approximate arbitrary complex numbers. A similar, but more familiar problem is that of approximating ordinary real numbers with fractions of integers or rationals. In this problem, there may be practical limits on the sizes of the integers in the approximating fraction (such as the size of a computer word). And so, one might ask, if I can only use integers up to a certain size, say N, how well can an arbitrary real number between 0.0 and 1.0 be approximated? It turns out that although the average error is about $1/N^2$, there are certain places on the real line where the error can be much worse, as large as $1/(2N)$, for example. These areas-of-bad-approximation occur near fractions with small denominators. For each fraction of the form a/b there is an interval around a/b of length at least $1/(bN)$ where no other fractions occur. Thus, there are relatively large gaps around fractions with small denominators like 0/1, 1/2, 1/3 and 2/3.

The arithmetic of Gaussian integers and Gaussian fractions is very similar to that of the usual integers and usual fractions and the same comments as above apply except for the novelty of the space of Gaussian fractions being in two dimensions instead of one. I was curious to know how the fractions with smaller denominators were distributed relative to the other fractions. To this end I made the size of the dot representing the fraction a/b to be inversely proportional to the size of b $(1/(2\ Size(b))$ to be exact). This made the fractions with smaller denominators stand out. You can see around each of these larger dots a halo of white showing the dearth of neighbors next to these fractions. The size of the dot was also chosen so that the dots would just barely touch in areas where approximation would be good. The resulting pattern gives a good visual indication of areas of good approximation and areas of bad approximation.

The pattern was generated by a program in the Postscript page description language (included below). The program is fairly straightforward but some comments are in order.

The algorithm for plotting Gaussian fractions is brute force. The procedure /fracmap enumerates all possible denominators whose size is less than the size limit passed in. The procedure /doDenom enumerates all possible numerators for a given denominator. For a given size, N, there are on the order of N^4 such fractions so the time to run the program goes up quickly in N. The pattern shown took about an hour to generate.

Multiple fractions plotted by the program may fall in the same place, but the one with the smallest denominator makes the largest circle (corresponding to the reduced fraction at that position) which covers the smaller ones at the same spot. One could design a more efficient algorithm that only plotted reduced fractions.

I was originally only interested in plotting those fractions with absolute value less than one. (These would all lie in a disc of radius one centered at the origin.) The program actually generates and plots fractions outside of that range. This is because I am taking all pairs of Gaussian integers whose size is less than N and forming their fractions and many of those numbers have sizes greater than one. To cut down on the time it took to plot the pattern, I put in some simple checks in the program to eliminate most of the fractions outside of the unit disc. For that reason you can see that the density of the dots falls off as one gets outside of the unit disc. When I first plotted the pattern, I liked the visual effect of these stray dots, so I left them in the program.

```
%! gaussian fraction coverage: gauss.ps

% Passed on the stack is the maximum abs value of denominators
% that are mapped.
```

```
/fracmap {
  gsave
  10 10 translate
  72 8 mul dup scale
  0 setlinewidth
  1 setlinecap
  /maxD exch def
  /maxDsqrd maxD dup mul def
  0 1 maxD {
    /b.re exch def
    0 -1 maxD neg {
      /b.im exch def
      /denom b.re dup mul b.im dup mul add def
      denom maxDsqrd gt {exit} if
      denom 0 gt {doDenum} if
      } for
    } for
  grestore
  } def

/doDenum {
  0 1 maxD {
    /a.re exch def
    0 -1 maxD neg {/a.im exch def doFrac {exit} if} for
    1 1 maxD {/a.im exch def doFrac {exit} if} for
    } for
  } def

/doFrac {
  /re a.re b.re mul a.im b.im mul add def
  /im a.im b.re mul a.re b.im mul sub def
  /size 0.5 denom sqrt maxD mul div def
  /x re denom div def
  /y im denom div def
  Dot
  x 0 lt x 1 gt or y 0 lt y 1 gt or or    % return true if out of range
  } def

/Dot {x y size 0 360 arc fill} def

25 fracmap
showpage
```

Ian O. Angell

Lattice Design 1

The lattice pattern here was produced by a computer and demonstrates various types of repetition and symmetry. To draw a lattice pattern, the computer takes an elementary set of line segments and arcs, and manipulates them using a space group (a sequence of reflections, rotations, and translations) into a tile. These tiles are then stacked in a regular lattice in two-dimensional space, thus initiating further symmetries. Even random, nonsymmetric starting sets of lines and arcs produce beautiful symmetrical patterns.

Reference

1. I. Angell, *Computer Geometric Art* (Dover, 1985).

Henry F. Fliegel and Douglas S. Robertson

Goldbach's Comet

Shown here is the number of prime pairs $G(E)$ which can be found to sum to a given even number E, plotted as a function of E. (For example, for $E = 10$, $G(E)$ is 2, since 10 can be expressed by just two prime pairs — as 3 + 7 or as 5 + 5.) This pattern is connected with the famous Goldbach Conjecture, named after the Russian mathematician, Christian Goldbach, who speculated in a letter to Leonhard Euler that every even number greater than 4 can be expressed in at least one way as the sum of two odd primes. Yes — at least. J. J. Sylvester was apparently the first to show that "irregularities" should appear in the function $G(E)$ because, if E is divisible by distinct prime factors $p1$, $p2$, $p3$, ..., then $G(E)$ will be increased by the multiplier

$$\frac{(p1-1)(p2-1)(p3-1)\ldots}{(p1-2)(p2-2)(p3-2)}$$

over even numbers which have no such prime factors. That multiplier splits $G(E)$ into bands, corresponding to values of E divisible by 3, 5, 7, $3 * 5$, $3 * 7$, etc. What we call $G(E)$ was analyzed extensively by Hardy and Littleton [1]. Nevertheless, to this day, there is no strict proof from basic axioms of what appears so clearly in the graph shown here — that Goldbach was right.

References

1. G. H. Hardy and J. E. Littlewood, "Some problems of 'Partitio Numerorum', III: On the expression of a number as a sum of primes", *Acta Mathematica* **44** (1922) 1–70.
2. H. F. Fliegel and D. S. Robertson, "Goldbach's Comet: The numbers related to Goldbach's conjecture", *J. Recreational Mathematics* **21**, 1 (1989) 1–7.

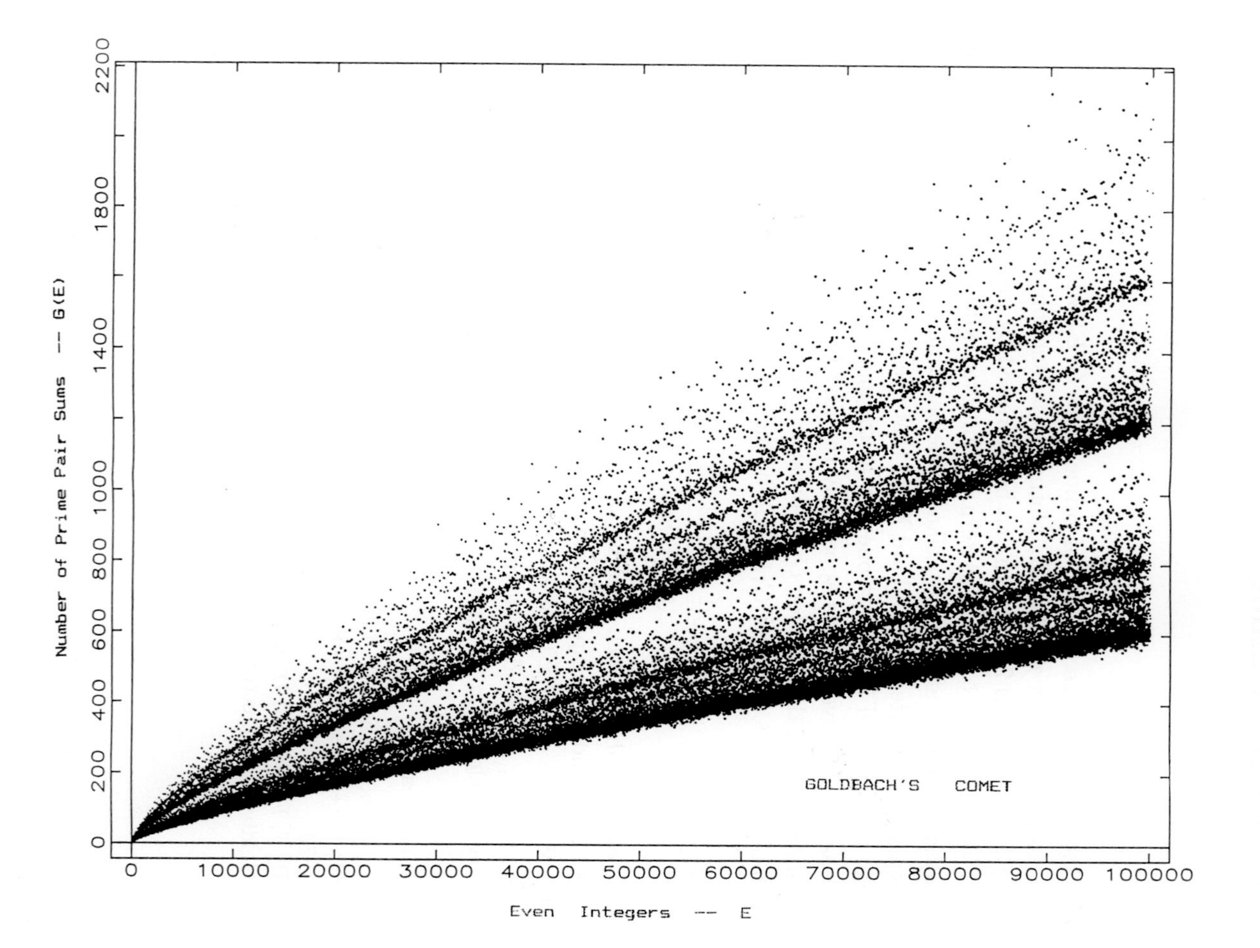
GOLDBACH'S COMET
Number of Prime Pair Sums -- G(E)
Even Integers -- E
0
200
400
600
800
1000
1400
1800
2200
0
10000
20000
30000
40000
50000
60000
70000
80000
90000
100000

Georg Jagoda

J-Curve (OLFRAC) (14 Iterations)

Described here is a pattern showing a curve that I call J-curve. Increasing the recursion depth by one looks like adding a second J-curve to the first one, rotated at an angle of 90°.

The pattern is created with OLFRAC by Ton Hospel from LISTSERV at BLEKULL11.

Datafile contains:

```
          axiom: ----F
number of rules: 1
       1st rule: F      +F--F+
          angle: 1 4                \ π  * 1/4 = 45°
recursion depth: 14
```

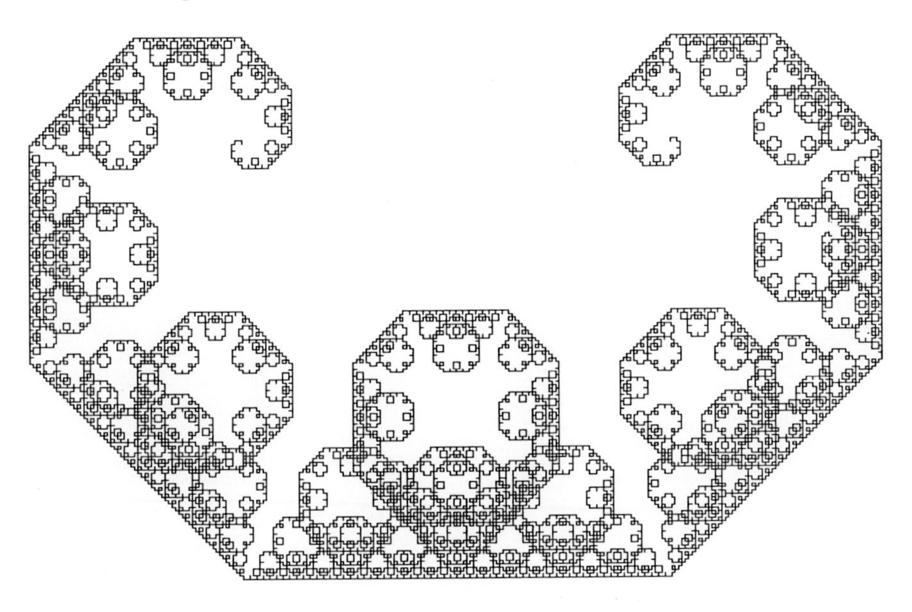

Georg Jagoda

Carpet (13 Iterations)

Described here is a pattern showing 4 J-curves.

The pattern is created with OLFRAC by Ton Hospel from LISTSERV at BLEKUL11.

Datafile contains:

```
          axiom:  F++F++F++F
number of rules: 1
      1st rule: F        +F--F+
          angle: 1 4              \ π  * 1/4 = 45°
recursion depth: 13
```

Georg Jagoda

3D-Cubes

Described here is a pattern showing 3 cubes. The three cubes are the pattern 3D-CUBE. The 180 "+" performs the rotation of $180 * \pi * 97^\circ/360^\circ = 180 * 180^\circ * 97^\circ/360^\circ = 180 * 48.5^\circ = 8730^\circ = 90^\circ + 24 * 360^\circ$. The digits in the axiom select the color: 1 (blue), 2 (red), 4 (green).

The pattern is created with OLFRAC by Ton Hospel from LISTSERV at BLEKUL11.

Datafile contains:

```
           axiom: ++++++++++++++++++++++++++++++++++++++++++\
                  ++++++++++++++++++++++++++++++++++++++++++\
                  ++++++++++++++++++++++++++++++++++++++++++\
                  ++++++++++++++++++++++++++++++++++++++++++\
                  ++++++++++++4c+ds[+ddl--1c]-ddl2c
number of rules: 20
       1st rule: c       |A+|B+|C|--
       2nd rule: s       sdsds
       3rd rule: l       lddlddl
       4th rule: d       e
       5th rule: e       f
       6th rule: A       FFV++FFH|--FFV++ffh|--
       7th rule: B       FW|-FFI+FW|-ffi+
       8th rule: C       FFX|-FJ+FFX|-fj+
       9th rule: H       HFFH[|--ffv++FFH|--ffv|ffv++ffh]FFH
      10th rule: h       hffh[|--ffv++FFH|--ffv|ffv++ffh]ffh
      11th rule: V       VFFV[++ffh|--FFV++ffh|ffh|--ffv]FFV
      12th rule: v       vffv[++ffh|--FFV++ffh|ffh|--ffv]ffv
      13th rule: I       IFFI[+fw|-FFI+fw|fw|-ffi]FFI
      14th rule: i       iffi[+fw|-FFI+fw|fw|-ffi]ffi
      15th rule: W       WFW[|-ffi+FW|-ffi|ffi+fw]FW
      16th rule: w       wfw[|-ffi+FW|-ffi|ffi+fw]fw
      17th rule: J       JFJ[+ffx|-FJ+ffx|ffx|-fj]FJ
      18th rule: j       jfj[+ffx|-FJ+ffx|ffx|-fj]fj
      19th rule: X       XFFX[|-fj+FFX|-fj|fj+ffx]FFX
      20th rule: x       xffx[|-fj+FFX|-fj|fj+ffx]ffx
           angle: 97 360            \ π * 97/360 = 48.5°
recursion depth: 6
sides                  A B C
      DRAW horizontal  H I J
don't draw horizontal  h i j
      DRAW vertical    V W X
don't draw vertical    v w x
   +------+
  /  C   /|
+------+ |
|       |B|
|  A    | +
|       |/
+------+
```

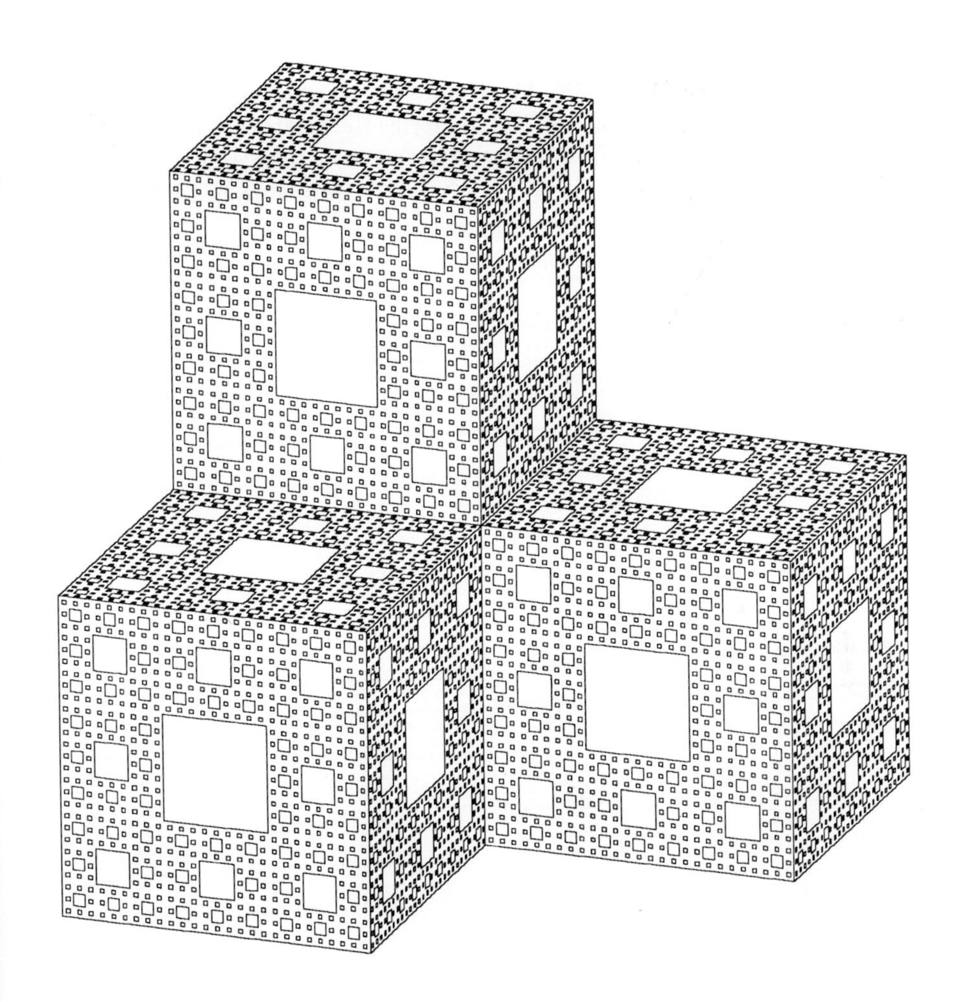

Georg Jagoda

The-End

Described here is a pattern showing 95 cubes building words. The 95 cubes look like the pattern 3D-CUBE. Three sides are shown but here you can see the six sides of the cubes. The 180 "+" performs the rotation of $180 * \pi * 97^\circ / 360^\circ = 180 * 180^\circ * 97^\circ/360^\circ = 180 * 48.5^\circ = 8730^\circ = 90^\circ + 24 * 360^\circ$.

The pattern is created with OLFRAC by Ton Hospel from LISTSERV at BLEKUL11.

Datafile contains:

```
+++++++++++++++++++++++++++++++++++++++++++++++++++++++++++\
+++++++++++++++++++++++++++++++++++++++++++++++++++++++++++\
+++++++++++++++++++++++++++++++++++++++++++++++++++++++++++\
+++[*r*r*d*d*d*d*d*d*uuuuuur*r*rr*d*d*d*d*d*d*uuur*r*r*r*d*\
d*d*uuuu*u*u*rr*r*r*r*lddd*l*l*u*u*ddd*d*d*r*r*r*]ddddddd*\
r*r*r*lddd*l*l*u*u*ddd*d*d*r*r*r*rr*u*u*u*u*u*u*rd*d*rd*rd*\
d*rd*u*u*u*u*u*u*rr*r*d*d*d*d*d*dl*r*r*r*ru*u*u*u*u*ul*l*
                                                    \ axiom
37                         \ number of rules
*  ABC[GLD|ABC]            \ cube (6 sides)
A  a                       \ delayed front side a
B  b                       \ delayed right-hand side b
C  c                       \ delayed upper side c
S  SeeSeeS                 \ S tep (big)
s  seses                   \ s tep (small)
e  f                       \ delayed f
U  eeS                     \ U p  /|\
u  zzT                     \ u p   |              +------+
D |eeS|                    \ D own   |           /  C   /|
d |zzT|                    \ d own  \|/         +------+ |
R  ++eeS--                 \ R ight  -->        |      |B|
r  ++zzT--                 \ r ight             |  A   | +
L |++eeS--|                \ L eft  <--         |      |/
l |++zzT--|                \ l eft    /         +------+
N |+es-|                   \ N ear  |/
n |+zt-|                   \ n ear  +-
G  +es-                    \ G o  -+        ┌──horizontal──┐
g  +zt-                    \ g o  /|        │              │
t  tztzt                   \     /          │ ┌─vertical──┐│
T  TzzTzzT                 \                │ │           ││
z  e                       \       don't draw          DRAW
a  |FFV++FFH|--FFV++FFH--  \ front side a: h v           V H
b  +FW|-FFI+FW|-FFI        \ right side b: i w           W I
c  ++|FFX|-FJ+FFX|-FJ|-    \ upper side c: j x           X J
H  HFFH[|--ffv++FFH|--ffv|ffv++ffh]FFH \   DRAW horizontal a
h  hffh[|--ffv++FFH|--ffv|ffv++ffh]ffh \  ¬draw horizontal a
V  VFFV[++ffh|--FFV++ffh|ffh|--ffv]FFV \   DRAW  vertical  a
v  vffv[++ffh|--FFV++ffh|ffh|--ffv]ffv \  ¬draw  vertical  a
I  IFFI[+fw|-FFI+fw|fw|-ffi]FFI    \         DRAW horizontal b
i  iffi[+fw|-FFI+fw|fw|-ffi]ffi    \ don't draw horizontal b
W  WFW[|-ffi+FW|-ffi|ffi+fw]FW     \         DRAW  vertical  b
w  wfw[|-ffi+FW|-ffi|ffi+fw]fw     \ don't draw  vertical  b
J  JFJ[+ffx|-FJ+ffx|ffx|-fj]FJ     \         DRAW horizontal c
j  jfj[+ffx|-FJ+ffx|ffx|-fj]fj     \ don't draw horizontal c
X  XFFX[|-fj+FFX|-fj|fj+ffx]FFX    \         DRAW  vertical  c
x  xffx[|-fj+FFX|-fj|fj+ffx]ffx    \ don't draw  vertical  c
97 360              \ angle = π * 97/360 = 48.5°
4                   \ recursion depth
up   down  right  left  near  go
/|\   |    -->    <--     /    -+
 |   \|/                 |/    /|
                         +-   /
```

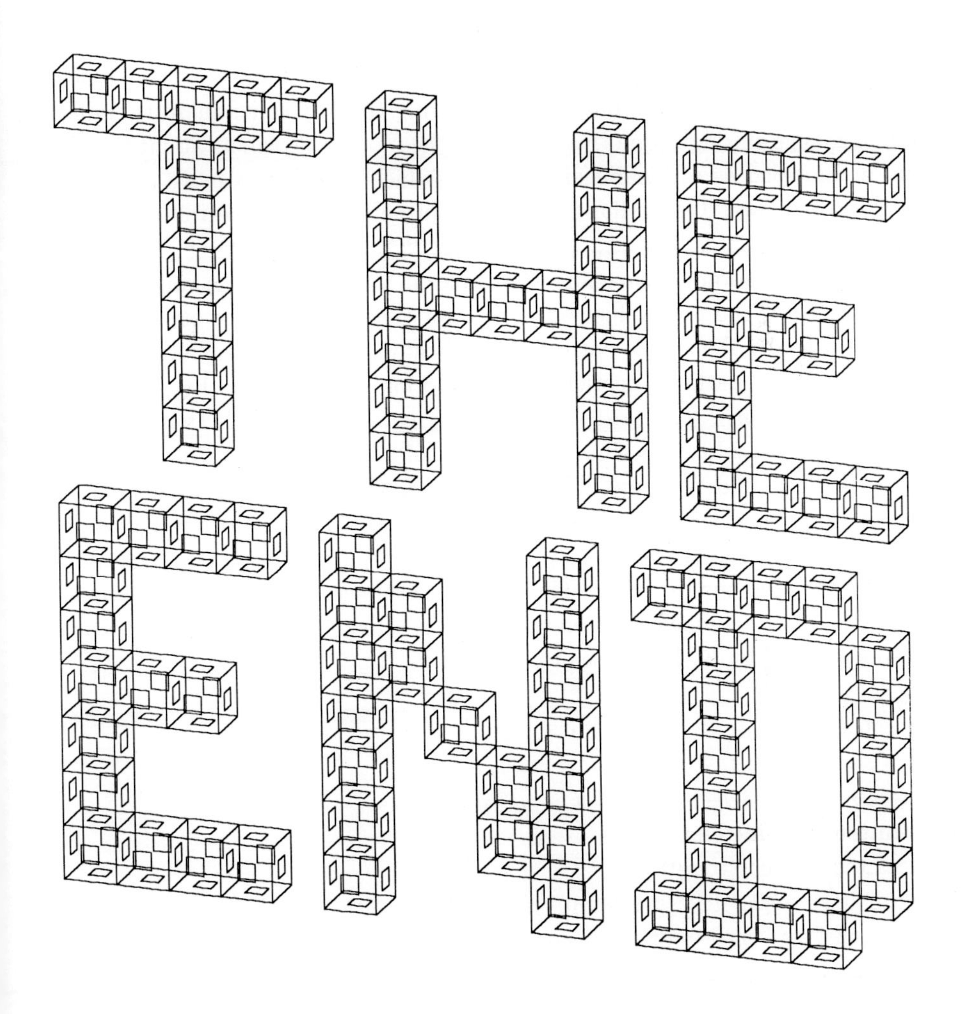
THE
END

Clifford A. Pickover

The Ikeda Attractor

A deep reservoir for striking images is the *dynamical system.* Dynamical systems are models containing rules describing the way some quantity undergoes a change through time. For example, the motion of planets about the sun can be modeled as a dynamical system in which the planets move according to Newton's laws. Generally, the pictures presented in this section track the behavior of mathematical expressions called *differential equations.* Think of a differential equation as a machine that takes in values for all the variables and then generates the new values at some later time. Just as one can track the path of a jet by the smoke path it leaves behind, computer graphics provides a way to follow paths of particles whose motion is determined by simple differential equations. The practical side of dynamical systems is that they can sometimes be used to describe the behavior of real-world things such as planetary motion, fluid flow, the diffusion of drugs, the behavior of inter-industry relationships, and the vibration of airplane wings. The pseudocode below describes how to produce the Ikeda pattern. Simply plot the position of variables j and k through the iteration. The variables scale, $x$off, and $y$off simply position and scale the image to fit on the graphics screen. The Ikeda attractor has been described by K. Ikeda (see references).

```
c1 = 0.4, c2 = 0.9, c3 = 6.0, rho = 1.0;
for (i = 0, x = 0.1, y = 0.1; i <= 3000; i++) (
  temp = c1 - c3 / (1.0 + x * x + y * y);
  sin_temp = sin(temp);
  cos_temp = cos(temp);
  xt = rho + c2 * (x * cos_temp - y * sin_temp);
  y = c2 * (x * sin_temp + y * cos_temp);
  x = xt;
  j = x * scale + xoff;
  k = y * scale + yoff;
)
```

References

1. K. Ikeda, "Multiple-valued stationary state and its instability of the transmitted light by a ring cavity system", *Opt. Commun.* **30** (1979) 257.
2. I. Stewart, "The nature of stability", *Speculations in Science and Tech.* **10**, 4 (1987) 310–324.
3. R. Abraham and C. Shaw, *Dynamics — The Geometry of Behavior, Part 3: Global Behavior* (Aerial Press 1985).*

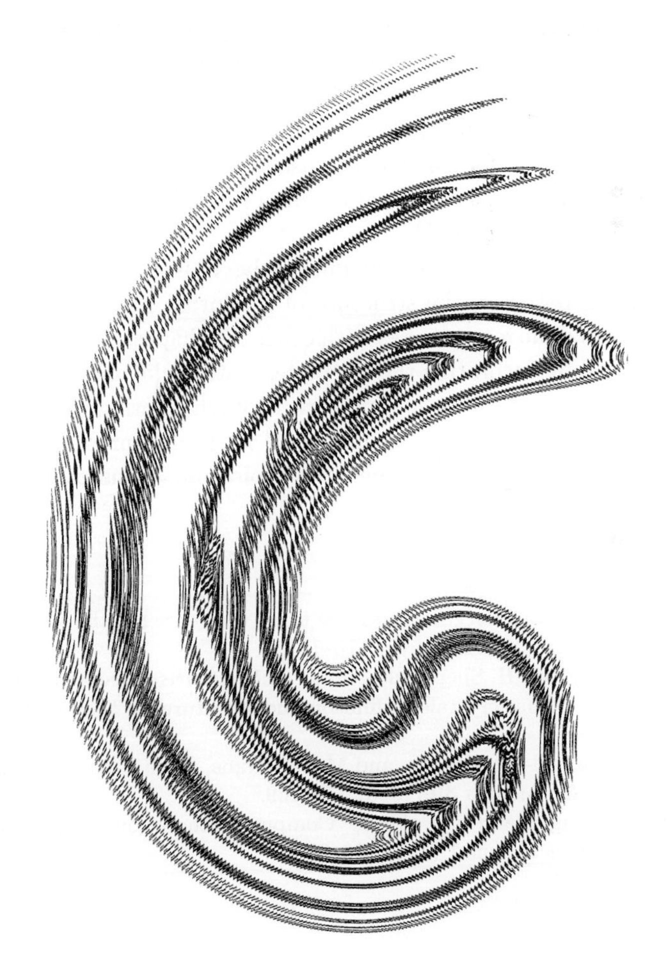

*Actually, the entire book collection of Aerial Press, including the Visual Math Series, is an educational wonderland.

J. Rangel-Mondragon and S. J. Abas

Star-NLP1-5m

Depicted here is a pattern obtained from a non-linear transformation applied to a Penrose tiling. Such tilings are generated using two different tile shapes — a "Kite" and a "Dart" and through imposing certain matching rules on their edges [1]. The properties of Penrose tilings are of wide-ranging interest. They give rise to non-periodic, self-similar patterns with fivefold symmetry which have a host of fascinating mathematical properties [2]. They offer a theoretical model for some newly discovered materials which have been called "Quasi-crystals" [3, 4]. They also offer a new structure on which to base aesthetically pleasing designs [5].

By decorating the tile pieces imaginatively, it is possible to generate a variety of designs based on the structure of a Penrose pattern in exactly the way that has been done for centuries with other tile shapes. The pattern shown here was obtained by decorating the tile pieces suitably [5] and applying a non-linear transformation to the whole pattern through altering the value of the golden ratio phi which arises in a number of ways in Penrose tiles [1]. This particular pattern is obtained through the choice phi=1 instead of the usual value phi=(1+sqrt(5))/2.

References

1. M. Gardner, "Extraordinary non-periodic tiling that enriches the theory of tiles", *Scientific American* (January 1977) 110–121.
2. B. Grunbaum and W. H. Shepherd, *Tilings and Patterns* (Freeman & Co., 1987).
3. P. J. Steinhardt, "Quasicrystals, definition and structure", *Physical Review* **34**, 2 (1986) 596–616.
4. S. J. Abas, J. Rangel-Mondragon, and M. W. Evans, "Quasi-crystals and penrose tiles", *J. Molecular Liquids* **39** (1988) 153–169.
5. J. Rangel-Mondragon and S. J. Abas, "Computer generation of penrose tilings", *Computer Graphics Forum* **7** (1988) 29–37.

Figure 1. A complex pattern with fivefold symmetry obtained through a non-linear transformation applied to a Penrose tiling.

Ilene Astrahan

Whirlpools

The pattern depicted here is an example of manipulation of mathematically derived material to produce a design. I used a commercial program ("Doug's Math Aquarium" — for the Amiga) to produce a fractal pattern. This software allows the user to change values in a basic equation. Fortunately, several equations are supplied with the software, which is helpful for innumerates like me. It takes a bit of experimentation to arrive at a result I like and can use. I can then save the file in IFF format and load it into my paint program. I can then manipulate it and use it in any manner. In this case, I printed a 32-color image in black and white. I used a region of the following equation:

```
Val: fa3(x,y,0)
FA: (a1*a1+a2*a2)>1?fa3(.5*(a2-frac(a1^2+a2^2)),-.5*a1+.5*a2,
a3+1):a3+a1
```

This is how it appears in this particular program.

Hardware: Commodore Amiga - Xerox 4020 inkjet printer.
Software: Deluxe Paint (Electronic Arts) Doug's Math Aquarium (Seven Seas Software).

Reference

1. B. B. Mandelbrot, *The Fractal Geometry of Nature* (W. H. Freeman and Co., 1977).

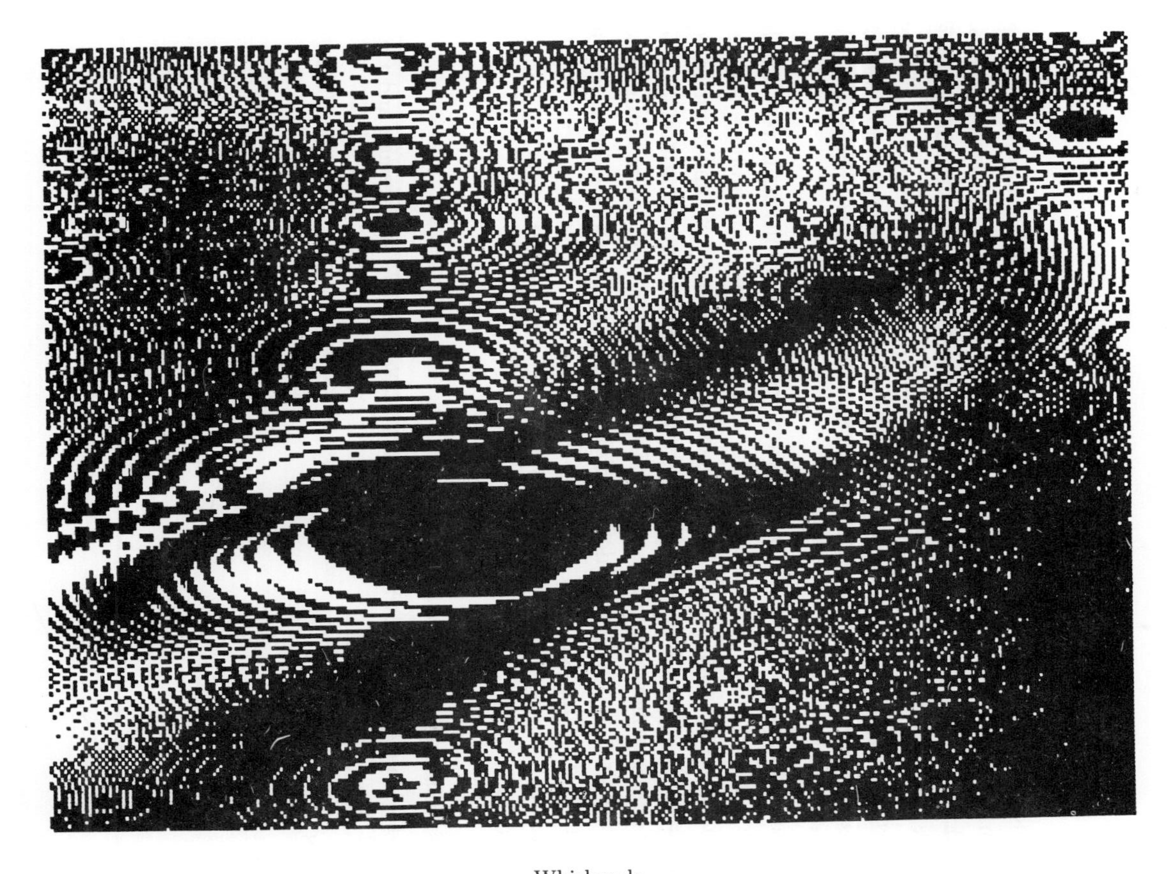

Whirlpools

Ilene Astrahan

Fern-Wolf (*Filicinae Lupus*)

The patterns depicted here are examples of fractal horticulture, showing life forms grown from user-defined seeds. A type of fractal genetics, perhaps. I used the public domain program called "Fractal Generator" written by Doug Houck. It allows you to draw simple line segments to create a shape. These segments continue to be replaced with small copies of the entire shape following the principles of self-similarity. The Koch Snowflake is the best known example of this type of shape.

The Principle of Sensitive Dependence on Initial Conditions is very much at work here. Slight alterations in the initial drawing can produce radical alterations in the final shape (not to mention the amount of time required to generate it). I have also done some drawing by hand on the printout. The initial shape is shown below — actual size — as traced from the monitor. The hardware used was a Commodore Amiga, a Xerox 4020 inkjet printer, and a Pigma .01 drawing pen.

Figure 1 shows the original printout. Figure 2 shows it after I did some additional artwork.

References

1. B. B. Mandelbrot, *The Fractal Geometry of Nature* (W. H. Freeman and Co., 1977) p. 43.
2. W. A. KcWorter and A. Tazelaar, "Creating fractals", *Byte Magazine* (August 1981) 125.

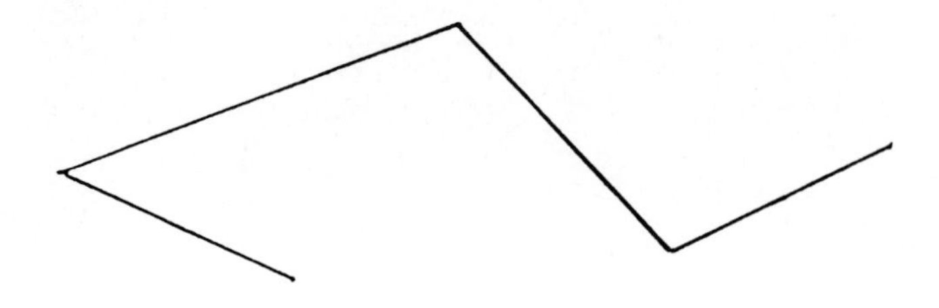

Figure 1.

Figure 2.

John Robert Hendricks

Magic Tessaract

Described herein is a pattern showing a projection onto a piece of paper of a four-dimensional magic hypercube of order three. It is merely one out of 58 possible ones of this size, each of which may be shown in 384 different aspects due to rotations and/or reflections. Larger magic hypercubes, both in size and dimension, have also been constructed.

You will notice in the figure that each horizontal row of three numbers, such as 1, 80, and 42, sums 123. The vertical columns, such as 1, 54, and 68, sum 123. Each oblique line of three numbers such as 1, 72, and 50, sums 123. A fourth linear direction shown by 1, 78, and 44 sums 123.

Pairs situated symmetrically opposite the center number 41 sum 82. Thus, 1, 41, and 81 sum 123; or 44, 41, and 38 sum 123, etc. The tessaract is bounded by cubes which are sketched. The corners of the cubes define the corners of the tessaract. "Opposite corners" define the four-dimensional diagonals.

The figure was first sketched in 1949. The pattern was eventually published in Canada in 1962, later in the United States of America. Creation of the figure dispelled the notion that such a pattern could not be made and it has taken many years for this representation to become generally accepted.

References

1. J. R. Hendricks, "The five- and six-dimensional magic hypercubes of Order 3", *Canadian Mathematical Bulletin* 5-2 (May 1962) 171–189.
2. J. R. Hendricks, "The magic tessaracts of Order 3 complete", *Journal of Recreational Mathematics* **22**, 1 (1990) 15–26.

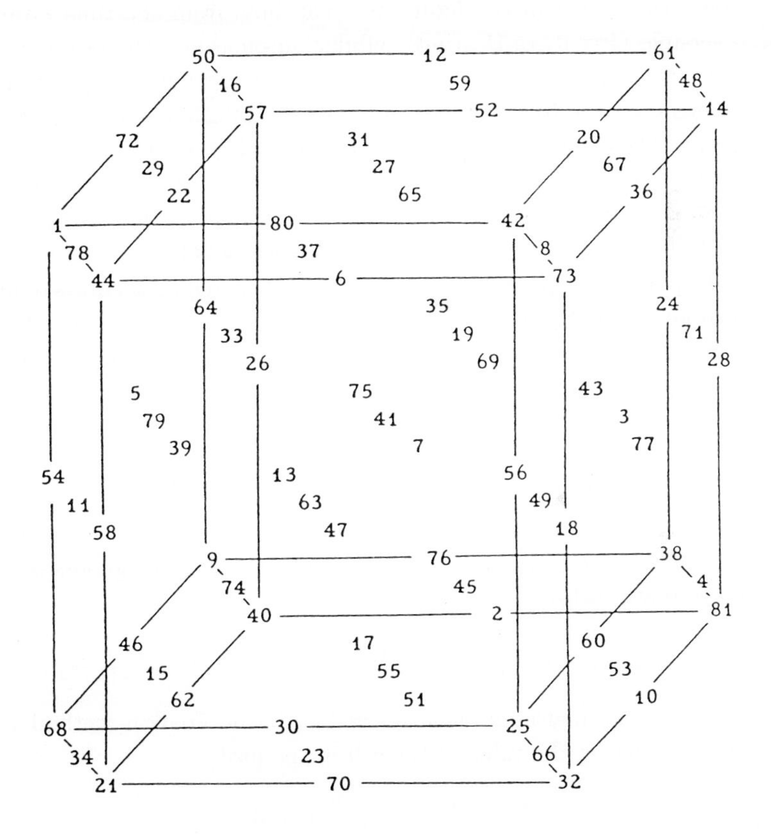

50 12 61
16 59 48
57 52 14
72 31 20
29 27 67
22 65 36
1 80 42
78 37 8
44 6 73
64 35 24
33 19 71
26 69 28
5 75 43
79 41 3
39 7 77
54 13 56
11 63 49
58 47 18
9 76 38
74 45 4
40 2 81
46 17 60
15 55 53
62 51 10
68 30 25
34 23 66
21 70 32

Pablo Tamayo and Hyman Hartman

The Reversible Greenberg-Hastings Cellular Automaton

The patterns shown in the figure are snapshots from the time evolution of the reversible Greenberg-Hastings cellular automaton. The original (non-reversible) Greenberg-Hastings cellular automaton was studied by Greenberg and co-workers [1] as a discrete model for reaction-diffusion in excitable media. Each cell can take one of three states: resting or quiescent (0), active (1), and refractory (2). A cell remains in the resting state until it is activated by an active neighbor. Once active, it will become refractory at the next time step independent of the neighborhood. Once refractory, it will become resting at the next time step independent of the neighborhood. The model attempts to model diffusion in an excitable medium composed of discrete chemical oscillators. It produces rotating spirals reminiscent of the Belousov-Zhabotinsky reaction [2]. In the two-dimensional von Neumann neighborhood, the rule can be described as:

$$a_{i,j}^{t+1} = F(t) = \begin{cases} a_{i,j}^t + 1 \bmod 3\,, & \text{if } a_{i,j}^t \neq 0\,; \\ \mathrm{A}_{\text{neig}}\,, & \text{if } a_{i,j}^t = 0\,; \end{cases} \tag{1}$$

where $a_{i,j}^t$ is the state of the cell at site (i,j) at time t, and A_{neig} is a Boolean activation function which is equal to 1 if at least one of the neighbors is active and equal to 0 otherwise,

$$\mathrm{A}_{\text{neig}} = [a_{i-1,j}^t = 1] \text{ or } [a_{i+1,j}^t = 1] \text{ or } [a_{i,j-1}^t = 1] \text{ or } [a_{i,j+1}^t = 1]\,. \tag{2}$$

This cellular automaton is made reversible by the Fredkin method which consists of subtracting the value of the cell in the past,

$$a_{i,j}^{t+1} = F(t) - a_{i,j}^{t-1} \bmod 3\,. \tag{3}$$

In this way the rule becomes reversible and displays a different behavior from the original non-reversible rule. It has a rich phenomenology which includes soliton-like structures and chemical turbulence [3]. Depending on the initial conditions it develops into three different regimes: (*i*) *regular regime* with short recurrence times, (*ii*) *chemical-turbulent regime*, and (*iii*) *disordered random regime*. The most interesting one is the chemical-turbulent regime, examples

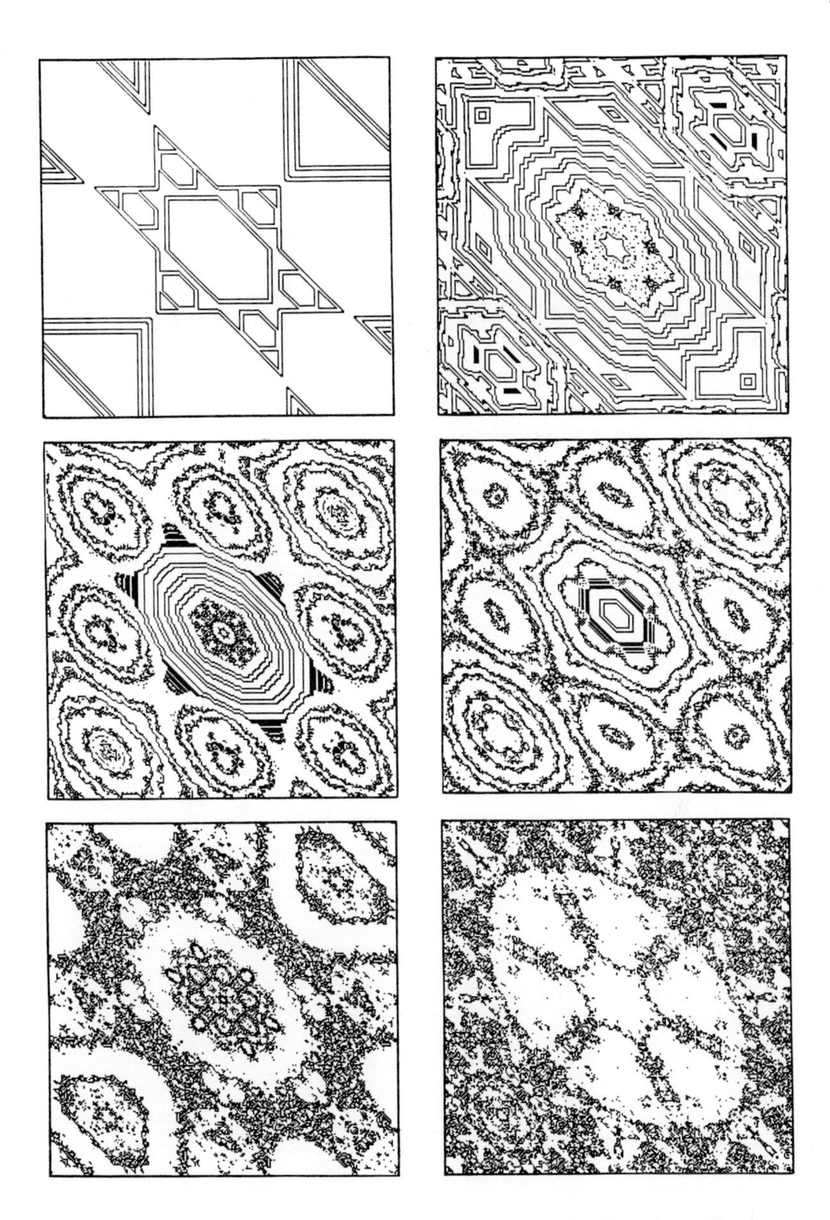

Figure 1. Patterns from the time evolution of the reversible Greenberg-Hastings model in the two-dimensional hexagonal lattice. Only active (black) and resting (white) cells are shown. The sequence shown in left-to-right, top-to-bottom order correspond to times: $t = 343$, $t = 500$, $t = 735$, $t = 1004$, $t = 2043$, $t = 15064$. The skewness is due to the fact that the hexagonal lattice was simulated using 6 neighbors in the square lattice.

of which are shown in the figure (for a hexagonal two-dimensional lattice). Only active (black) and resting (white) cells are shown. The initial condition consisted of one single active cell in a resting background. These patterns were generated using a CAM-6 Cellular Automata simulator [4]. For more information on the reversible Greenberg-Hastings model, see reference 5.

References

1. J. M. Greenberg and S. P. Hastings, *Siam J. Appl. Math.* **34** (1978) 515; J. M. Greenberg, B. D. Hassard, and S. P. Hastings, *Bull. of the Am. Math. Soc.* **84** (1978) 1296. See also: A. Winfree, *Physica* **17D** (1985) 109.
2. H. Meinhardt, *Models of Biological Pattern Formation* (Academic Press, 1982).
3. Y. Kuramoto, *Chemical Oscillations, Waves and Turbulence* (Springer-Verlag, 1984).
4. T. Toffoli and N. Margolus, *Cellular Automata Machines* (MIT Press, 1987).
5. P. Tamayo and H. Hartman, "Reversible cellular automata and chemical turbulence", *Physica D* **45** (1990) 293.

Ilene Astrahan

Cybernetic Rapids

The pattern depicted here is an example of manipulation of mathematically derived material to produce a design. I used a commercial program ("Doug's Math Aquarium" — for the Amiga) to produce a fractal pattern. This software allows the user to change values in a basic equation. Fortunately several equations are supplied with the software, which is helpful for innumerates like me. It takes a bit of experimentation to arrive at a result I like and can use. I can then save the file in IFF format and load it into my paint program. I can then manipulate it and use it in any manner. In this case, I printed a 32-color image in black and white. I used a region of the following equation:

```
Val: fa3(x,y,0)*10
FA: a3<1000?fb4(a1*a1,a2*a2,2*a1*a2,a3+1)*(a1-a2)+1 : 0
FB: a1+a2<4?fa3(a1-a2+x,a3+y,a4):0
```

This is how it appears in this particular program.

Hardware: Commodore Amiga - Xerox 4020 inkjet printer.
Software: Deluxe Paint (Electronic Arts) Doug's Math Aquarium (Seven Seas Software).

Reference

1. B. B. Mandelbrot, *The Fractal Geometry of Nature* (W. H. Freeman and Co., 1977).

Cybernetic Rapids

C. William Henderson

Pinwheels

Described here is a pattern produced on an AMIGA 1000, a 512K computer, and output to a Xerox 4020 color ink jet printer. The software is Doug's Math Aquarium by Seven Seas Software, a program that permits the artist to create images based solely on mathematical equations. The AMIGA graphics dump can simulate in print nearly all of the computer's 4,096 color range, though only 32 can be displayed on screen at one time. However, in the case of the image shown here, because of the black and white limitations, the graphics dump was set to reproduce in a range of 16 levels of gray.

Since the images are generated purely by mathematics, appearance is determined by the artist's selection and manipulation of various formulas. No paint system nor image touch-up is involved. From the formulas the software calculates a three-dimensional surface by equating Z to the computed combinations of X and Y, then projects the three-dimensional surface onto the screen as a two-dimensional contour map (though 3D displays are possible). The colors represent the contour intervals. To the degree that the contours exceed 32 (in this case 16), the colors repeat themselves.

As an example, the simplest formula might be just X. Thus, for every combination of X and Y, Z would be equal to X only. For the entire height of the screen, Z would be at a distance from the screen of X. This would result in a computed vertical plane going into the screen from the right. Its contours would be vertical lines. However, if the colors grade from dark to light to dark, the illusion on screen would be a series of vertical cylinders. Thus, judicious selection of color and sequence can produce illusions of 3D bas-reliefs, the actual three-dimensional surface having little resemblance to the 2D projection.

The $X - Y$ formula used to generate the Pinwheels image here involves simple trigonometric functions. The equation that resulted in this unusual image is simply ... $\sin(X + \cos Y)/\cos(Y + \sin X)$.

Pinwheels

C. William Henderson

Circlefest

Described here is a pattern produced on an AMIGA 1000, a 512K computer, and output to a Xerox 4020 color ink jet printer. The software is Doug's Math Aquarium by Seven Seas Software, a program that permits the artist to create images based solely on mathematical equations. The AMIGA graphics dump can simulate in print nearly all of the computer's 4,096 color range, though only 32 can be displayed on screen at one time. However, in the case of the image shown here, because of the black and white limitations, the graphics dump was set to reproduce in a range of 16 levels of gray.

Since the images are generated purely by mathematics, appearance is determined by the artist's selection and manipulation of various formulas. No paint system nor image touch-up is involved. From the formulas the software calculates a three-dimensional surface by equating Z to the computed combinations of X and Y, then projects the three-dimensional surface onto the screen as a two-dimensional contour map (though 3D displays are possible). The colors represent the contour intervals. To the degree that the contours exceed 32 (in this case 16), the colors repeat themselves.

As an example, the simplest formula might be just X. Thus, for every combination of X and Y, Z would be equal to X only. For the entire height of the screen, Z would be at a distance from the screen of X. This would result in a computed vertical plane going into the screen from the right. Its contours would be vertical lines. However, if the colors grade from dark to light to dark, the illusion on screen would be a series of vertical cylinders. Thus, judicious selection of color and sequence can produce illusions of 3D bas-reliefs, the actual three-dimensional surface having little resemblance to the 2D projection.

The following formula for CIRCLEFEST involves an IF-THEN-ELSE statement and MODULO (an integer remainder, i.e., 11 mod 3 $=$ 2). Some of the smaller circles are a moire effect, a special feature of the program.

If $35 \times (X\hat{\ }2 + Y\hat{\ }2)\hat{\ }.1 \bmod 4 = 1$ then $(FA \bmod 8) + 16$ else $(FB \bmod 16) + 8$, where $FA = Y - X$ and $FB = X\hat{\ }2 + Y\hat{\ }2$.

Circlefest

C. William Henderson

XOR Size

Described here is a pattern produced on an AMIGA 1000, a 512K computer, and output to a Xerox 4020 color ink jet printer. The software is Doug's Math Aquarium by Seven Seas Software, a program that permits the artist to create images based solely on mathematical equations. The AMIGA graphics dump can simulate in print nearly all of the computer's 4,096 color range, though only 32 can be displayed on screen at one time. However, in the case of the image shown here, because of the black and white limitations, the graphics dump was set to reproduce in a range of 16 levels of gray.

Since the images are generated purely by mathematics, appearance is determined by the artist's selection and manipulation of various formulas. No paint system nor image touch-up is involved. From the formulas the software calculates a three-dimensional surface by equating Z to the computed combinations of X and Y, then projects the three-dimensional surface onto the screen as a two-dimensional contour map (though 3D displays are possible). The colors represent the contour intervals. To the degree that the contours exceed 32 (in this case 16), the colors repeat themselves.

As an example, the simplest formula might be just X. Thus, for every combination of X and Y, Z would be equal to X only. For the entire height of the screen, Z would be at a distance from the screen of X. This would result in a computed vertical plane going into the screen from the right. Its contours would be vertical lines. However, if the colors grade from dark to light to dark, the illusion on screen would be a series of vertical cylinders. Thus, judicious selection of color and sequence can produce illusions of 3D bas-reliefs, the actual three-dimensional surface having little resemblance to the 2D projection.

The formula for this image involves trigonometry, XOR (bitwise exclusive OR function of integer values) and an IF-THEN-ELSE statement.

Thus, if $X > Y$ then $FA \times FB$, else $FC \times FB$.

Where $FA = (2X \text{ XOR } 1) + (Y/2 \text{ XOR } 1) \times (Y \text{ XOR } X)$

$FB = \sin(X + Y) + \sin(X - Y)$

$FC = (X/2 \text{ XOR } 5) + (2Y \text{ XOR } 5) \times (Y \text{ XOR } X)$

XOR Size

C. William Henderson

Trig Gem

Described here is a pattern produced on an AMIGA 1000, a 512K computer, and output to a Xerox 4020 color ink jet printer. The software is Doug's Math Aquarium by Seven Seas Software, a program that permits the artist to create images based solely on mathematical equations. The AMIGA graphics dump can simulate in print nearly all of the computer's 4,096 color range, though only 32 can be displayed on screen at one time. However, in the case of the image shown here, because of the black and white limitations, the graphics dump was set to reproduce in a range of 16 levels of gray.

Since the images are generated purely by mathematics, appearance is determined by the artist's selection and manipulation of various formulas. No paint system nor image touch-up is involved. From the formulas the software calculates a three-dimensional surface by equating Z to the computed combinations of X and Y, then projects the three-dimensional surface onto the screen as a two-dimensional contour map (though 3D displays are possible). The colors represent the contour intervals. To the degree that the contours exceed 32 (in this case 16), the colors repeat themselves.

As an example, the simplest formula might be just X. Thus, for every combination of X and Y, Z would be equal to X only. For the entire height of the screen, Z would be at a distance from the screen of X. This would result in a computed vertical plane going into the screen from the right. Its contours would be vertical lines. However, if the colors grade from dark to light to dark, the illusion on screen would be a series of vertical cylinders. Thus, judicious selection of color and sequence can produce illusions of 3D bas-reliefs, the actual three-dimensional surface having little resemblance to the 2D projection.

The $X - Y$ formula used to generate the TRIG GEM image here involves simple trigonometric functions. The equation that resulted in this exotic image is simply: $[\sin(X+Y)/\sin(X-Y)] + [\sin(X-Y) \times (X+Y)]$. The image field is offset from the intersection of the X and Y axes.

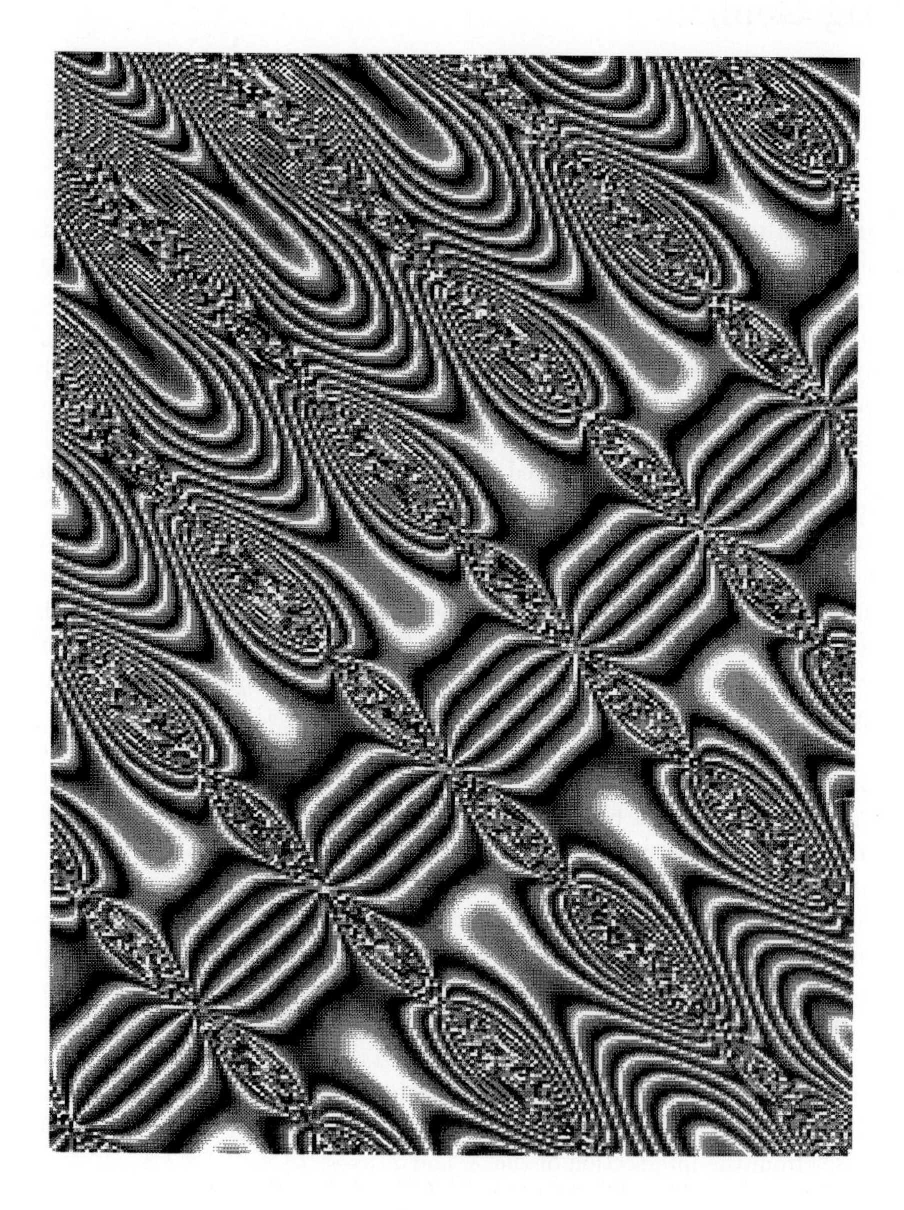

Trig Gem

Gary Ricard

Locked Links

Described here is a pattern generated by a program which draws "locked links". The links are always regular polygons and they are "locked" together by weaving their edges over and under each other. This is best illustrated in Fig. 1. Input to the program which generates the locked links includes the total number of sides for the polygon in the center of each figure. Even numbered inputs provide figures with two separate intertwined links, each with half the number of sides originally requested. Odd numbered inputs generate a single link with the input number of sides that intertwines with itself, basically forming a knot. Figure 1 contains examples for initial inputs of 3, 4, 5, 6, and 8 sides. Other inputs to the program include the length of a side of the polygon in the center of the figure, thickness of a link, center of the figure, and angle at which to begin drawing. The program is an intense exercise in algebra and trigonometry but is actually quite compact when completed. (My version is about 100 lines of code.)

Figure 2 is a tiling using a 4-sided locked link. Each locked link is placed at the center of a gothic cross. Thus, the base geometry for the tiling is given below:

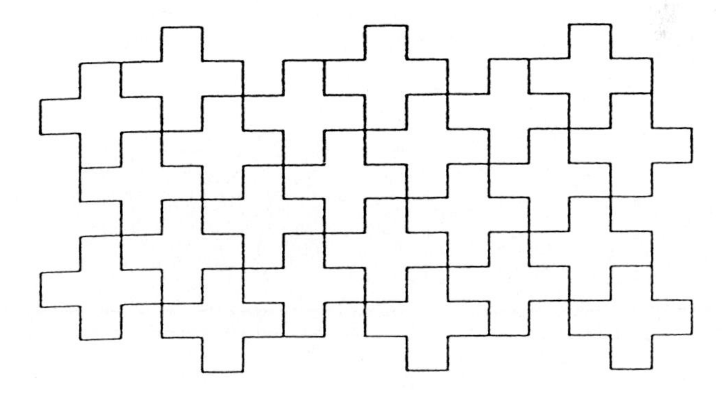

What makes Fig. 2 interesting is that it is a minor optical illusion. The figures along the diagonals in the pattern seem to bend and curve as one attempts to follow a diagonal with the eye. If you stare at a particular region of the pattern, the diagonals on the periphery of your vision seem to curve to the boundary of the figure. The effect is certainly caused by the positioning of the locked links relative to one another. The mind wants to force the center of the locked links to be on a 45-degree line even if the straight diagonal must be bent to do it!

Figure 1.

Figure 2.

Gary Ricard

Exponential Tunnel

Described here is a pattern generated by incrementing variable "s" by 1 from 0 to 750 in the following 2 pairs of parametric equations:

1) $x = \exp(-.012s)\cos(1.24s)$
   $y = \exp(-.012s)\sin(1.24s)$

2) $x = \exp(-.012s)\cos(1.24s + .5236)$
   $y = \exp(-.012s)\sin(-1.24s - .5236)$

The first pair of equations generates a design with 5 arms spiraling clockwise from the center. The second pair generates the same design but with the 5 arms spiraling counterclockwise from the center. (If you are examining the color figure these are the blue and red spirals respectively.) The result, especially in color, looks like an exponential tunnel with spiral arms descending into infinity. Alternatively, some see the result as a 5-pelated flower. This is actually the more common interpretation when both spirals are the same color. Another interesting point is that the spirals are generated using only straight lines.

The intertwined border surrounding the spirals is generated using the "locked links" program as described in the article by that name.

A program which will generate the spirals on the VGA screen using Turbo Pascal follows. Unfortunately the screen version of the design does not look nearly as impressive as the plotter version.

```
PROGRAM scrspiral;
  USES graph;
  VAR m, b, phase: REAL;
      grdriver, grmode: INTEGER;

  FUNCTION x(s: INTEGER; phase: REAL): REAL;
    BEGIN
      x:=exp(-0.012 * s) * cos(1.24 * s+phase)
    END;
```

```
FUNCTION y(s, sign: INTEGER; phase: REAL): REAL;
   BEGIN
      y:=exp(-0.012 * s) * sin(sign * (1.24 * s+phase))
   END;

FUNCTION scale(z:REAL): INTEGER;
   BEGIN
      scale:=round(z * m+b)
   END;

PROCEDURE spiral(phase: REAL; sign: INTEGER);
   VAR s: INTEGER;
   BEGIN
      s:=0;
      moveto(scale(x(s, phase)), scale(y(s, sign, phase)));
      WHILE s<=750 DO
          BEGIN
             s:=s+1;
             lineto(scale(x(s, phase)), scale(y(s, sign, phase)))
          END
   END;

BEGIN
   m:=239.5;
   b:=m;
   initgraph(grdriver, grmode, 'C:\PASCAL\GRAPHICS');
   setwritemode(xorput);
   setbkcolor(lightgray);
   setcolor(blue);
   spiral(0.0,1);
   setcolor(red);
   spiral(0.5235987756,-1);
   readln;
   closegraph
END.
```

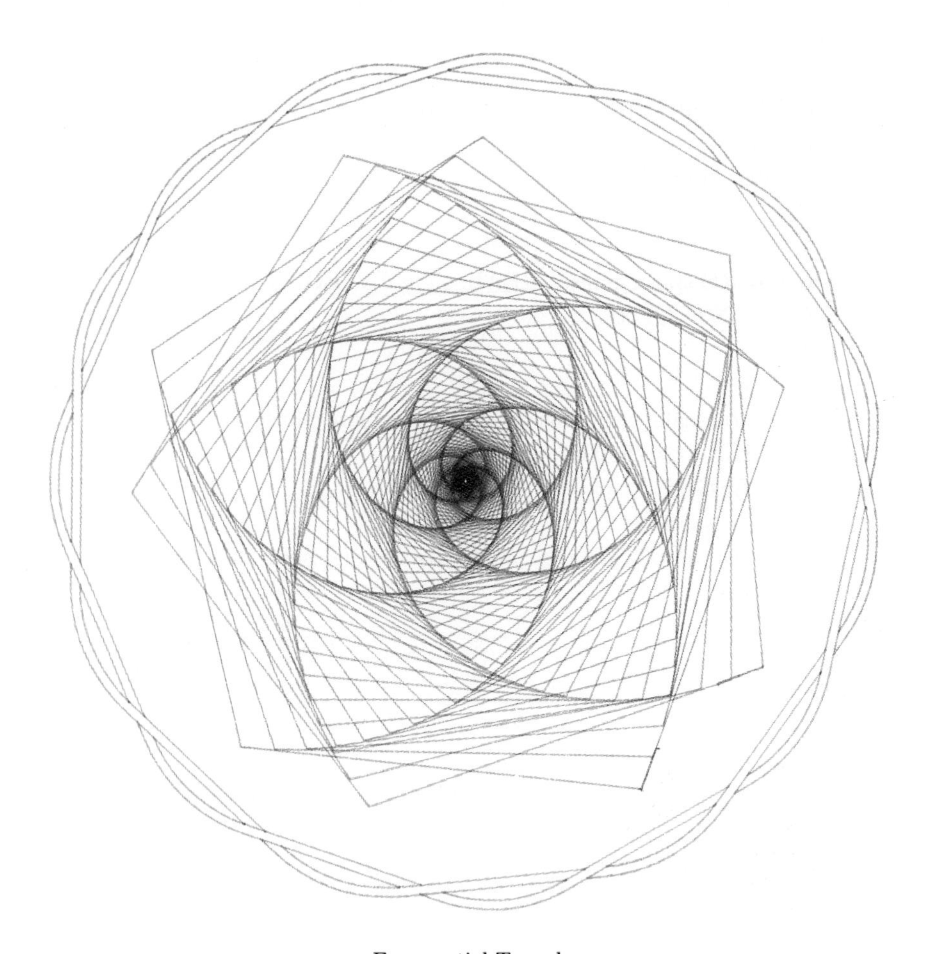

Exponential Tunnel

Hiroshi Okumura

A Generalization of the Regular Tiling (4^4)

Described here is a generalized pattern of (4^4), the regular tiling by congruent squares. Figure 1 shows a configuration consisting of two points X, Y and lines x_i and y_i (i and j are integers) passing through X and Y respectively, such that the quadrangle determined by x_i, x_{i+1}, y_j, y_{j+1} has an incircle for any i and j [1]. Figure 2 shows a special case where both X and Y are points of infinity, and it also shows that Fig. 1 can be regarded as a generalized pattern of regular tiling (4^4).

Our generalized configuration has the following properties:

(1) For any two circles, one of the centers of similitude (two intersections of internal common tangents and external common tangents) lies on the line XY.

(2) If we denote the curvature (reciprocal of the radius) of the circle touching x_i, x_{i+1}, y_j, y_{j+1} by $[i,\, j]$ then

$$[i,\, j] + [i+m,\, j+n] = [i+m,\, j] + [i,\, j+n]\,,$$
$$[i,\, j][i+m+n,\, j-m+n] = [i+m,\, j-m][i+n,\, ,j+n]\,,$$

provided we make the convention that we regard the circles below the line XY as having negative radii.

(3) For any integers i, j, k, the quadrangle determined by x_i, x_{i+k}, y_j, y_{j+k} has an incircle (Rigby [2], Theorem 4.5).

References

1. H. Okumura, "Configurations arising from the three-circle theorem", *Mathematics Magazine* **63**, 2 (1990) 116–121.
2. J. F. Rigby, "Cycles and tangent rays, circles and tangent lines", *Mathematics Magazine* **64**, 3 (1991) 155–167.

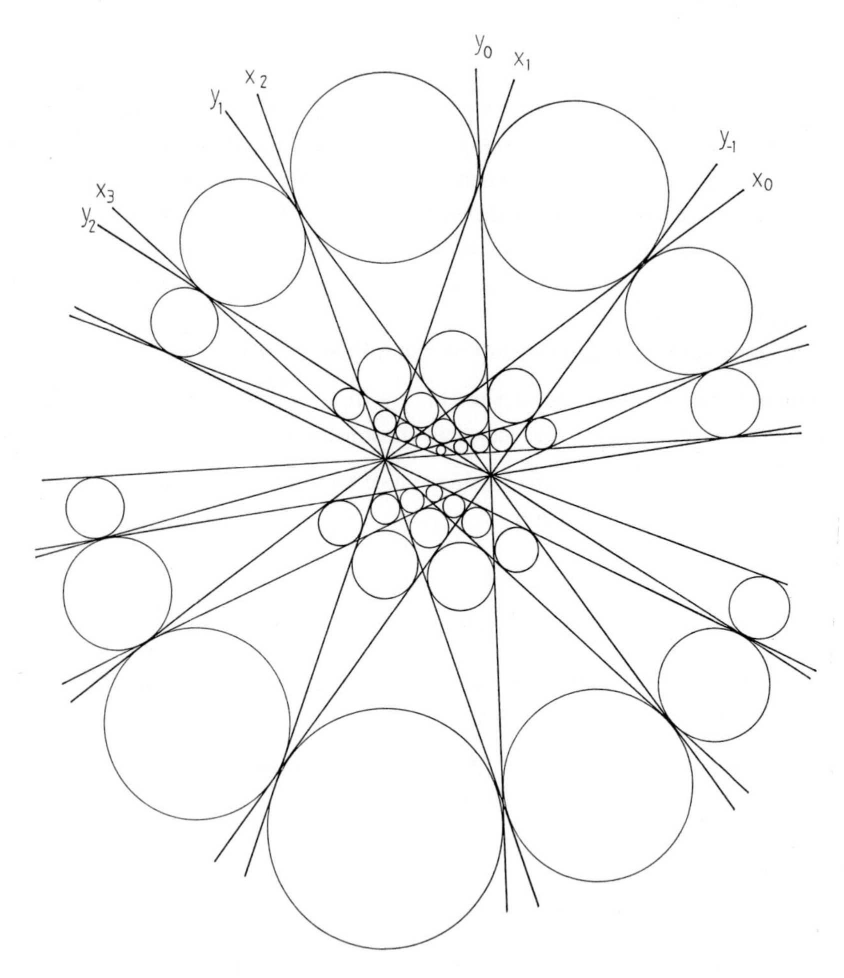

Parallel lines intersect at a point of infinity.

Figure 1.

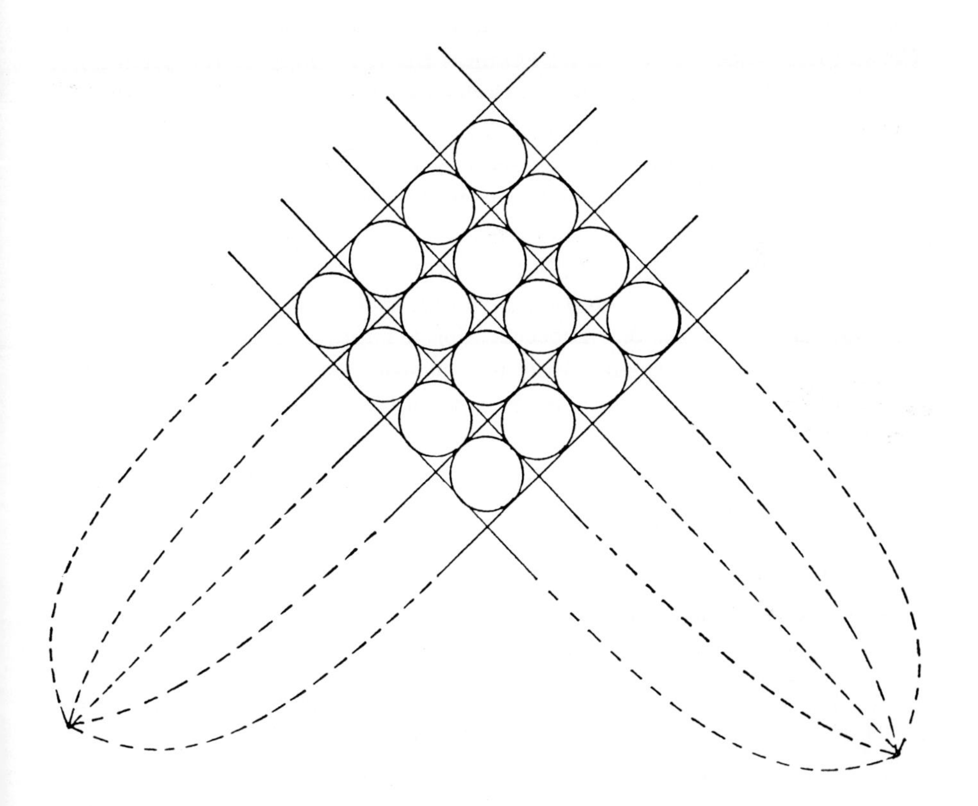

Figure 2.

Hiroshi Okumura

A Pattern by Fujita Configurations

Described here is a pattern by Fujita Configurations. Let ABCD be a parallelogram; E, a point on the segment CD; F and G, points on the segments CE and DA, respectively, and let FG meet BE at H. Suppose the quadrangles ABHG, BCFH, and DGHE have incircles (see Fig. 1). This figure seems to have been first considered by one of the Wasan mathematicians S. Fujita for the case where ABCD is a square (see Fig. 2) [1]. (Wasan refers to Japanese mathematics developed independently of Western science between the 17th and 19th centuries.) And we call this figure a Fujita configuration.

The four incircles in the Fujita configuration (including the incircle of the triangle FEH) have an interesting property, that is, the sums of the opposite radii are equal [2, 3]. Especially for a Fujita configuration, in which its parallelogram is a square, FEH is a right-angled triangle with the sides EF:FH:HE=5:4:3 and the ratio of the radii of the four incircles is 1:2:3:4 (see Fig. 3). Using this figure, we can construct a pattern in the plane (see Fig. 4). In this pattern, we can superpose similar ones. Two similar patterns which are double and four times the size are drawn on the figure.

References

1. S. Fujita, *Seiyô Sampô* **3** (1781).
2. H. Okumura, "Four circles in a parallelogram", *J. Recreational Mathematics* **19**, 3 (1987) 224–226.
3. H. Okumura, "Fujita configurations", *J. Recreational Mathematics* **21**, 1 (1989) 29–34.

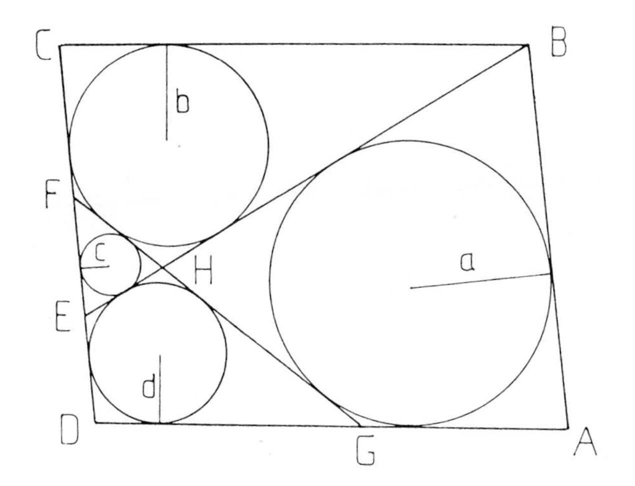

Figure 1. $a + c = b + d$.

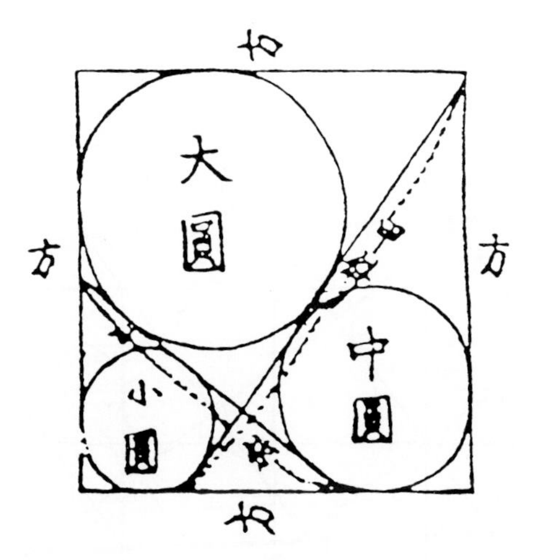

Figure 2.

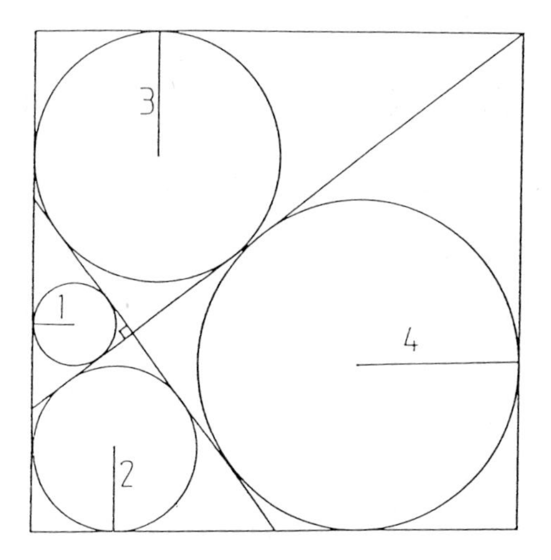

Figure 3.

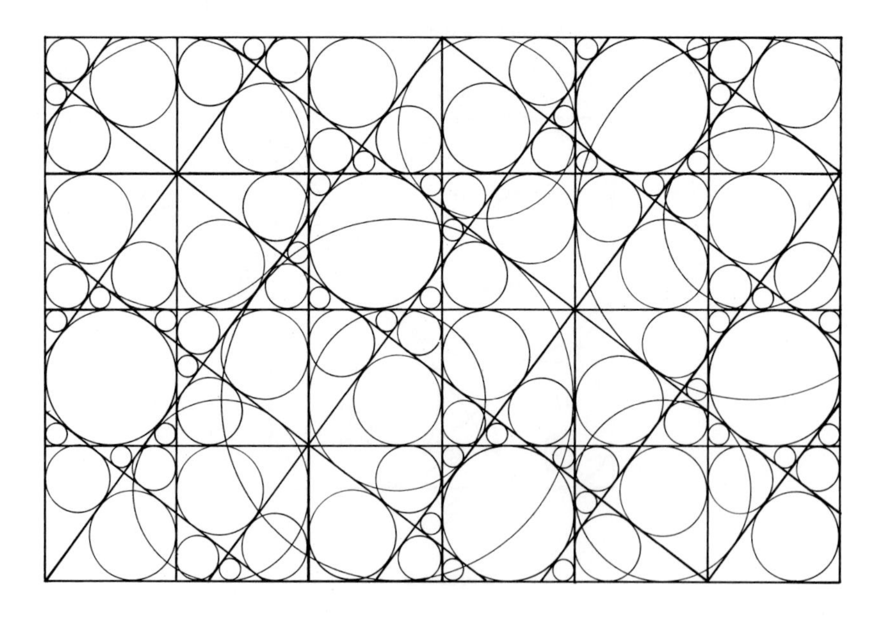

Figure 4.

Kenneth J. Hooper

Star Trails

Described here is a pattern showing one example of the structures contained within the Mandelbrot Set [1]. This area is sometimes ignored by traditional computer graphics methods, which produce images from just outside the Mandelbrot Set [2, 3].

Computer-generated images of the Mandelbrot Set are created by repeatedly solving the following two equations:

$$x = x^2 - y^2 + a \quad \text{and} \quad y = 2xy + b\,.$$

Variables a and b typically range in value from -2 to 2. Initially, x and y are set to zero but subsequently take on the values from the previous calculation. Each x-y pair defines the position of a point on a complex plane and every repetition of the calculations yields a new position. If a starting location inside the Mandelbrot Set is chosen, the x-y points remain bounded within a small region, sometimes tracing a path that repeats itself. Figures 1 and 2 were generated by counting the number of repetitions it took to trace a path before it repeats.

An algorithm for counting the number of repetitions makes use of the ratio y/x. Each ratio is then stored for later comparison with the first y/x value to determine when the sequence repeats. The following pseudocode for an inner loop details this method (the outer loops supply values for a, b, and pixel column and row):

```
x <- 0, y < - 0
for i <- 0 to 15
  xx <- x * x - y * y + a
  yy <- 2 * x * y + b
  tan(i) <- yy / (xx + .000001) Note: Prevents divide by zero.
  x <- xx
  y <- yy
endloop
for j <- 1 to 15
  error <- tan(0) - tan(j)
```

```
  if( |error| < .05) then          Note: 1. |error| = absolute value
    setpixel(column, row)                2. The loop counter can be
    exit loop                               used to define pixel
endloop                                  shade of color.
```

Figure 1 shows a starting location near $a = -0.42, b = 0.63$, which is near the boundary of the Mandelbrot Set. Self similar patterns are apparent as the stars seem to trail away in the distance. Actually, the patterns cross over to outside the Mandelbrot Set but the ratio y/x still repeats. The difference is that the points are no longer bounded. Figure 2 is located near $a = 0.44, b = 0.34$, and shows similar patterns crossing the boundary of the Mandelbrot Set.

References

1. C. A. Pickover, "Inside the Mandelbrot Set", *Algorithm* **1**, 1 (1989) 9–12.
2. B. B. Mandelbrot, *The Fractal Geometry of Nature* (W. H. Freeman, 1983).
3. A. K. Dewdney, "Computer recreations", *Scientific American* (September, 1986).

Figure 1. Star patterns within the Mandelbrot Set seem pulled toward the boundary.

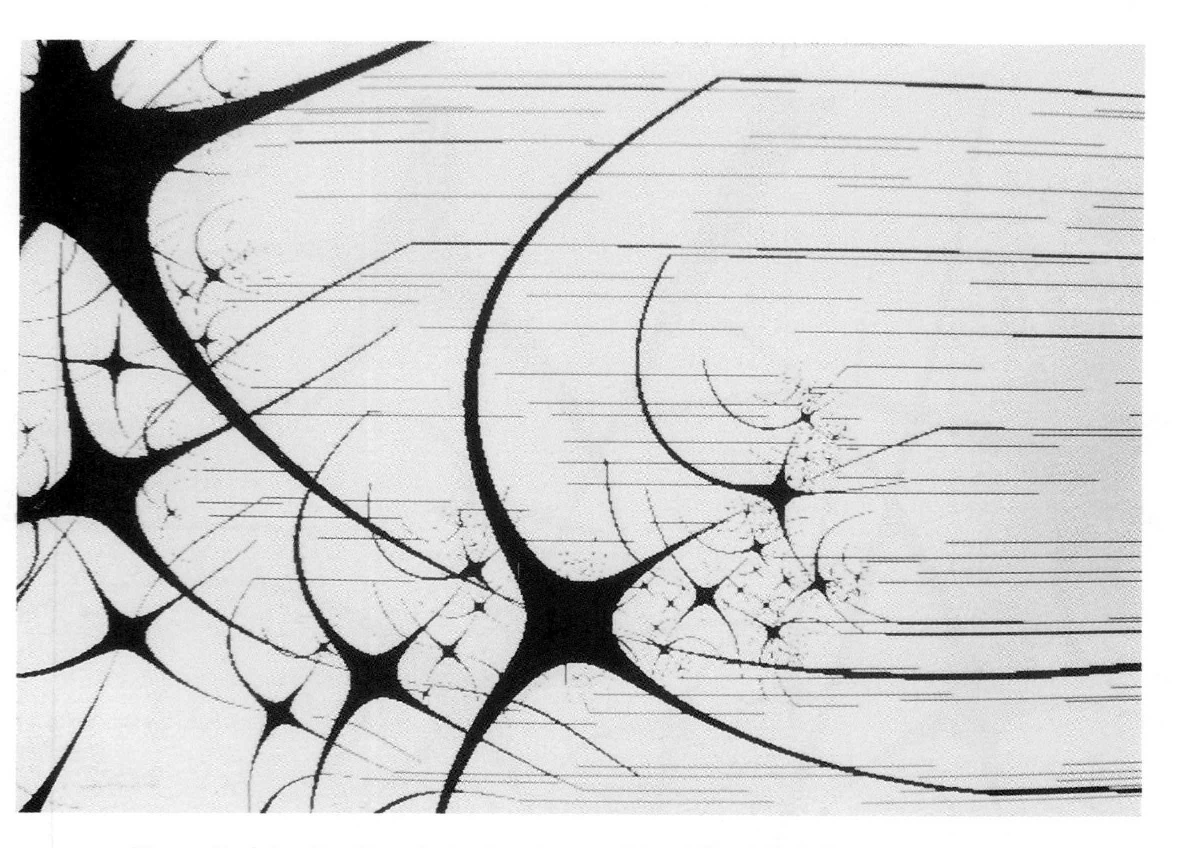

Figure 2. A familiar “flame” structure from traditional Mandelbrot Set images is instead suggested from the star patterns.

David Scruton

An Iteration Map

Shown here is an image derived from complex dynamics. In particular, the figure represents a complex variable recursion produced by iterating $z = z^3 + \mu$ forty times. μ is a complex constant. Diverging points whose magnitude is between 2 and 3 are plotted. The picture was computed on a micro-VAX and plotted on a Talaris laser printer.

Mieczyslaw Szyszkowicz

A Self-Similar Structure Generated by a Broken-Linear Function

Consider the broken-linear transformation with the parameter

$$f(x, r) = r(1 - 2|x - 0.5|), \quad 0 < r \leq 1,$$

which maps

$$f(x, r) : [0, 1] \longrightarrow [0, 1].$$

Figure 1 shows the behavior of the sequences $\{x_n\}$ and $\{y_n\}$ produced by this transformation used iteratively with $r = 1$

$$x_{n+1} = 1 - 2|x_n - 0.5|,$$
$$y_{n+1} = 1 - 2|y_n - 0.5|.$$

The iterative process is realized with the stopping criterion $(x_{n+1} - x_n)^2 + (y_{n+1} - y_n)^2 < \tau$, where τ is the assumed threshold. Starting with the initial point $(x_0, y_0) \in [0, 1] \times [0, 1]$, the sequence of the points (x_n, y_n) is generated. If the stopping criterion is satisfied, then the iterative process is stopped and the number of executed iterations modulo 2 is mapped on the point (x_0, y_0). Figure 1 represents a black (0) and white (1) self-similar pattern constructed by the above algorithm.

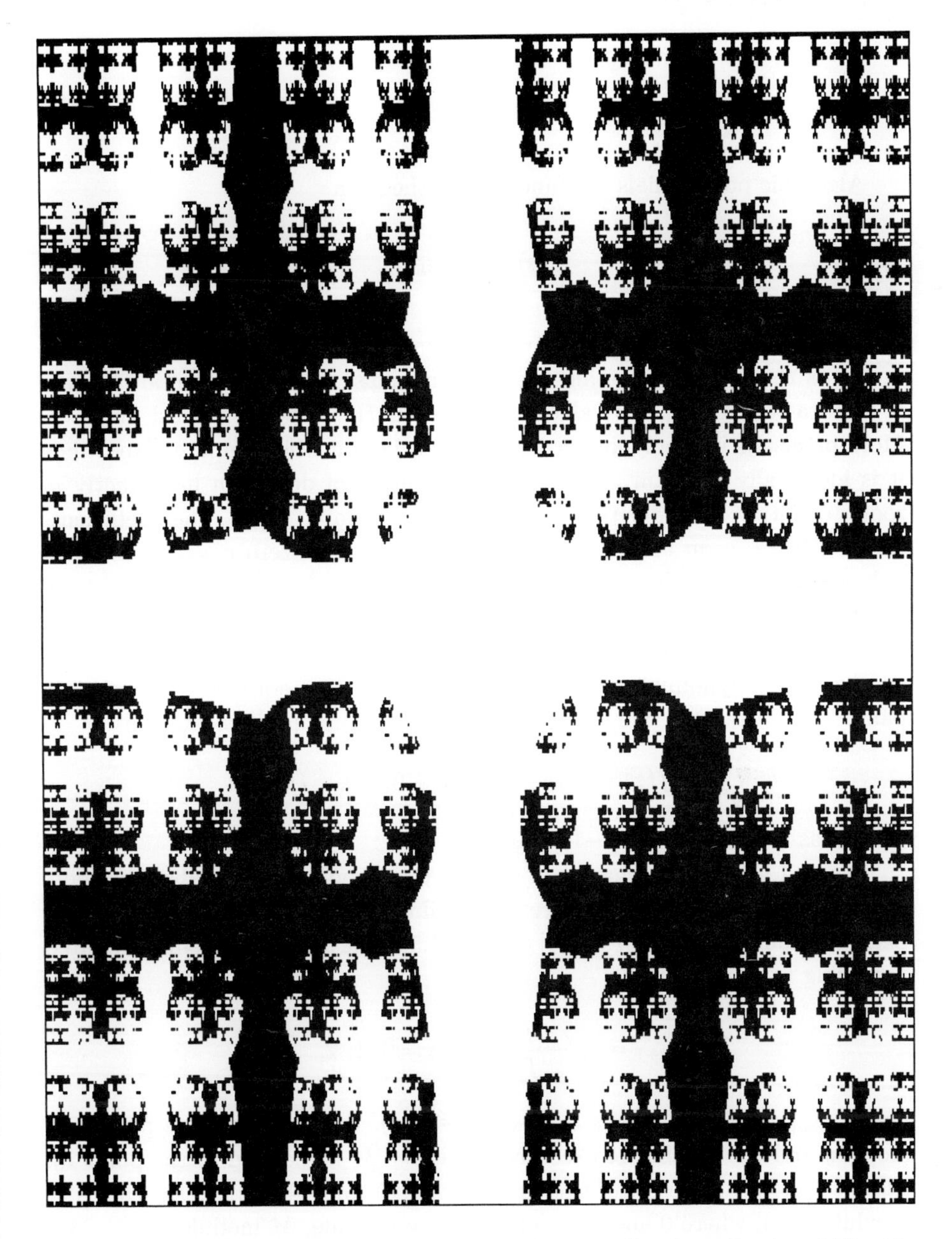

Figure 1. Self-similar structure generated by the function $f(x, r) = r(1 - |x - 0.5|)$ with $r = 1$.

Mieczysław Szyszkowicz

Roots of Algebraic Polynomials

Algebraic polynomials are functions of the form

$$f(z) = \sum_{k=0}^{m} a_k z^k ,$$

where a_k, $k = 0, 1, \ldots, m$ are fixed real or complex numbers. The number $m \geq 1$ is the degree of the algebraic polynomials. The fundamental theorem of algebra states that every algebraic equation $f(z) = 0$ has at least one root. More precisely, if a polynomial vanishes nowhere in the complex plane, then it is identically constant. An algebraic equation of degree m has exactly m roots, counting multiplicities. If the real polynomial f (i.e., the coefficients a_k, $k = 0, 1, \ldots, m$ are real numbers) has a complex zero $r = \alpha + i\beta$, then it has also the complex root $r = \alpha - i\beta$. Here, r denotes the conjugate of r, α and β are real numbers, and $i = \sqrt{-1}$. In other words, the complex zeros of a real algebraic polynomial always appear in conjugate pairs. The zeros of an algebraic polynomial with complex coefficients may not occur in conjugate pairs. As an example, the following algebraic polynomial is considered

$$f(z) = z(z - 0.25i)(z + 0.5)(z + i) ,$$

which has the zeros 0, $i/4$, $-1/2$, and $-i$.

Figure 1 shows the behavior of Newton's method used to localize the roots of the equation $f(z) = 0$, where f is defined above. The initial values z_0 belong to the quadrant $[-1.5, 1.5] \times [-1.5, 1.5]$. The sequence $\{z_n\}$ is generated according to Newton's rule

$$z_{n+1} = z_n - \frac{f(z_n)}{f'(z_n)} \qquad n = 0, 1, 2, \ldots ,$$

and if $|z_{n+1} - z_n| < \tau$ (τ is a fixed threshold), then the iterative process is stopped and the number of executed iterations M is mapped to the point z_0. Figure 1 shows the result of this map as a black and white pattern: (black = 0, white = 1), where 0 and 1 are obtained by the rule, M modulo 2.

Figure 1. Newton's method used to localize the complex algebraic polynomial.

Rastislav Telgársky

Mosaics

Described here are patterns derived from the chessboard, where black and white squares become rectangles, and their uniform colors are replaced by other patterns.

Each mosaic is a repetition, in an M by N matrix, of a mosaic element. This element is generated by a single vector, whose coordinates are increasing. I have used arithmetic, geometric, Fibonacci, and also random generated sequences to control the rate of growth, regularity, etc. The mosaic element has four axes of symmetry. The drawing of each rectangle is repeated four times, except for the middle cross, where the number of repetitions is two — one at the nodes of the cross, and one at the middle of the cross.

The following fragment of C code describes a mosaic element which is repeated M by N times:

```
for (i = 0; i < m - 1; i++)
  for (j = 0; j < m - 1; j++) {
    k = ((i + j) % t) + 1 + w;
    pat(sel, a + v[i], b + v[j], a + v[i + 1] - 1, b + v[j + 1] - 1, k);
    if (i! = m - 2)
       pat(sel, a + sm - v[i + 1] + 1, b + v[j], a + sm - v[i], b + v[j + 1] - 1, k);
    if (j! = m - 2)
       pat(sel, a + v[i], b + sm - v[j + 1] + 1, a + v[i + 1] - 1, b + sm - v[j], k);
    if (i + j! = 2*m - 4)
       pat(sel, a + sm - v[i + 1] + 1, b + sm - v[j + 1] + 1, a + sm - v[i], b + sm - v[j], k);
}
```

where k is the color, sel is the selected pattern pat (such as drawing two lines between some of the four points), v is the control vector of dimension m (for example, representing an arithmetic sequence, like $v[0] = 0$, $v[1] = 3$, for $(i = 2; i < m; i++)v[i] = v[i-1] + i + 1$, $sm = v[m-2] + v[m-1] - 1)$, and (a, b) are coordinates of the left top of the mosaic element.

Further modifications are obtained by using a random increment in the definition of the control vector. Different mosaic patterns that result from warping onto curved surfaces, and carefully projected onto a view plane, for example, are achieved by using Painters Algorithm. Finally, color mosaics, displayed on a high resolution screen, are a real feast for the eyes.

Figure 1. Mosaic based on arithmetic sequence with two uniform fillings.

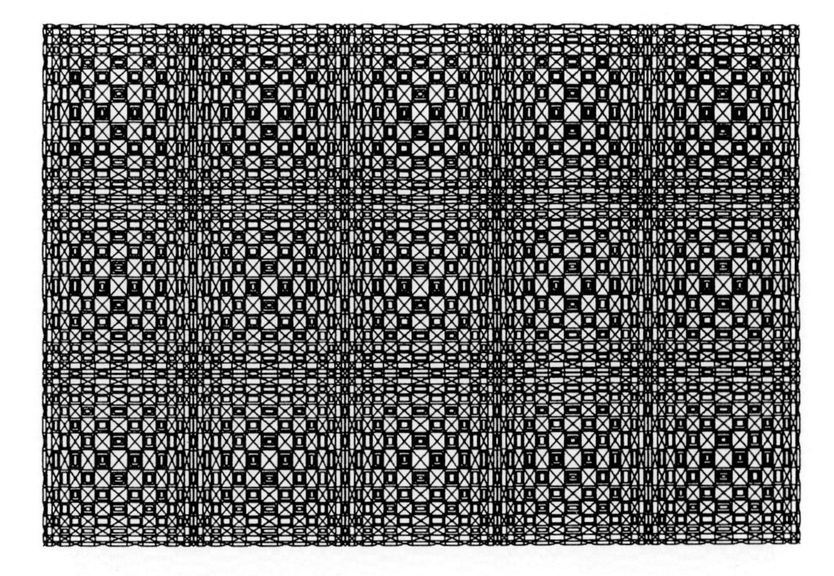

Figure 2. Mosaic based on arithmetic sequence and two alternating patterns: two diagonal line segments and a recursive box.

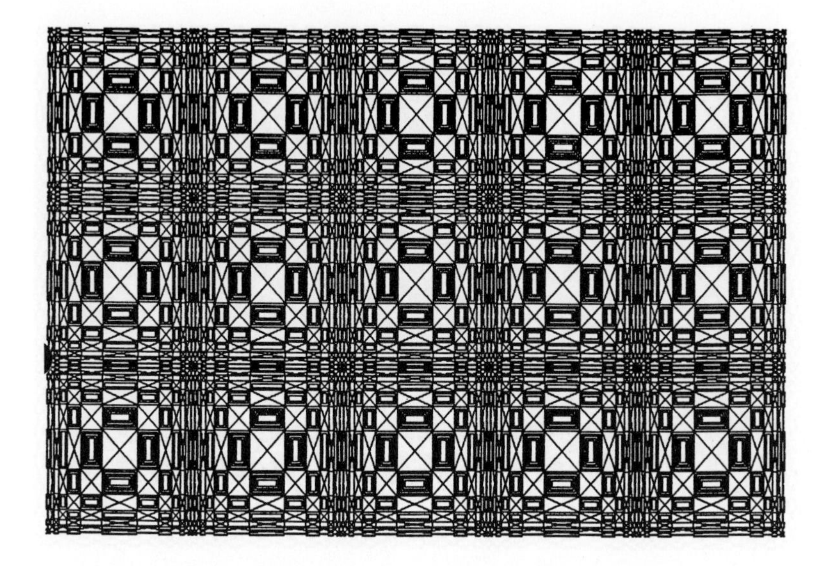

Figure 3. Mosaic based on Fibonacci sequence and two alternating patterns: two diagonal line segments and a recursive box.

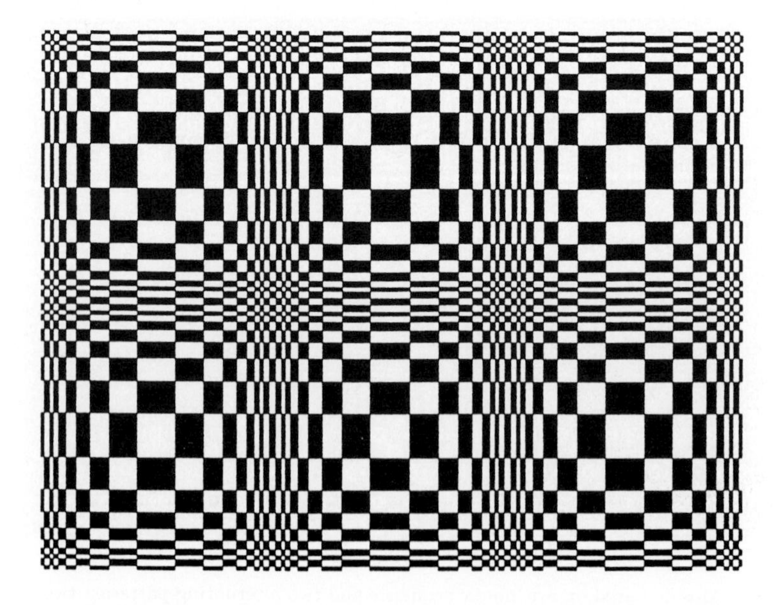

Figure 4. Mosaic based on exponential sequence with two uniform fillings.

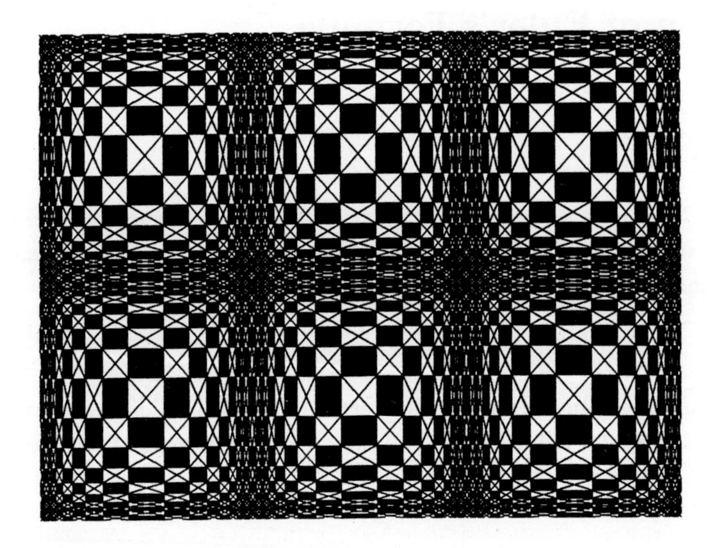

Figure 5. Mosaic based on exponential sequence and two alternating patterns: two diagonal line segments and a recursive box.

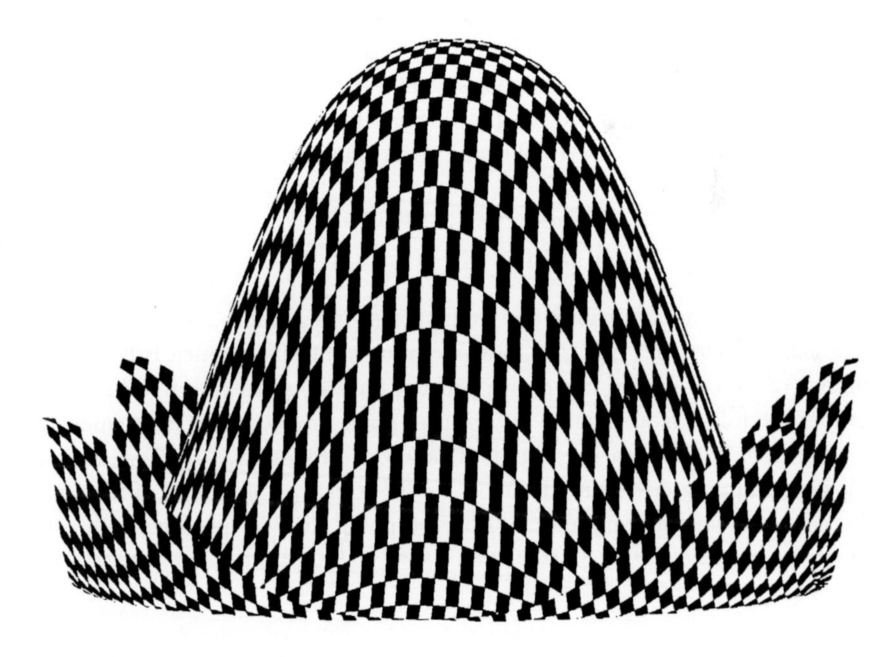

Figure 6. Mosaic drawn on hat-like surface and projected onto a view plane.

Mieczysław Szyszkowicz

Pattern of Euler's Formula

Much computer graphics art is created as a result of the illustration of the behavior of the sequence $\{z_n\}$ through some kind of iterative process. Usually this sequence is generated by the iteration of the form

$$z_{n+1} = \Phi(z_n), \quad n = 0, 1, 2, \ldots,$$

where z_0 is a given initial value. The iterative function Φ defines the recurrence relation. Often the behavior of the sequence $\{z_n\}$ is studied as a function of the initial values z_0. The same approach is used here. Consider the following initial value problem defined as the system of ordinary differential equations

$$dx/dt = \sin(y),$$
$$dy/dt = \sin(x),$$

with the initial values x_0, y_0 and $x_0, y_0 \in [-5, 5]$. The problem is to find the function $x(t)$ and $y(t)$. t is an independent variable. One of the simplest numerical methods able to solve this problem is Euler's method. In this situation, Euler's formula generates the following recurrence processes

$$x_{n+1} = x_n + h\sin(y_n),$$
$$y_{n+1} = y_n + h\sin(x_n),$$

where h is a constant called the step of the integration. $x_n \simeq x(t_n)$ and $y_n \simeq y(t_n)$, where $x(t)$ and $y(t)$, are the exact solutions. Figure 1 displays the number of executions of the above process for the given initial values. This number modulo 2 is associated with initial point (x_0, y_0), and consequently, a black and white pattern is created. The above process was realized with the following stopping criterion: IF $(x_{n+1} - x_n)^2 + (y_{n+1} - y_n)^2 < \tau$, THEN stop, where τ is a constant number. Other stopping criteria produce interesting pictures.

Figure 1. Euler's formula as a function of the initial values.

Mieczyslaw Szyszkowicz

Self-Accelerating Version of Newton's Method

Traub [1] presented the following self-accelerating recursion to solve the nonlinear equation $f(z) = 0$

$$z_{n+1} = z_n - \frac{f_n}{\gamma_n}, \quad n = 0, 1, 2, \ldots,$$

$$\gamma_n = \frac{f(z_n + C_n f_n) - f_n}{C_n f_n},$$

$$C_n = -\frac{1}{\gamma_{k-1}},$$

where z_0 is given and $f_n = f(z_n)$. This method was proposed for use in the complex plane. Presented here is a pattern showing the behavior of the sequence $\{z_n\}$ generated by the above iterative process. The sensitivity of $\{z_n\}$ to the initial values z_0 is displayed. Figure 1 shows the number of executed iterations (modulo 2) by alternating colors. The equation $f(z) = z^4 - 1 = 0$ is solved in the complex domain. This equation has four roots, which are $1, -1, i$ and $-i$, $i = \sqrt{-1}$. $C_0 = -1/f'(z_0)$. The stopping criterion of this self-accelerating algorithm is as follows: IF $|z_{n+1} - z_n| < \tau$ THEN stop, where τ is a constant number ($\tau = 10^{-6}$). Initial values are between -1.5 and 1.5 in the real and imaginary directions.

Reference

1. J. F. Traub, *Iterative Methods for the Solution of Equations* (Prentice Hall, 1964).

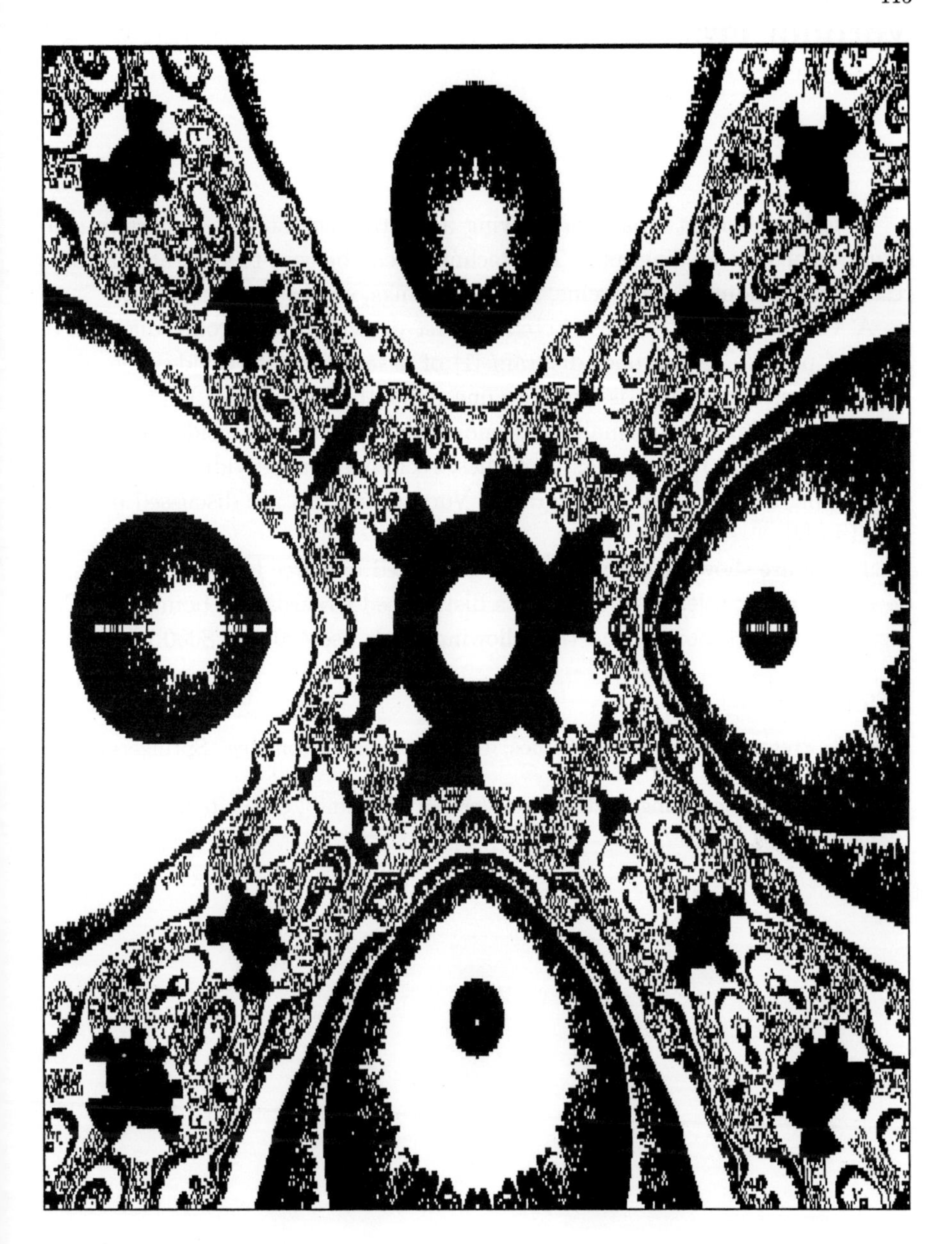

Figure 1. Self-accelerating methods used to localize the roots.

Ken Shirriff

Voronoi Fractal

Described here is a pattern showing a fractal generated by recursive construction of Voronoi diagrams. This technique can be used to generate patterns resembling roadmaps, leaf veins, butterfly wings, or abstract patterns.

A Voronoi fractal is constructed by first drawing the Voronoi diagram of a set of points. (A Voronoi diagram [1] of a set of points divides the plane into regions; each region is closer to one point than any other.) Then, using a larger set of points, a smaller Voronoi diagram is drawn inside each of the original regions. This process continues, recursively subdividing each region by drawing smaller Voronoi diagrams. Voronoi fractals are discussed in more detail in [2].

The figure shows a Voronoi fractal generated by three levels of subdivision on points in a circle, with the points distributed towards the boundary. The first level uses 20 points, and the following levels use 400 and 8000 points.

References

1. F. P. Preparata and M. I. Shamos, *Computational Geometry* (Springer-Verlag, 1985).
2. K. Shirriff, "Generating fractals from Voronoi diagrams", *Computers and Graphics* **17**, 2 (1993) 165–167.

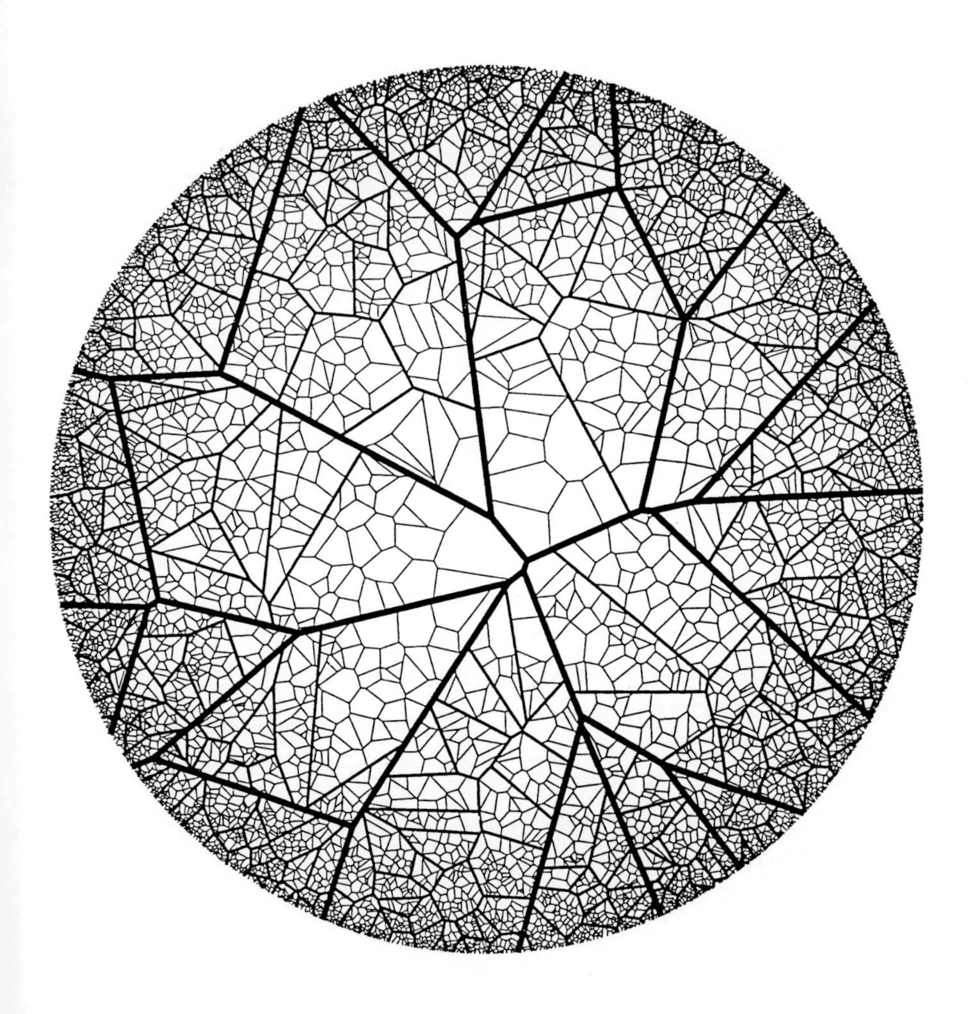

Mieczyslaw Szyszkowicz

The Logistic Map in the Plane

Presented here is a pattern obtained by repeating a simple mathematical operation, such as

$$x_{k+1} = rx_k(1 - x_k), \quad k = 0, 1, 2, \dots,$$

where x_0 is a given initial value and r is a parameter between 0 and 4. The behavior of the sequence $\{x_k\}$ is very sensitive to the value of the parameter r (see [1]). The obtained sequence may be attracted (converge) to one point, a group of points, or an infinite number of points composed by a fractal structure called a strange attractor. Figure 1 was generated by the above iteration executed for two variables x and y:

$$\begin{aligned} x_{k+1} &= rx_k(1 - x_k), \\ y_{k+1} &= sy_k(1 - y_k), \\ k &= 0, 1, 2, \dots, \end{aligned}$$

where $r = s$ or $r \neq s$. The above process was realized with $r = s = 3.7$ for the points (x_0, y_0), where x_0 and y_0 have values between 0 and 1. For each initial point the sequences $\{x_k\}$ and $\{y_k\}$ were produced with the following stopping criterion:

$$\text{IF } abs(x_k - x_{k+1}) < \tau \quad \text{OR} \quad abs(y_k - y_{k+1}) < \tau \text{ THEN stop}.$$

Here, τ is a constant value. Figure 1 was generated with $\tau = 0.45$. The number of executed iterations modulo 2 is represented.

Reference

1. R. May, "Simple mathematical models with very complicated dynamics", *Nature* **261** (1976) 359–467.

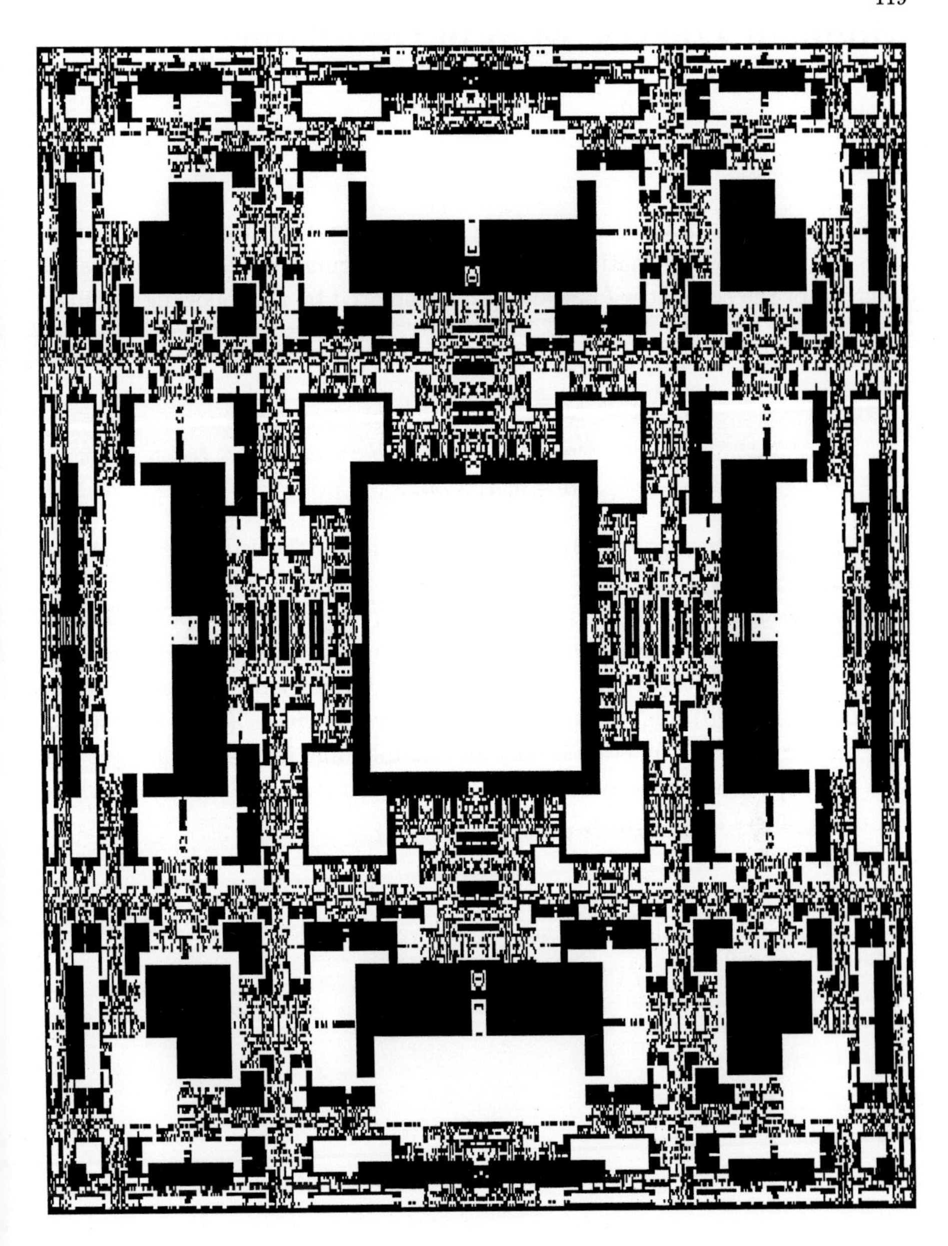

Figure 1. Pattern generated by the logistic parabola $f(x) = rx(1-x)$.

Mieczyslaw Szyszkowicz

Modified Logistic Map in the Plane

Presented here is a pattern resulting from the iteration, or repeated application, of a simple mathematical operation. In particular, the pattern shows the behavior of the sequences $\{x_n\}$ and $\{y_n\}$ generated by the following formulae

$$
\begin{aligned}
x_{n+1} &= R_n x_n (1 - x_n)\,, \\
y_{n+1} &= R_n y_n (1 - y_n)\,, \\
R_n &= r + (x_n + y_n)/2\,, \\
n &= 0,\, 1, =, 2, \ldots,\, N\,.
\end{aligned}
$$

If $r \in [0,\, 3]$, and $x_0,\, y_0 \in [0,\, 1]$, then the elements of the sequences $\{x_n\}$ and $\{y_n\}$ are between 0 and 1 [1]. Figure 1 shows the pattern obtained by the above iterations with $r = 3$ and $0 < x_0,\, y_0 < 1$. For each start point $(x_0,\, y_0)$ the iterative process is executed and after seven iterations ($N = 6$), the value $x_7 + y_7$ is used to determine the color of the initial point $(x_0,\, y_0)$. In our situation, black (0) and white (1) was used by the rules $5(x_7 + y_7)$ mod 2. The logistic parabola function is only an example function, others are good as well. The pattern produced by reduction of the interval $[0,\, 1]$ to $[0,\, a]$, where $0 < a < 1$ shows more details.

Reference

1. R. May, "Simple mathematical models with very complicated dynamics", *Nature* **261** (1976) 359–467.

Figure 1. Modified logistic function in the plane.

Mieczyslaw Szyszkowicz

Newton's Method in the C-Plane

Presented here is a pattern showing the behavior of the sequence $\{z_n\}$ as a function of the variable c in the following nonlinear equation:

$$f(z) = z^4 - c = 0 .$$

This sequence is generated by Newton's method which is used to localize a root of the above equation. The function f is considered in the complex plane. Newton's method is realized by the iteration

$$z_{n+1} = z_n - \frac{f(z_n)}{f'(z_n)} \quad n = 0,\ 1,\ 2, \ldots . \tag{1}$$

This equation is well known and described in [1] for fixed values of c (usually $c = 1$), and z_0 is changed over the complex plane. In this situation, the Julia sets are created (see [1]). We propose the realization of the iteration (1) with a fixed z_0 and variable c. This is something similar to the approach used to create the Mandelbrot set [1]. Denote by Ω_c a subset of the complex plane, and assume that Ω_c is the region of the variability of c. The iteration (1) is executed with the stopping criterion (the test $|z_{n+1} - z_n| < \tau$, where $\tau = 10^{-6}$), and if this criterion is satisfied, then the iteration is stopped. The number of executed iterations is expressed by black and white contours. Figure 1 is an example pattern for $z_0 = 1 + i$. For the uniquely defined Ω_c there is infinite number of patterns. Each of these is specified by an initial z_0. The reader may wish to study the behavior of $\{z_n\}$ simultaneously with variable c and z_0 values.

Reference

1. H.-O. Peitgen and D. Saupe, eds., *The Science of Fractal Images* (Springer-Verlag, 1988).

Figure 1. Newton's method applied to the equation $f(z) = z^4 - c = 0$, with the initial value $1 + i$, and c from the region $[-0.5, 20.5] \times [-15.0, 15.0]$.

Mieczysław Szyszkowicz

Iterations with a Limited Number of Executions

Iterative processes are used very often to create computer graphics arts. The classic example is the iteration realized in the complex plane

$$z_{n+1} = z_n^2 + c\,, \quad n = 0,\, 1,\, 2, \dots\,,$$

where $z_0 = 0$ and c is varied over the complex plane. This iteration is the basis for the construction of the Mandelbrot set [1, 3]. In the above iterative process, the number of executions is controlled by the value of the norm z, i.e., $|z| = \sqrt{x^2 + y^2}$ where $z = x + iy$. These numbers are often displayed as color patterns. Another approach is to realize the iterative process for a fixed number of times and to show the obtained value for the sequence. As an example, consider the iteration generated by the logistic map [2, 4]

$$x_{k+1} = r x_k (1 - x_k)\,, \quad k = 0,\, 1,\, 2\,, \quad \{x_k\} \subset [0,\, 1]\,,$$

where x_0 is a given initial value. The behavior of the sequence $\{x_k\}$ is very sensitive to the value of the parameter $r, r \in [0,\, 4]$. Figure 1 shows the values $x_K + y_K$, represented by black and white, and obtained after $K = 6$ iterations

$$\begin{aligned} x_{k+1} &= r x_k (1 - x_k)\,, \\ y_{k+1} &= s y_k (1 - y_k)\,, \end{aligned}$$

where $r = s$ or $r \neq s$. Figure 1 was obtained with $r = s = 3.9$ and with the initial values x_0, y_0, where x_0 and y_0 have values between 0 and 1. The point with the coordinates $(x_0,\, y_0)$ was colored by scaled value $x_6 + y_6$.

References

1. A. Dewdney, *Computer Recreation* (Scientific American, 1985).
2. B.-L. Hao, *Chaos* (World Scientific, 1984).
3. B. Mandelbrot, *The Fractal Geometry of Nature* (W. H. Freeman and Co., 1983).
4. R. May, "Simple mathematical models with very complicated dynamics", *Nature* **261** (1976) 359–467.

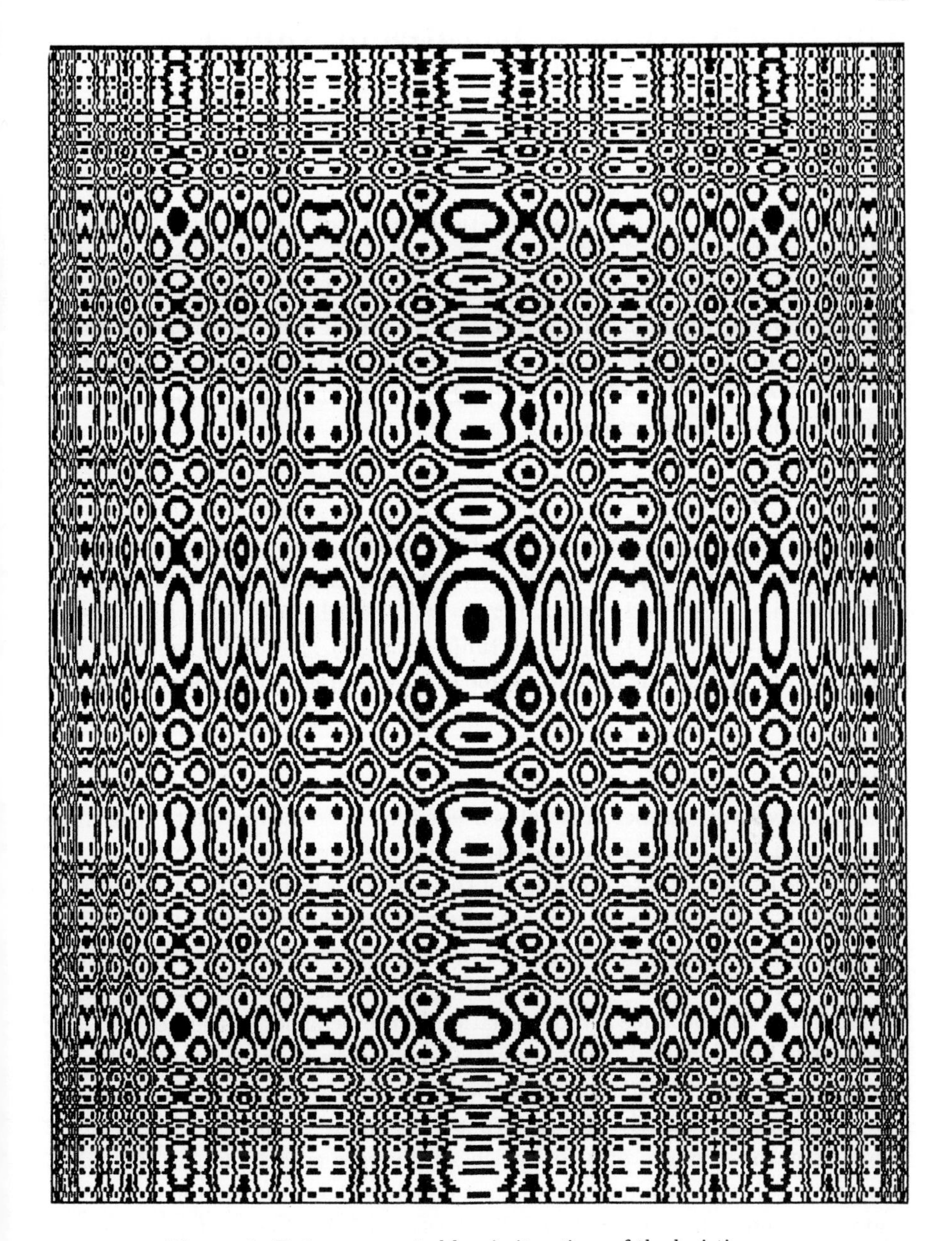

Figure 1. Pattern generated by six iterations of the logistic map.

Stephen D. Casey

The Starry Night — Iterates of Tan(z)

"The Starry Night" is a "mathematical Van Gogh", giving us a glimpse of a nebular cluster which has been generated by mathematical, as opposed to natural, forces. The study of this pattern falls under the iteration theory of Fatou and Julia, which examines discrete dynamical systems on the Riemann Sphere $\widehat{\mathbf{C}}$ with a dynamic given by a non-linear holomorphic or meromorphic mapping from $\Omega \subset \widehat{\mathbf{C}}$ to $\widehat{\mathbf{C}}$ (see [1–3]). This particular pattern was generated by the meromorphic function

$$f(z) = \lambda \cdot \tan(z)$$

in λ or parameter space in a neighborhood of the point $\sqrt{2}/2 + i(\sqrt{2})/2$. At each pixel position $(x, y) = x + iy = \lambda$, the orbit of the initial value $\pi/4$ was calculated, i.e., $z_{n+1} = \lambda \cdot \tan(z_n)$, $z_0 = \pi/4$, which generated the sequence $z_1 = \lambda$, $z_2 = \lambda \cdot \tan(\lambda)$, $z_3 = \lambda \cdot \tan(\lambda \cdot \tan(\lambda)), \ldots$. This sequence was generated until either the absolute value of z_n was greater than or equal to 100 (which guarantees that z_n diverges to ∞) or the number of iterates exceeded 256. In the picture, the black regions correspond to points λ for which $|z_{256}| < 100$, while the brighter regions correspond to divergent z_n. The brighter the tone, the fewer the number of iterates needed for z_n to escape. Note that there is a hot spot in the center of each "star".

The mathematical interpretation of the pattern tells us many things. First, we should note that the pattern is actually a bifurcation diagram for the family of dynamical systems $f(z) = \lambda \cdot \tan(z)$. From the picture, we can see that $g(z) = \tan(z)$ is not structurally stable (as defined in [3]) because a slight variation in λ can produce a wide variation in the behavior of the iterates. For fixed λ, the function $f(z)$ has an essential singularity at ∞. Therefore, by the Great Picard Theorem, in every neighborhood of ∞, $f(z)$ often takes on every complex number infinitely. By an argument similar to that in [3], it can be shown that those values of λ for which z_n diverges correspond to dynamical systems $\lambda \cdot \tan(z)$ which have a Julia set equal to the entire complex plane. It is also of interest to note that the distribution of the divergent points forms a fractal pattern, which is seen by observing that the larger "stars" generate smaller ones in a quasi-self-similar pattern.

Figure 1. The Starry Night — Iterates of Tan(z). A bifurcation diagram for the family of dynamical systems $f(z) = \lambda \cdot \tan(z)$.

The pattern was developed in the spring of 1986. The generating algorithm was a variation of the "escape-time algorithm" discussed in [4], modified so as to handle iterates of the tangent. The author was influenced by the work of [3] on iterates of $\lambda \cdot e^z$. However, this study of iterates of $f(z) = \lambda \cdot \tan(z)$ differs in many respects from [3]. Since $|e^z| = e^{\Re z}$ and e^z is entire, iterates of $\lambda \cdot e^z$ go to infinity with the growth of $\Re z_n$. However, $\tan(z)$ has simple poles at odd multiples of $\pi/2$ on the real axis, and is bounded by $|\coth(y)|$ off the axis. Thus, an iterate diverges by first being drawn into a sufficiently small neighborhood of some pole. The "stars" in the pattern represent how iterates of $\lambda \cdot \tan(z)$ spread the original poles throughout the complex plane.

Calculations for the pattern were performed on a Digital VAX 11/750, with code written in VAX-11 FORTRAN (DEC). The pattern was then produced on a Raster Technologies Model ONE/80, with the hardcopy produced by taking a photograph of the screen. The equipment was made available by the signal processing branch of Harry Diamond Labs. The author wishes to thank the branch for access to the equipment and Rob Miller of the Labs for his assistance with the Raster Technologies equipment.

References

1. P. Blanchard, "Complex analytic dynamics", *Bull. Amer. Math. Soc. (N.S.)* **11**, 1 (1986) 85–141.
2. S. D. Casey, "Formulating fractals", *Computer Language* **4**, 4 (1987) 28–40.
3. R. L. Devaney, "Julia sets and bifurcation diagrams for exponential maps", *Bull. Amer. Math. Soc. (N.S.)* **11**, 1 (1986) 167–171.
4. A. K. Dewdney, "Computer recreations", *Scientific American* **253**, 2 (1985) 16–24.

Stephen D. Casey

From Asymmetry to Symmetry — Iterates of Exp(z) and Exp(z^2)

The two patterns in "From Asymmetry to Symmetry" are a study in contrast. Figure 1 is a window into an asymmetric world in which one has no sense of orientation, whereas Fig. 2 is an extremely complex pattern exhibiting symmetry with respect to both the x and y axes. The study of these patterns falls under the iteration theory of Fatou and Julia, which examines discrete dynamical systems on the Riemann Sphere $\widehat{\mathbf{C}}$ with a dynamic given by a non-linear holomorphic or meromorphic mapping from $\Omega \subset \widehat{\mathbf{C}}$ to $\widehat{\mathbf{C}}$ (see [1–4]). Figure 1 was generated by the function

$$f_1(z) = \lambda \cdot e^z ,$$

while Fig. 2 was generated by the function

$$f_2(z) = \lambda \cdot e^{z^2} ,$$

in λ or parameter space. At each pixel position $(x, y) = x+iy = \lambda$, the orbit of the initial value 0 was calculated, i.e., for $j = 1, 2$, $z_{n+1} = f_j(z_n)$, $z_0 = 0$. This procedure was repeated until either $\Re(z_n)$ was greater than or equal to 100, which guarantees that z_n diverges to ∞, or the number of iterates exceeded 256. (Since $|e^z| = e^{\Re z}$ and e^z is entire, iterates of $\lambda \cdot e^z$ go to infinity with the growth of $\Re(z_n)$.) In the picture, the black regions correspond to points λ for which $\Re(z_{256}) < 100$, while the bands in the other regions correspond to the number of iterates (modulo 2) needed for $\Re(z_n) \geq 100$.

Devaney [4] gives us a precise mathematical interpretation of the patterns. They are bifurcation diagrams for the families of dynamical systems generated by $f_1(z) = \lambda \cdot \exp(z)$ and $f_2(z) = \lambda \cdot \exp(z^2)$. From the pictures, we can see that neither $\exp(z)$ nor $\exp(z^2)$ is structurally stable (as defined in [4]), because a slight variation in λ can produce a wide variation in the behavior of the iterates. Devaney also shows that those values of λ for which z_n diverges correspond to dynamical systems with dynamics f_1, f_2 having a Julia set equal to the entire complex plane. It is also of interest to note that the distribution of the divergent points forms a fractal pattern, which is seen by observing that the larger "bends" generate smaller ones in a quasi-self-similar pattern.

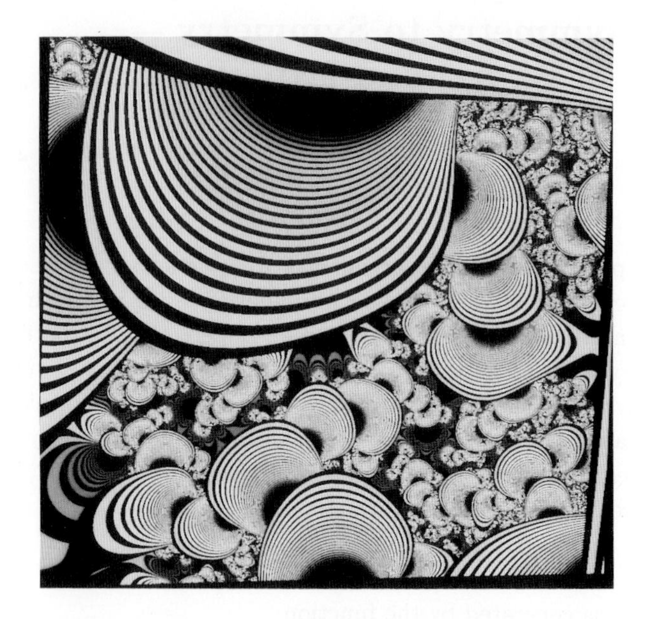

Figure 1. A bifurcation diagram for the family $f_1(z) = \lambda \cdot e^z$.

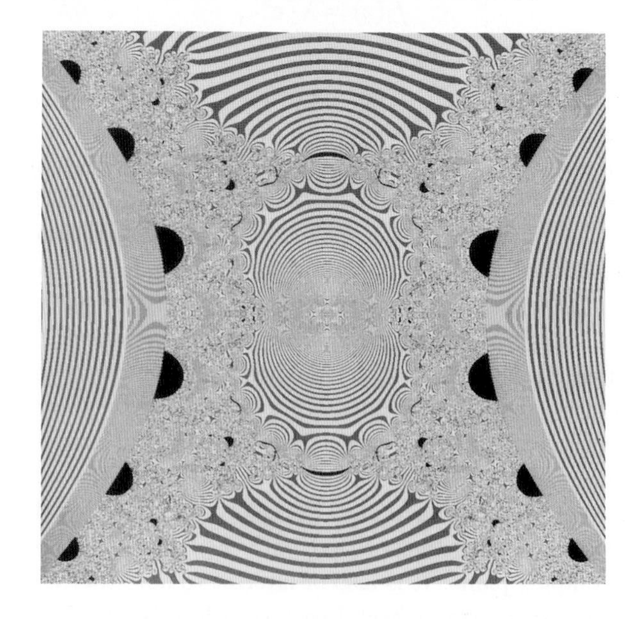

Figure 2. A bifurcation diagram for the family $f_2(z) = \lambda \cdot e^z$.

The pattern was developed in the spring of 1986. The generating algorithm was a variation of the "escape-time algorithm" discussed in [5], modified so as to handle iterates of exponentials. The author was reproducing the experiments of Devaney with iterates of $\lambda \cdot e^z$. The symmetry in Fig. 2 was expected. After producing Fig. 2, the author produced a "three-fold" symmetric pattern using $\lambda \cdot \exp(z^3)$. (See [2, 3] for additional patterns, pseudocode for generating the patterns, and further discussion.)

Calculations for the pattern were performed on a Digital VAX 11/750, with code written in VAX-11 FORTRAN (DEC). The pattern was then produced on a Raster Technologies Model ONE/80, with the hardcopy produced by taking a photograph of the screen. The equipment was made available by the signal processing branch of Harry Diamond Labs. The author wishes to thank the branch for access to the equipment, Rob Miller of the Labs for his assistance with the Raster Technologies equipment, and Joseph Comick for his assistance in the preparation of the prints. He also wishes to thank Professors Alan Brownstein and Richard Kreminski for helpful mathematical discussions.

References

1. P. Blanchard, "Complex analytic dynamics", *Bull. Amer. Math. Soc. (N.S.)* **11**, 1 (1986) 85–141.
2. S. D. Casey, "Formulating fractals", *Computer Language* **4**, 4 (1987) 28–40.
3. S. D. Casey, "Fractal images: procedure and theory", HDL Technical Report HDL-TR-2119 (1987) p. 42.
4. R. L. Devaney, "Julia sets and bifurcation diagrams for exponential maps", *Bull. Amer. Math. Soc. (N.S.)* **11**, 1 (1986) 167–171.
5. A. K. Dewdney, "Computer recreations", *Scientific American* **253**, 2 (1985) 16–24.

Stephen D. Casey and Nicholas F. Reingold

A Variation on a Curve of Mandelbrot

The pattern described here was generated by an efficient recursive algorithm developed by the authors which produces approximations of self-similar fractal sets (see [3]). Such sets are constructed by a repeated scaling, translation, reflection, and/or rotation of a fixed pattern or set of patterns. The procedure is a "pattern rewriting system" in which a given geometric pattern is drawn repeatedly after suitable mappings. The pattern used to generate the fractal set by the rewriting system will be called a *seed.* The *base* is the initial configuration. This particular pattern was produced by using the "monkey's tree" as a seed with the Gosper curve as a base.[a] The procedure was iterated 4 times. Using their algorithm, the authors duplicated the self-similar fractal sets in Mandelbrot [4], which constitute approximately 45% of the graphics plates in the book.

The curve C produced by repeating this particular generating process infinitely is often a self-similar fractal. It can be mapped onto itself by appropriate similarity transforms, and thus is "self-similar". And, if $\mathcal{N}(r)$ equals the *minimal* number of closed line segments of length r needed to cover C, then C has the non-integer fractal dimension $D = \lim_{r\to 0} \frac{\log(\mathcal{N}(r))}{\log(1/r)} \sim 1.8687$, which, by definition, makes it a fractal (see [2–4]).

The basic idea of the pattern rewriting system is to simply replace each segment of the pattern at each level of recursion with an appropriately scaled and rotated copy of the seed. The routine is a cousin of the string rewriting or L-systems introduced by A. Lindenmayer and developed by P. Prusinkiewicz (see [5]). which are themselves cousins of LOGO, the language developed at MIT under the direction of S. Papert (see [1]). However, it differs from these systems in some important ways. The generating schemes for patterns are designs with some built-in information on how to orient patterns in later levels of iteration. In L-systems and in LOGO, a pattern is generated by "teaching the turtle" where to move based on some fundamental set of instructions. In pattern rewriting, if you can draw a piece-wise linear pattern on a piece of paper, identifying its vertices and its orientation scheme, then you can produce

[a]See [4], plates 31 and 146 for the former, and plates 46, 47, and 70 for the latter.

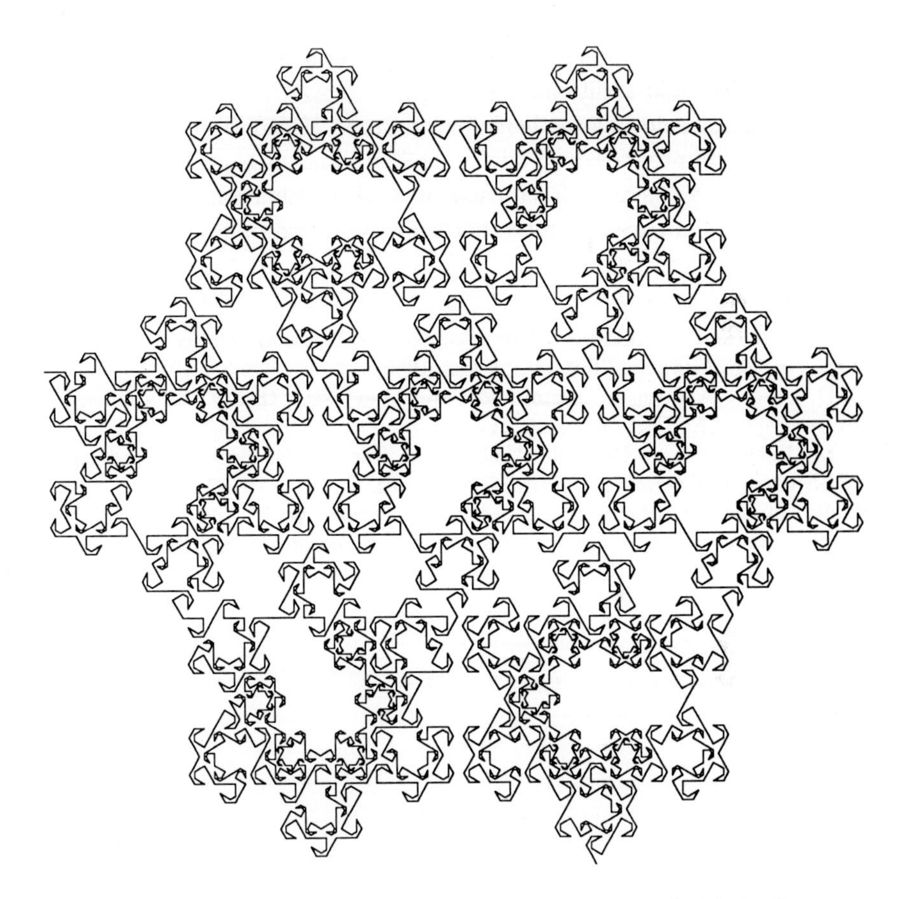

Figure 1. Pattern generated by using the "monkey's tree" as a seed with the Gosper curve as a base.

it on the computer. The system is also efficient, in that the only overhead between levels of recursion is a single boolean variable. In L-systems, the higher the level of iteration, the longer the string produced. Pattern rewriting gave the authors an interactive *Fractal Geometry* "text", with which they created and explored approximations to classic fractal sets (e.g., Cantor sets, Peano curves), and reproduced many of the patterns in [1] and [4]. (See [2, 3] for further details and additional patterns.)

The authors would like to thank Larry Crone of American University for mathematical and programming advice, and our student Nora Wade, who catalogued all of the self-similar fractal sets in [1, 4]. Calculations were performed on an AST Premium/386, with the hardcopy produced on a QMS-PS 410.

References

1. H. Abelson and A. diSessa, *Turtle Geometry: The Computer as a Medium for Exploring Mathematics* (The MIT Press, 1980).
2. S. D. Casey, "Analysis of fractal and Pareto-Levy sets: theory and application", *Proceedings of EFTF* (1990) 205–211.
3. S. D. Casey and N. F. Reingold, "Self-similar fractal sets: theory and procedure", *IEEE Computer Graphics and Applications* **14**, 3 (1994) 73–82.
4. B. B. Mandelbrot, *The Fractal Geometry of Nature* (W. H. Freeman, 1983).
5. H.-O. Peitgen and D. Saupe, eds. *The Science of Fractal Images* (Springer-Verlag, 1988).

Stephen D. Casey and Nicholas F. Reingold

An Asymmetric Sierpinski Carpet

The pattern described here was generated by an efficient recursive algorithm developed by the authors which produces approximations of self-similar fractal sets (see [3]). Such sets are constructed by a repeated scaling, translation, reflection, and/or rotation of a fixed pattern or set of patterns. The procedure is a "pattern rewriting system" in which a given geometric pattern is drawn repeatedly after suitable mappings. The pattern used to generate the fractal set by the rewriting system will be called a *seed.* The *base* is the initial configuration. The seed for this particular pattern consisted of three components. The base was a square drawn counter-clockwise. The procedure was iterated 5 times to produce the figure. Using their algorithm, the authors duplicated the self-similar fractal patterns in Mandelbrot's *The Fractal Geometry of Nature* [4], which constitute approximately 45% of the graphics plates in the book. This particular pattern is an original configuration.

The set S_C produced by repeating this particular generating process infinitely is often a self-similar fractal. It can be mapped onto itself by appropriate similarity transforms, and thus is "self-similar". And, if $\mathcal{N}(r)$ equals the *minimal* number of closed disks of radius r needed to cover S_C, then S_C has the non-integer fractal dimension $D = \lim_{r \to 0} \frac{\log(\mathcal{N}(r))}{\log(1/r)} = \frac{\log(5)}{\log(3)} \sim 1.4649$, which, by definition, makes it a fractal (see [2–4]).

The basic idea of the authors' pattern rewriting system is to simply replace each segment of the pattern at each level of iteration with an appropriately scaled and rotated copy of the seed. The routine is a cousin of the string rewriting or L-systems introduced by A. Lindenmayer and developed by P. Prusinkiewicz (see [5]), which are themselves cousins of LOGO, the language developed at MIT under the direction of S. Papert (see [1]). However, it differs from these systems in some important ways. The generating schemes for patterns are designs with some built-in information on how to orient patterns in later levels of iteration. In L-systems and in LOGO, a pattern is generated by "teaching the turtle" where to move based on some fundamental set of instructions. In pattern rewriting, if you can draw a piece-wise linear pattern on a piece of paper, identifying its vertices and its orientation scheme, then you can produce it on the computer. The system is also efficient, in that the only

Figure 1. An original design generated by the pattern rewriting system.

overhead between levels of recursion is a single boolean variable. In L-systems, the higher the level of iteration, the longer the string produced.

The routine gave the authors an interactive *Fractal Geometry* "text", with which they created and explored approximations to classic fractal sets (e.g., Cantor sets, Peano curves), and reproduced many of the patterns in [1] and [4] (see [2, 3] for further details and additional patterns). Motivation for developing the routine was provided by the figures in [1, 4] and the simplicity and elegance of LOGO.

The authors would like to thank Larry Crone of Amercian University for mathematical and programming advice, and our student Nora Wade, who catalogued all of the self-similar fractal sets in [1, 4]. Calculations were performed on an AST Premium/386, with the hardcopy produced on a QMS-PS 410.

References

1. H. Abelson and A. diSessa, *Turtle Geometry: The Computer as a Medium for Exploring Mathematics* (The MIT Press, 1980).
2. S. D. Casey, "Analysis of fractal and Pareto-Levy sets: theory and application", *Proceedings of EFTF* (1990) 205–211.
3. S. D. Casey and N. F. Reingold, "Self-similar fractal sets: theory and procedure", *IEEE Computer Graphics and Applications* **14**, 3 (1994) 73–82.
4. B. B. Mandelbrot, *The Fractal Geometry of Nature* (W. H. Freeman, 1983).
5. H.-O. Peitgen and D. Saupe, eds., *The Science of Fractal Images* (Springer-Verlag, 1988).

Stephen D. Casey

The "Computer Bug" as Artist — Opus 3

The pattern described here was generated by an efficient recursive algorithm developed jointly by the author and N. F. Reingold. The algorithm is explained in [1].

The procedure used to develop the pattern duplicated the self-similar fractals sets in Mandelbrot's *The Fractal Geometry of Nature* [2], which constitute approximately 45% of the graphics plates in his book. This particular pattern is the result of a mistake by the author, and it reminded the author of graphics plates 246 and 293 in Mandelbrot, which are entitled *The Computer "Bug" as Artist, Opus 1 (2)*, respectively. It is of interest to note that this "computer bug" (or, more appropriately, "programmer bug") actually produced (something which looks like) a bug.

Calculations for the pattern were performed on an AST Premium/386, with the hardcopy produced on a QMS-PS 410.

References

1. S. D. Casey and N. F. Reingold, "Self-similar fractal sets: theory and procedure", *IEEE Computer Graphics and Applications* **14**, 3 (1994) 73–82.
2. B. B. Mandelbrot, *The Fractal Geometry of Nature* (W. H. Freeman, 1983).

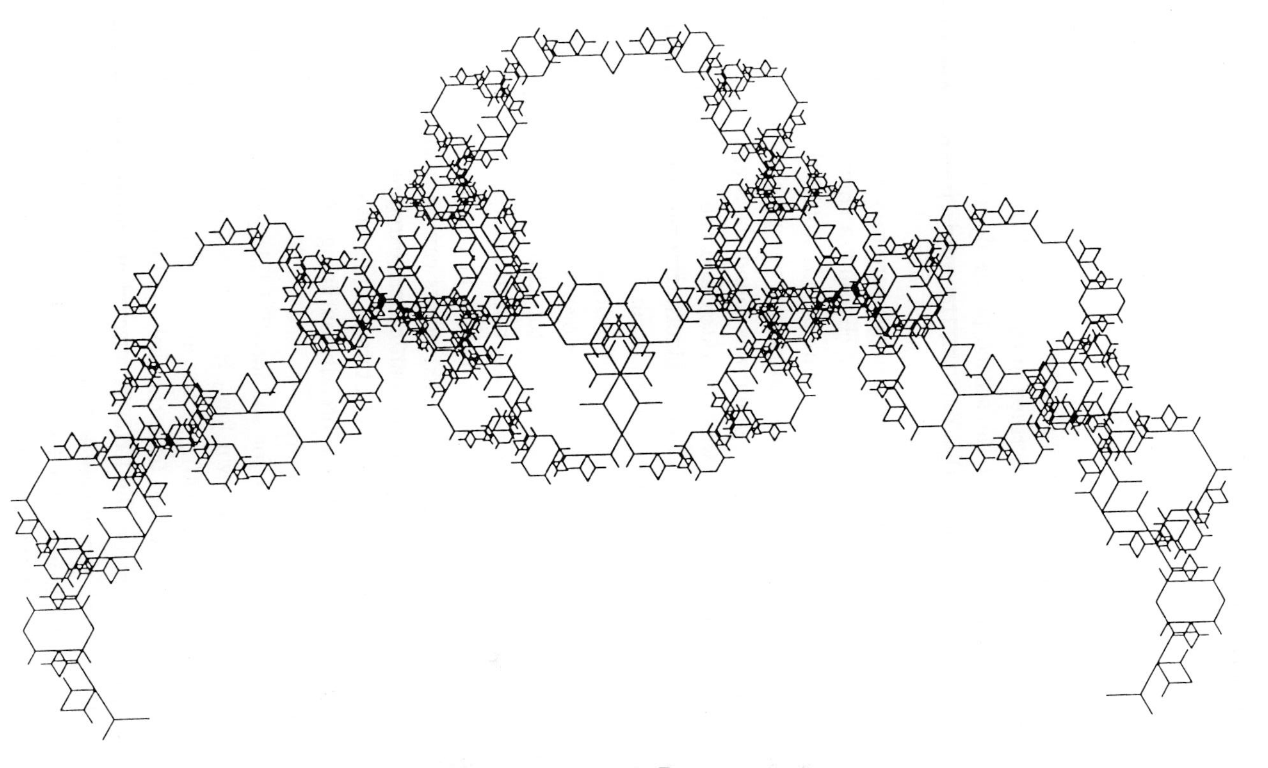

Computer Bug

Edward M. Richards

The Last Word in Fractals

Described here is a pattern which grows fonts or words or fractal shapes by random walk techniques as described in the next TOPSY-TURVY pattern. This pattern however shows that the seeded points do not have to be regular geometric figures. To generate this pattern, the word "fractals" was written using a drawing package. For effect, the word was also underscored with a wiggly line. The word "fractals" and its underscore were then used as seeded points. Each random walk was started by choosing a column (0, 1, ..., 639) and a row (0, 1, ..., 479) at random. The search would then begin in any one of the four major directions, the distance of one pixel at a time, looking for seeded points. When a seeded point was found the searching pixel stuck to it and another starting point was generated and the search repeated. To prevent overcrowding as more and more pixels aggregated on the screen, the program only allowed a starting point to occur if it was more than 9 pixels away from any other pixel on the screen. If overcrowding was detected, a new potential starting was generated. Using this overcrowding protection prevents the random starting points from occurring too close to the points which are already aggregated. This helps to maintain the fractal quality of the pattern. Without the overcrowding DENSITY function in the program, the growth could have lumpy and solid spots. Seeded points can be made with different colors and an ABSORB color function can cause searching pixels to take on the color of the seeded pixel. Thus, clusters of different colors may be grown on the screen. With seeded pixels broadly outlining anything from fonts to countries, interesting patterns emerge.

I have to thank C. Pickover for the idea for this pattern. The approach shown here, while growing its own font, is different from his approach given in *Algorithm* (July/August 1990). In the article "Growing your own fonts", Pickover actually grows the font from the skeleton form of the letter outwards. In his approach, a point on the letter is selected at random and then grown outwards. The pattern shown here sort of sneaks in from the outside and sticks to whatever it finds. It was after I read his article that I thought of applying my FRACTAL AGGREGATION program to a similar task.

References

1. J. Feder, *Fractals* (Plenum Press, 1988).
2. C. A. Pickover, *Computers, Pattern, Chaos, and Beauty — Graphics From an Unseen World* (St. Martin's Press, 1990).
3. C. A. Pickover, "Personal programs: growing your own fonts", *Algorithm* **1**, 5 (1990) 11–12.

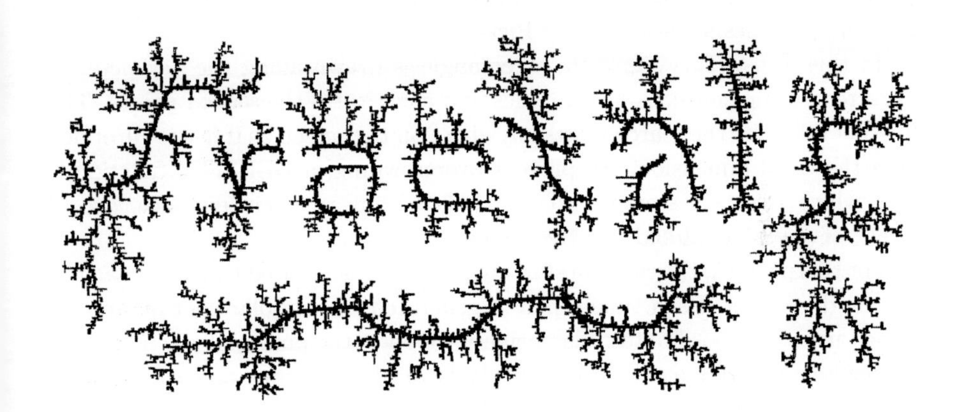

Edward M. Richards

Topsy-Turvy Fractal Growth

Described here is a pattern which was grown using a program developed by the author. The program generates fractal shapes by random walks of "sticky" diffusing particles as described by Pickover [2].

In this Topsy-Turvy growth, a rectangle is drawn along the perimeter of the screen using conventional graphics and constitutes the seeded points. The starting point for the search starts anywhere on the screen as if it were dropped from the third dimension. To prevent overcrowding, a DENSITY function is used which rejects a starting point if it is within a certain distance, e.g., 9 pixels, of any pixel already on the screen.

For this pattern, the random starting points are generated by random angles and random radii. Each time a pair is generated, these polar coordinates are turned into XY coordinates. The neighborhood of the potential starting point is then searched for a distance of 9 pixels in all directions and the point is selected if there are no neighbors. If there is a neighbor, a new potential starting point is calculated. The probability function used to determine the radius for this pattern was the uniform distribution used to select a column (e.g., $0, 1, \ldots, 639$ for a 640-column screen). The XY coordinates were tested to see if they fell within the range of the screen. If not, they were folded back into it. Obviously, other probability functions can be used to determine the radius.

The pattern shown here is technically not a diffusion-limited aggregation (DLA) such as those shown in Feder [1] since the density of the growth would have gradually increased to a solid as new random starting points were selected. This overcrowding was prevented by the density function.

References

1. J. Feder, *Fractals* (Plenum Press, 1988).
2. C. A. Pickover, *Computers, Pattern, Chaos and Beauty* (St. Martin's Press, 1990).

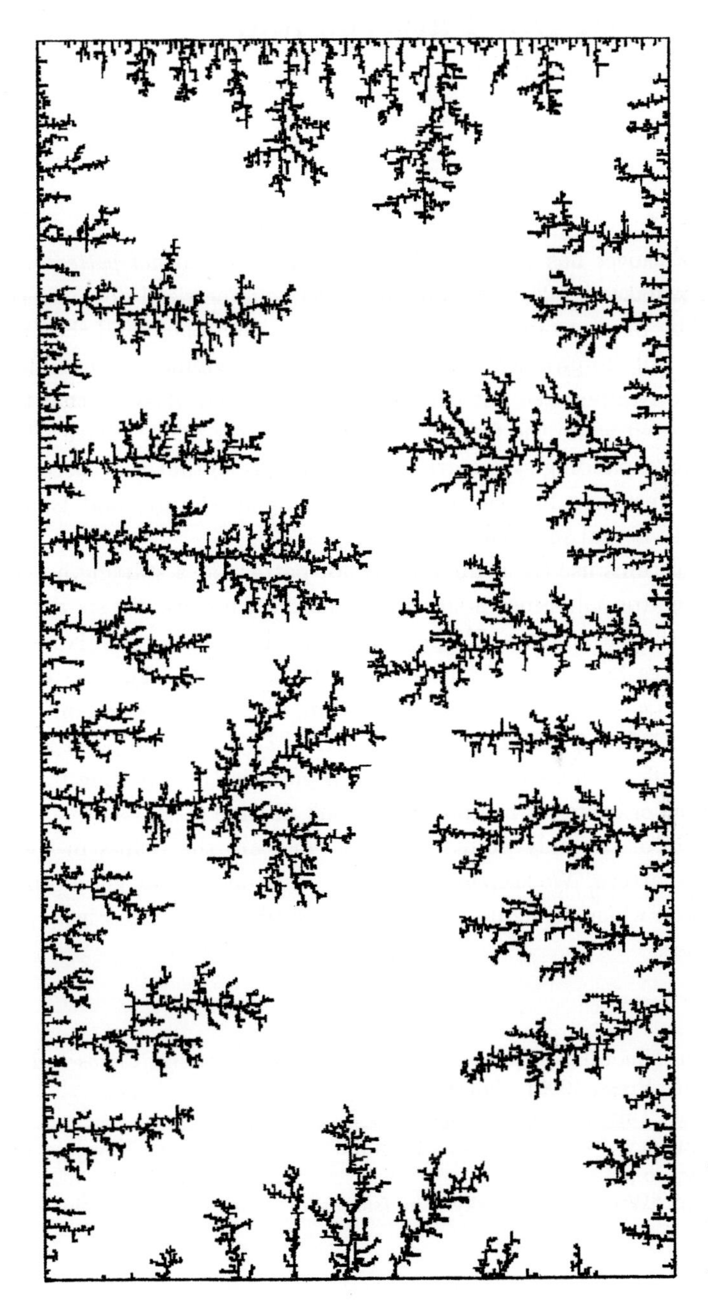

(*Landscape*)

William K. Mason

Symmetrized Dot-Patterns and Cellular Automata

Clifford Pickover has experimented with *symmetrized dot-patterns*, which resemble snowflakes [1, 2]. Generally speaking, symmetrized dot-patterns are constructed by placing randomly positioned dots on a plane and then mirror-reflecting the dots to produce a pattern with sixfold symmetry. One difference between these symmetrized dot-patterns and actual snowflakes is that the latter are connected while the former consist of discreet dots. I have tried two algorithms for expanding symmetrized dot-patterns so that they become connected. The resulting figures are often pleasing to the eye and have some resemblance to actual snowflakes.

Both algorithms use the symmetrized dot-pattern as a starting point for a cellular automaton [3]. Since a snowflake has six arms, we use six-dot neighborhoods for the automatons. If a point has coordinates (x, y) its six neighbors have coordinates $(x+1, y+1)$, $(x+1, y-1)$, $(x-1, y+1)$, $(x-1, y-1)$, $(x+2, y)$ and $(x-2, y)$. In the first algorithm, a point is "turned on" at time t if exactly one of its 6 neighbors is "on" at time $t-1$. Once a point is on it is never turned off. In the second algorithm, a point is "on" at time t and only if an odd number of its six neighbors are "on" at time $t-1$.

At time zero, the only points that are "on" are the points in the symmetrized dot-pattern. We keep cycling the automaton until we get a connected figure. We stop when the figure is most pleasing to the eye. See Figs. 1, 2, and 3 for examples.

References

1. C. Pickover, "Snowflakes from sound", in *Computers, Pattern, Chaos and Beauty* (St. Martin's Press, 1990) Chapter 4.
2. C. Pickover, "Unseen worlds", *Algorithm* **2**, 2 (1991).
3. A. K. Dewdney, "Wallpaper for the mind", *The Armchair Universe* (W. H. Freeman & Co., 1988).
4. I. Peterson, "Written in the sky", in *Islands of Truth* (W. H. Freeman & Co., 1990).

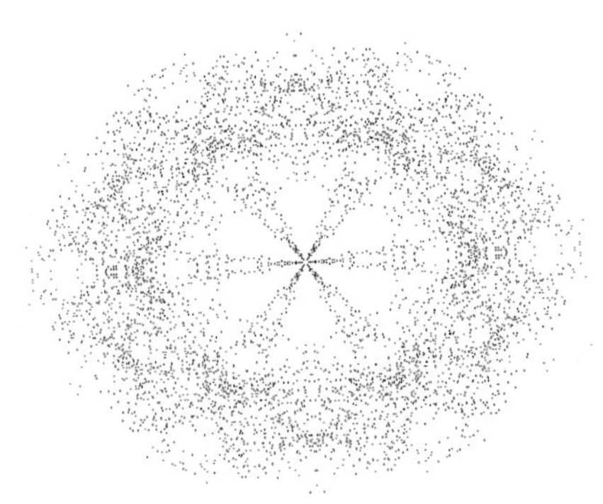

Figure 1. Original Pattern — 500 points.

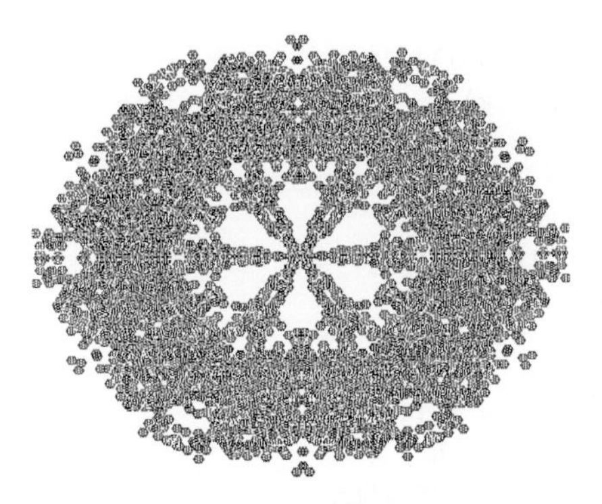

Figure 2. Type 1 Automaton — 3 cycles.

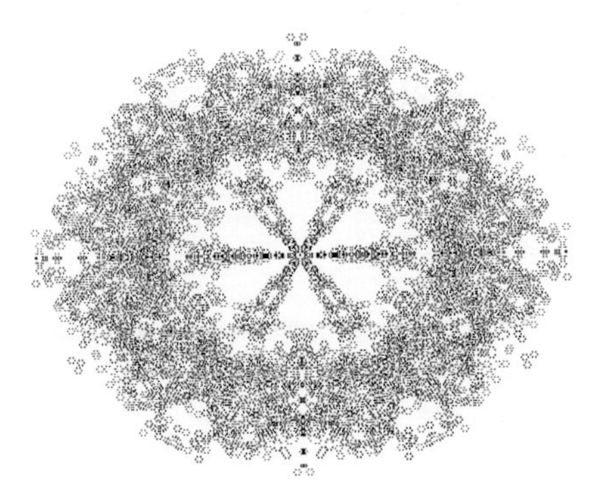

Figure 3. Type 2 Automaton — 2 cycles.

Snowflake Algorithm.

```
VARIABLES sflake, nxtflake are 2-dimensional arrays, 100x100 for example
   sum
   atype is 1 for type 1 algorithm, 2 for type 2 algorithm

Load symmetrized dot_pattern into array sflake(x,y) so that
     sflake(x,y)=1 if point (x,y) is a dot in the pattern
     sflake(x,y)=0 if point (x,y) is not in the pattern

UNTIL key pressed
   for x = 1 to 100
       for y = 1 to 100
           sum = sflake(x+1,y+1)+
             sflake(x+1,y-1)+
             sflake(x-1,y+1)+
             sflake(x-1,y-1)+
             sflake(x+2,y) +
             sflake(x-2,y)
           IF atype = 1 THEN
                IF sum = 1 THEN
                    nxtflake(x,y)=1
                ELSE
                    nxtflake(x,y)=0
                END IF
           [end for atype = 1]
           IF atype = 2 THEN
                IF sum is odd THEN
                    nxtflake(x,y)=1
                ELSE
                    nxtflake(x,y)=0
                END IF
           [end for atype = 2]
       [end for y]
   [end for x]
   for x = 1 to 100
       for y = 1 to 100
            IF atype = 1 THEN
                 IF nxtflake(x,y)=1 THEN sflake(x,y)=1
            IF atype = 2 THEN
                 sflake(x,y)= nxtflake(x,y)
       [end for y]
   [end for x]
   Plot all points (x,y) for which sflake(x,y)=1
   Erase all points (x,y) for which sflake(x,y)=0
```

Ian D. Entwistle

"Carpet of Chaos": Mapping the Bilateral Symmetric $z^z + c$ in the Complex z and c Planes

The patterns Figs. 1 and 2 are respectively the Mandelbrot and modified Julia sets for the function $z - z^z + c$.

For iteration of polynomials $z^n + c$ where n is an integer, the "Mandelbrot set" (M) consists of those complex numbers z such that the sequence $z = 0$, $z_1 = z$, $z_2 = z^2$, $z_3 = (z^2 + z)^2, \ldots$, never satisfies $|z| > 2$. A number of algorithms for carrying out this iterative process have been widely used to generate the M set mappings of many functions which were not polynomials [1]. Complex functions which are analytic and continuous would be expected to generate Julia sets on iteration. Those mapped from $z^z + c$ are rather unattractive. The M set for a function can also be defined as the collection of points for which the Julia set of f_c is connected. Julia sets derived from points c close to the origin would then be expected to have outlines close to the shape of the lemniscate controlled by the escape test. In the case of $z^z + c$, very few points in the plane are bounded for values of c close to $0 + 0i$. The Julia sets derived from such points are therefore not of interest. A reflection perhaps of the less attractive properties of the M set. Alternative escape tests [2] do however yield mappings from the iteration, which are visually exiting [3] such as Fig. 2. Figure 1 was first obtained by the standard Level Set Method [1] applied to $z^z + c$ using lines to plot the divergent set points only where adjacent points differ in divergence rate. Selection of the point c $(0.09 + 0i)$ allowed the speed of mapping to be halved as the pattern is symmetrical. Even using a language such as Fortran, which supports complex variables, the iteration process is much slower than that for simple polynomial iteration. The Pseudo code utilized is listed. This utilizes the "square" escape test previously discussed [2] and an additional test to control the point plotting. The data for producing both figures is appended.

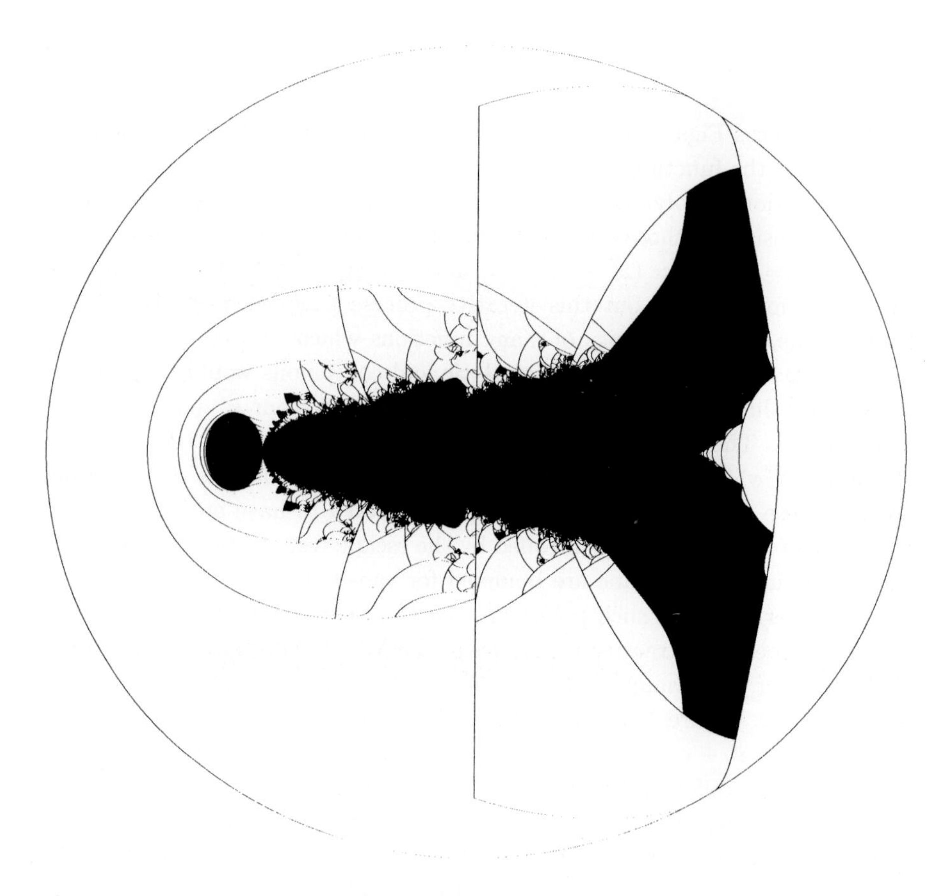

Figure 1.

Figure 2.

Appendix

	nx,	ny,	$x$min,	$x$max,	$y$min,	$y$max,	itermax,	c,	es,	ss
Fig. 1.	2400,	2400	−2.0	2.2	−2.0	2.0	25	$0.09 + 0i$	30	35
Fig. 2.	1920	1840	−3.1	1.1	−2.2	2.2	100			

```
Pseudo code for Julia set
        Variables: Integer sx, sy, iter   Real rz, iz, x, y
Complex c, z
        DO sy =1 to nx
        DO sx =1 to ny
            y=ymin+sy*(ymax−ymin)/ny
            x=xmin+sx*(xmax−xmin)/nx
            z = cplx(x, y)
        DO iter=1 to itermax
            z=z^z + c
            rz=real(z):iz=imag(z)
```

References

1. H.-O. Peitgen and D. Saupe, *The Science of Fractal Images* (Springer-Verlag, 1986).
2. I. D. Entwistle, "Julia set art and fractals in the complex plane", *Computers and Graphics* **13**, 3 (1989) 389–392.
3. I. D. Entwistle, " 'Entrapped Lepidoptera': An alternative mapping of a Julia Set for the function $z - z^4 - z + c$", in *The Pattern Book: Fractals, Art, and Nature*, ed. C. A. Pickover (World Scientific, 1995).
4. C. A. Pickover, *Computers, Pattern, Chaos, and Beauty — Graphics From an Unseen World* (St. Martis Press, 1990).

Ian D. Entwistle

"Spidermorphs": Mappings from Recursion of the Function $z \to z^z + z^4 - z + c$

The two patterns Figs. 1 and 2 result from studying the Julia set mappings of the dynamic $f(z) = z^z + z^4 - z + c$ and identifying areas of the maps which have disconnected fractal geometry resembling biological forms.

Two types of computer-generated geometries have introduced the term "Biomorph". Firstly, in the brilliant discourse on evolution, "The Blind Watchmaker", the spidery line graphical images generated by recursion of lines were used to represent evolution of biological forms [1]. At the same time, a collection of curious fractal patterns resembling low order biological forms resulting from iteration of algebraic equations in the complex plane were classed as "Biomorphs" [2]. For the purpose of describing Figs. 1 and 2, the term "Biomorph" encompasses organismic morphologies created by small changes to traditional Julia set calculations [3]. Of particular interest for such pattern generation are the escape radius tests utilized, as well as additional tests for determining which points to plot in the complex plane [4]. Among the many reported examples [2, 3] of biomorphs, those derived from iteration of dynamics containing the bilateral symmetric z^z appear to most closely resemble known biological shapes [3]. The patterns were generated using tests previously described [4, 5] and are outlined in the Pseudo code listed. This utilizes the magnitude of the side of a square lemniscate rather than the radius of the escape circle. In addition, a limit is put on the points to be plotted by fixing the maximum reached by $z$real or $z$imag. Figure 1 is symmetrical as its origin point in the c plane of the function lies on the imaginary axis about which the c plane mapping is symmetrical. Figure 2 can be observed if the window is extended to $-5, 5, -5, 5$. Data for generation of both patterns is appended.

Parameters

	nx,	ny,	$x$min,	$x$max,	$y$min,	$y$max,	itermax,	c,	es,	ss
Fig. 1.	2400,	3000	-4.1	4.1	-4.1	4.1	25	$-0.78 + 0i$	20	25
Fig. 2.	2400	2400	-1.31	-0.17	-1.35	-0.13	25	$0.1 + i0$	20	25

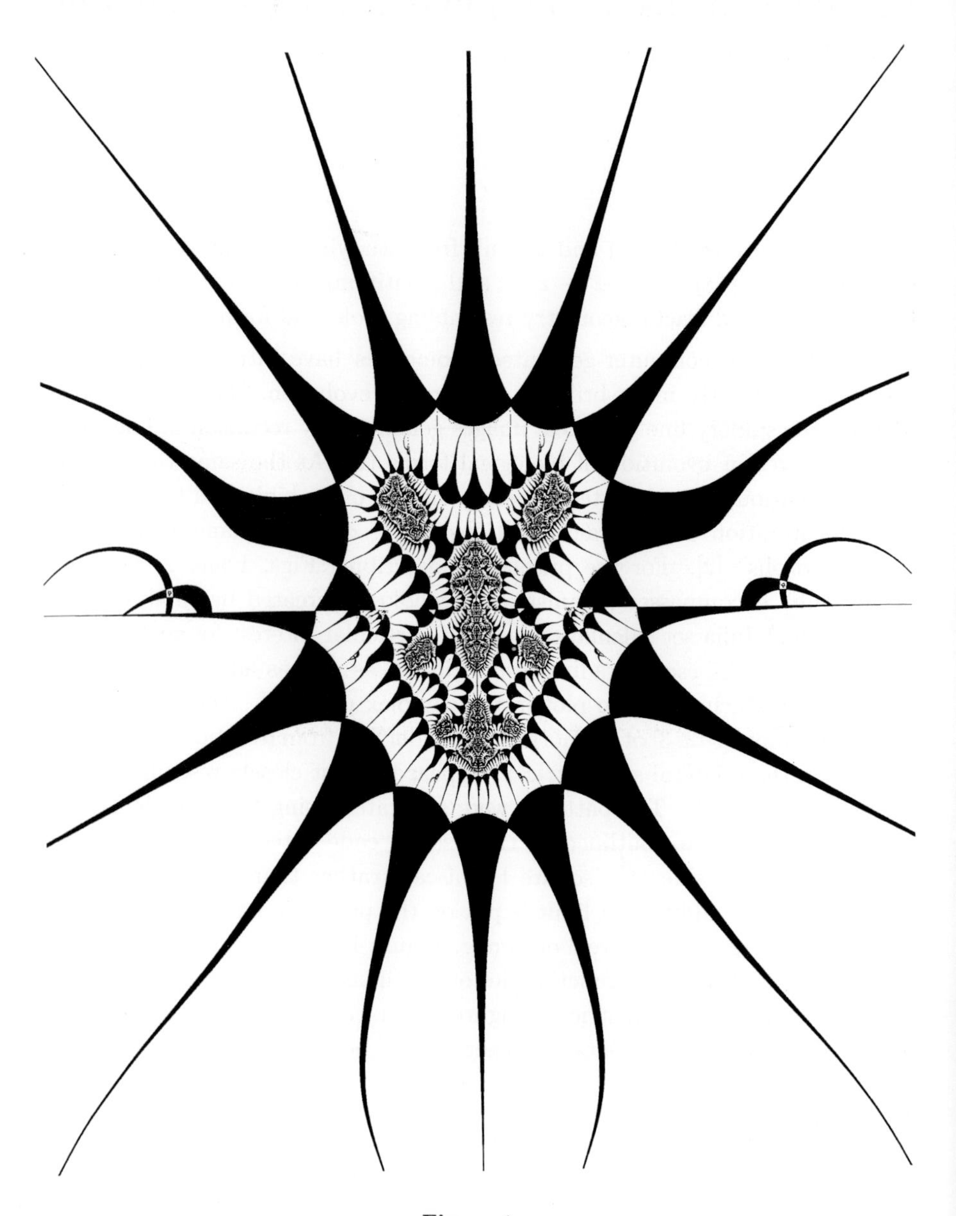

Figure 1.

Figure 2.

Pseudo code

```
Variables:Integer sx, sy, iter   Real rz, iz, x, y   Complex c, z
DO sy =1 to nx
DO sx =1 to ny
    y=ymin+sy*(ymax−ymin)/ny
    x=xmin+sx*(xmax−xmin)/nx
    z=cplx(x, y)
DO iter=1 to itermax
    z=z^z + z^4 − z + c
    rz=real(z):iz=imag(z)
    if rz^2 > es or iz^2 > es exit loop
END
    if rz^2 < ss or iz^2 < ss then print
END:END
```

References

1. R. Dawkins, *The Blind Watchmaker* (Longmans, 1986).
2. C. A. Pickover, "Biomorphs: Computer displays of biological forms generated from mathematical feedback loops", *Comp. Graphic Forum* **5** (1986) 313–316.
3. C. A. Pickover, *Computers, Pattern, Chaos, and Beauty — Graphics From an Unseen World* (St. Martis Press, 1990).
4. I. D. Entwistle, "Julia set art and fractals in the complex plane", *Computers and Graphics* **13**, 3 (1989) 389–392.
5. I. D. Entwistle, " 'Entrapped Lepidoptera': An alternative mapping of a Julia Set for the function $z - z^4 - z + c$", *The Pattern Book: Fractals, Art, and Nature*, ed. C. A. Pickover (World Scientific, 1994).

Ian D. Entwistle

"Fractal Turtle" and "Elephant Star": Multiple Decomposition Mappings of $f(z) : z \to z^5 + c$ in the Complex Plane

The patterns illustrate a Julia set derived by iteration of the function $z \to z^5 + c$ in the complex plane using a non-standard escape test and additional tests utilizing the values of z and $|z|$ at escape to control the point plotting.

Mappings in the complex plane derived from iteration of a wide range of algebraic functions have been explored by computer graphical methods in order to study their fascinating and often beautiful fractal geometry [1]. Of special interest is the way in which the iterative mathematics produce "chaos" patterns close to the divergent and bounded point boundaries. For higher polynomial functions such as $z \to z^5 + c$ the mappings are not visually very appealing since the area of chaos is small and the "quasi" circle outlines which illustrate the different divergence rates are close to the bounded set outline. The use of alternate divergence tests, especially those that make direct use of the values of z or $|z|$, have been investigated [2]. They frequently enhance the mappings of particular functions. For generation of the patterns illustrated in Figs. 1 and 2, a number of such non-standard tests were used. Binary decomposition has been described in detail [3]. In brief, a complex number can be considered to represent a vector in the complex plane which has a direction from the origin as well as magnitude. If the angle of this direction is determined for each point and the point plotted according to whether this angle is $>$ or $< 180°$ then a pattern of alternate black and white will result. Multiple decomposition will be affected if the $360°$ angle range is divided into numerous smaller ranges and more color control is applied to these. Effective use of this method has been described for illustrating alternative mappings of the Mandelbrot set [4]. A modified version of the algorithm Binary Decomposition Method for Julia Sets (BDM/J) [5] was utilized to generate Figs. 1 and 2. For computer languages which do not support complex variables, algebraic substitution [6] $z = x + iy$ and $c = a + ib$ into $z^5 + c$ results in the two equations (1) and (2) which are iterated using the escape test $(z\text{imag})^2 > 10$. The angle data is utilized by calculating the value (in radians) using equation (3). Color values

Figure 1.

Figure 2.

of black and white are then related to the integer angle number associated with each divergent point. Note that the ratio $z$imag/$z$real uses the data generated in the iteration loop. To obtain the pattern displayed by the bounded points, the minimum value of $|z|$ reached at maximum iteration is converted to an absolute integer using equation (4). Values of 0 and 1 are then assigned to these integers for printing. The data for generating the patterns is appended. Note that the "turtle" shape of Fig. 1 was achieved by elongating the imaginary axis.

Appendix

Equations
(1) $z\text{real} = x^5 - 10^* x^{3*} y^2 + 5^* x^{4*} y + \text{creal}$
(2) $z\text{imag} = y^5 - 10^* y^{3*} x^2 + 5^* x^* y^4 + \text{cimag}$
(3) Integer angle = INT(ABS(ARCTAN($z$imag/$z$real))) MOD 40
(4) Integer $|z|$ = INT(Min$|z|$*120)

Data	Pixels	$x$min,	$x$max,	$y$min,	$y$max,	Iterations,	creal,	cimag
Fig. 1.	2400*3300	−1.5	1.5	−1.5	1.5	50	0.4	0.68
Fig. 2.	2400*2400	−1.5	1.5	−1.5	1.5	50	−0.55	−0.6

References

1. C. A. Pickover, *Computers, Pattern, Chaos, and Beauty — Graphics From an Unseen World* (St. Martins Press, 1990).
2. H.-O. Peitgen and D. Richter, *The Beauty of Fractals* (Springer-Verlag, 1986).
3. I. D. Entwistle, "Julia set art and fractals in the complex plane", *Computers and Graphics* **13**, 3 (1989) 389–392.
4. I. D. Entwistle, "Julia sets: Alternative mapping of polynomial Julia sets", *Fractal Report* **15** (1991) 2–6.
5. [2], pp. 40–44, 64–76.
6. J. D. Jones, "Three unconventional representations of the Mandelbrot set", *Computers and Graphics* **14**, 1 (1990) 127, 129.
7. H.-O. Peitgen and D. Saupe, *The Science of Fractal Images* (Springer-Verlag, 1988).
8. F. J. Flanagan, *Complex Variables: Harmonic and Analytical Functions* (Dover Publications, 1972).

Ian D. Entwistle

"Floral Table": A Mapping of the Function $z \to z^4 - z + c$ in the Complex z Plane

The circular pattern in Fig. 1 illustrates a mapping of the polynomial $f_c(z) = z^4 - z + c$ in the complex z plane for which the divergent points in the z plane iteration of $z + c$ were identified by the iteration value for which $|z| < 2$. This definition of the escape radius implemented by Mandelbrot [1] determines that the curve for iteration value $k = 1$ is a circle. Alternative tests for divergence of points in similar mappings have been used [2]. All the points in the bounded set were iterated to a limit of 150.

Aesthetically pleasing patterns are generated by the mappings of linear polynomials of the type $z^n + c$ in the complex plane and their properties have been widely studied. Many algorithms for achieving variations in mappings have been reported [2]. Although the perimeters of the bounded sets for these mappings appear to increase in complexity with increase in n, the higher rates of divergence note for points close to the bounded circumference lessens the aesthetic appeal of the resulting images. This is also due to the tendency as n increases for the geometry of the outline to resemble more of a cirlce. One alternative approach to obtaining more appealing images from mappings of functions in the complex plane which has not been widely studied does produce attractive patterns [3]. In the present example, addition of $-z$ to the function $z^4 + c$ does produce a function which is readily mapped in both z and c planes and, as the pattern Fig. 1 illustrates, has a more attractive "Mandelbrot" set. Other examples of mappings produced by function summation appear elsewhere in this book. The algorithm described in the pseudo code below for Figs. 1 and 2 is a modified version of the Level Set Method [4]. In order to better differentiate the areas of the pattern mapped by divergent points close to the bounded areas in black and white, an alternative strategy for plotting the points was adopted. Bounded set points and those which diverge at rates differing from neighboring points are plotted. In addition, the circular black area was mapped by printing all the points which diverged after only three iterations. Figure 2, a magnification of the sea horse shapes in Fig. 1, illustrates clearly the fractal nature of the patterns. The input data is appended.

Figure 1.

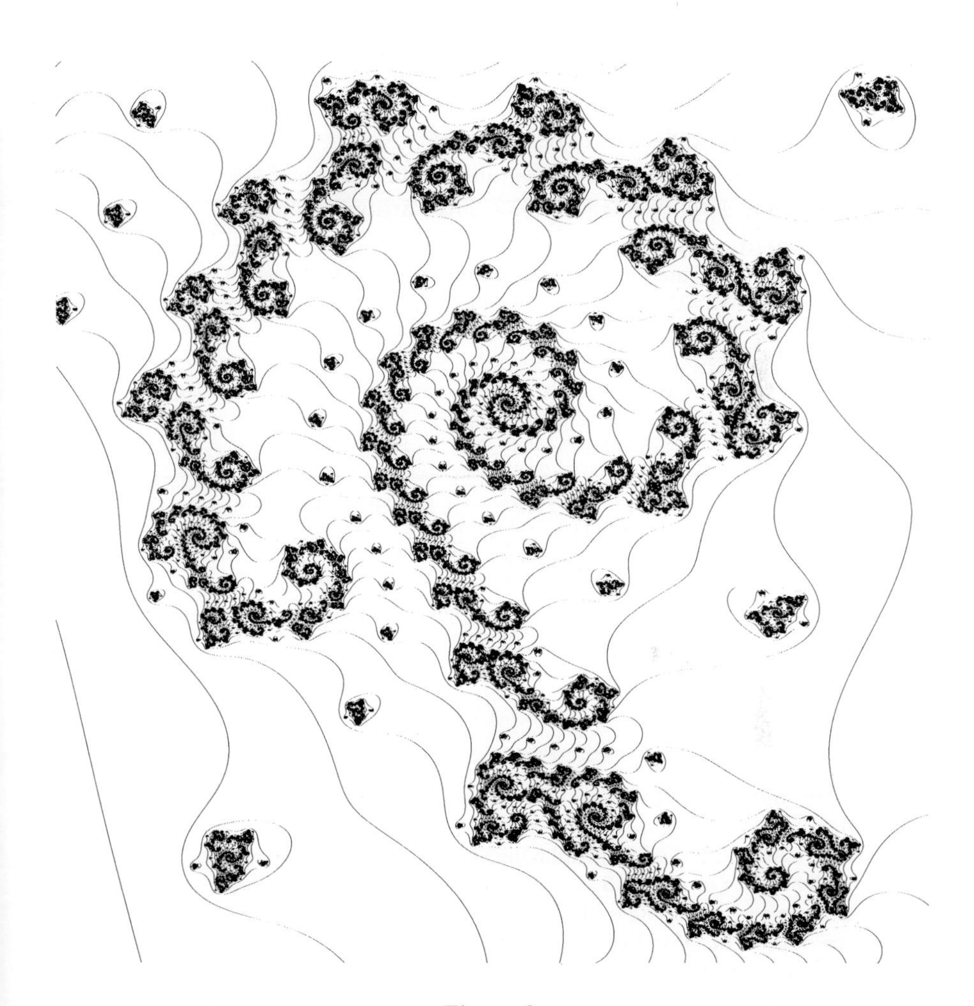

Figure 2.

Appendix

	$nx = ny$,	$x$min,	$x$max,	$y$min,	$y$max,	itermax
Fig. 1.	2400	−1.2	1.2	−1.2	1.2	150
Fig. 2.	2400	−.686	−.665	.3313	.356	150

Pseudo code for modified Level Set Method

```
Arguments: nx, ny, xmin, xmax, ymin, ymax, intermax
Variables: real x, y, cs integer: sy, sx, iter array: cc(nx)
BEGIN    FOR sy=1 To ny
                cy=ymin+sy*(ymax - ymin)/ny
         FOR sx=1 TO nx
                cx=xmin + sx*(xmax - xmin)/nx
                x=0:y=0
         FOR iter=1TO itermax
                      zx=x^4 - 6*x^2*y^2 + y^4 - x + cx
                      zy=4*x^3*y - 4*x*y^3 - y + cy
                      x=zx : y=zy : cs=x^2 + y^2
                      IF cs > 4 THEN exit LOOP to subroutine
                      END LOOP:PRINT
                      SUBROUTINE
                      cc(sx)=iter
                      IF cc(sx) - cc(sx - 1) <> 0 THEN PRINT:END ROUTINE
                      IF iter=3 THEN PRINT:END ROUTINE
```

References

1. P. Fisher and W. Smith, eds., *Chaos, Fractals, and Dynamics* (Marcel Dekker, 1985).
2. I. D. Entwistle, "Methods of displaying the behaviour of the mapping $z \to z^2 + c$", *Computers and Graphics* **13**, 4 (1989) 549–551.
3. C. A. Pickover, *Computers, Pattern, Chaos, and Beauty — Graphics From an Unseen World* (St. Martins Press, 1989).
4. M. Barnsley, *Fractals Everywhere* (Academic Press, 1988).
5. H.-O. Peitgen and D. Saupe, *The Science of Fractal Images* (Springer-Verlag, 1988).

Ian D. Entwistle

"Entrapped Lepidoptera": An Alternative Mapping of a Julia Set for the Function $z \rightarrow z^4 - z + c$

The pattern is an alternative mapping of a Julia set for the function $z \rightarrow z^4 - z + c$ which utilizes an escape test $|z_{\text{real}}|$ or $|z_{\text{imag.}}|$ <sqrt 20 to determine the divergent points in the complex plane.

Julia sets for the dynamic $f(z) = z^2 + c$ have been widely studied and many beautiful images resulting from a boundedness test $|z_n| > 2$ have been published [1]. Fewer studies have been made of the patterns which emerge from iteration of other dynamic functions when alternative tests or additional tests are applied. Of particular interest are mappings of transcendental functions [2] and also studies with alternative tests on iteration of higher polynomials and mixed algebraic functions [3]. Others have given rise to some remarkable "biomorph" images [4]. In the pattern illustrated, divergence is assessed by testing the value of either the real or imaginary values of z. The lemniscate for iteration value $k = 1$ is thus square-shaped rather than circular. Both the real and imaginary values of the function $z \rightarrow z^4 - z + c$ rise rapidly on iteration and very few points require more than 20 iterations to reach the escape value. Since this function is a simple sum of z^4 and $-z$, the standard Level Set Method algorithm was used to define a programme listing. A modification was used to control the coloring of the map. This involves the extra subroutine test indicated in the pseudo code listed. Thus, only points in the complex plane which have values of z at escape between 20 and 25 are printed. The pattern is symmetrical about the imaginary axis and so only half the points of the map need to be calculated. The input data required to generate the pattern is appended. The fractal geometry of the pattern is exemplified by the appearance of the "gothic butterfly biomorph" shapes at all scales.

Appendix

$nx = ny$,	$x$min,	$x$max,	$y$min,	$y$max,	es,	ss,	itermax,	cx,	cy
2400	-1.52	1.52	-1.52	1.52	20	25	25	$-.78$	0

Figure 1.

Pseudo code for modified Level Set Method

```
Arguments: nx, ny, xmin, xmax, ymin, ymax, itermax, cx,cy,cs,ss
Variables: real x, y integer sx, sy, iter
BEGIN    FOR sy=1 TO ny
         FOR sx=1 TO nx
              y=ymin + sy*(y max -y min)/ny
              x=xmin + sx*(x max -x min)/nx
         FOR iter=1 TO itermax
              zx=x^4 - 6*x^2*y^2 + y^4 - x + cx
              zy=4*x^3*y - 4*y^3 - y + cy
              x=zx:y=zy
              IF x^2 > es or y^2 > es THEN exit LOOP to SUBROUTINE

              END LOOP
              SUBROUTINE
              IF x^2 < ss or y^2 < ss THEN PRINT
              END ROUTINE
```

References

1. H.-O. Peitgen and P. H. Richter, *The Beauty of Fractals* (Springer-Verlag, 1986).
2. R. T. Stevens, *Fractal Programming in C* (M&T Books, 1989).
3. C. A. Pickover, "Chaotic behaviour of the transcendental mapping ($z \rightarrow$ Cosh(z)+μ), *The Visual Computer* **4** (1988) 243–246.
4. I. D. Entwistle, "Julia set, art and fractals in the complex plane", *Computers and Graphics* **13**, 3 (1989) 389–392.
5. C. A. Pickover, "Biomorphs: Computer displays of biological forms generated from mathematical feedback loops", *Comp. Graphics Forum* **5** (1986) 313–316.
6. M. Barnsley, *Fractals Everywhere* (Academic Press, 1988).

S. Dean Calahan and Jim Flanagan

Self-Mapping of Mandelbrot Sets by Preiteration

The images described here show how pictures of Mandelbrot sets of $f_c(z) = z^2+c$ and $f_c(z) = z^3+c$ change shape as the points constituting them undergo additional individual iterations. The canonical M-set computation colors a pixel based on the escape conditions of the associated point in the complex plane. "Self-mapping" by preiteration colors the pixel based on the escape conditions of forward points in the orbit of the point under consideration. That is, for a point $c \in \mathbf{C}$, iterate ck times, the result being $c_k \equiv f_c^k(0)$. Color c based on the escape conditions of c_k.

Let $M_n = \{c \in \mathbf{C}\colon f_c^k(0) \nrightarrow \infty \text{ as } k \to \infty\}$. $M_1(= M_0)$ is just the Mandelbrot set for the function under study (the first two sets in this sequence are the same when, as customary, the initial z point is considered to be 0+0i). M_2 maps the escape conditions of the first iterates of points in M_1 back to the original point. M_3 maps the escape conditions of the second iterates of each point in M_0 back to the original point. Alternately, M_3 maps the escape conditions of the first iterate of each point in M_2 back to the point in M_2. Thus, the sequence of maps $M_0, M_1, M_2, \ldots$, suggest the description that each map results from "folding" or "kneading" the points of the previous map by one iterative step.

The Figures

Figures 2 and 3 show sequences of preiteration maps for $z^2 + c$ and $z^3 + c$. Figure 3 depicts the topmost bud of Fig. 1 : M_{20} at higher magnification (corners at $-0.4 + 1.14$i, 0.14+0.6li). Figure 4 enlarges the entire map shown in Fig. 1 : M_{10}.

Commentary

An apparent feature of these sequences is that as the number of preiterations grows larger, the maps look more and more like M_0. At each step, groups of buds vanish, replaced by "ghosts" of the original, apparently a kind of dust suggesting the outline of the original bud in M_0 (note especially Fig. 1: M_{20}

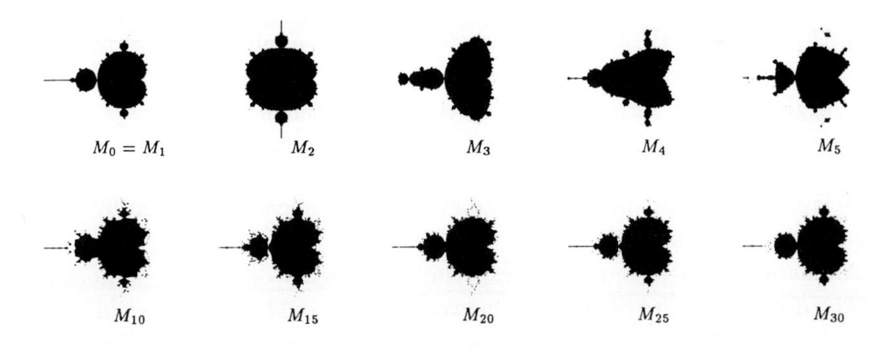

Figure 1.

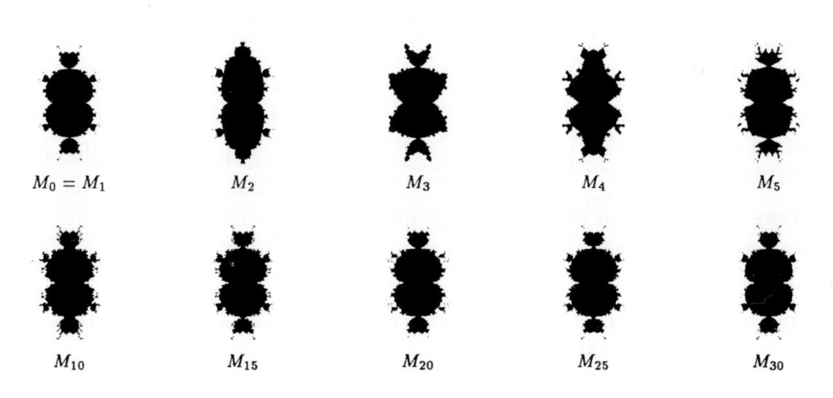

Figure 2.

Figure 3.

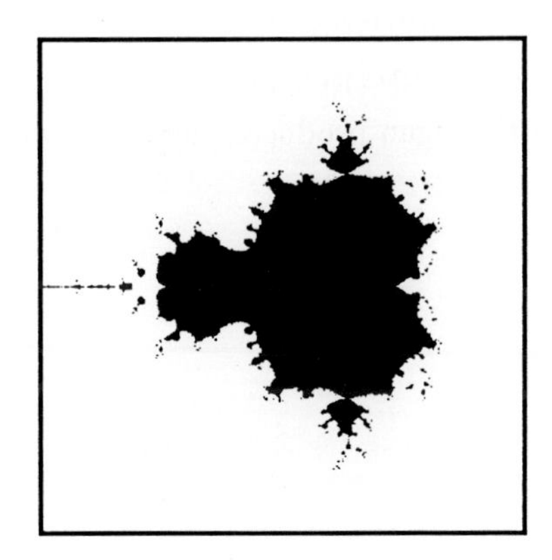

Figure 4.

and M_{30}, and Fig. 3). As the process described essentially throws away points of the original set which escape quickly, this is perhaps not surprising.

Suggestions for further study of this phenomenon immediately suggest themselves:

- Relating the "ghost-bud-dutsts" to external arguments generated by binary (or n-ary) decompositions.
- Animation of these sequences, especially for higher degrees of preiteration. The discrete nature of the step between maps would probably yield dissatisfying animations for a low degree of preiteration. Perhaps in the future, the concept of "partial iteration" will acquire meaning, or another model for this process will be discovered, that affords a smooth transition between frames.
- Study of the "ghosting" effect. The disappearance of parts of the set indicates that the forward orbits of some of the points inside a Mandelbrot set contain points outside the set. How else might these points be different?
- Preiterated Julia sets. Such sets should be interesting themselves, as well as helping to illuminate the "ghosting" effect.
- Comparison with vector field maps. This mapping technique emulates philosophically the study of vector fields, measuring a kind of "flow" in which points are carried to other points by the process under consideration. Further investigations might profitably compare and contrast pictures of the vector fields of iterative steps with pictures of the kind shown here.

Acknowledgments

A SPARCstation 2 running software under development by co-author Flanagan produced these images.

S. Dean Calahan

Unrolling the Mandelbrot Set

These images graph some measurements on components of a binary decomposition of the Mandelbrot set, M. For excellent renderings of binary decompositions, see [1] p. 74.

Figure 1, a schematic diagram of the binary decomposition, emphasizes its regular tree structure — a collection of fork bifurcations. The horizontal boundaries between black and white cells correspond to segments of boundaries between level sets of M. The vertical boundaries divide differences in the escape conditions of groups of points in a level set. For points in white cells, the orbit escapes above the real line; for points in black cells, the orbit escapes below the real line.

Within a level set the number of cells n is a function of the escape time e of points in the level set: $n = 2^e$. Two kinds of boundaries divide the cells within a level set, in an alternating fashion: the outer "tine" of a fork, crossing the level set and forking yet again, and the middle tine of a fork, crossing an infinite number of level sets and approaching the tip of a tendril. At a bifurcation, the outer tines (the "handles" of new forks) are connected to the left and right crossbars. The middle tine is always connected to the right crossbar, separated from the left crossbar by a gap. In a picture of M, few of the tines or crossbars are straight (actually only the outermost ones): indeed the curving of these forks exposes some of the dynamics of M.

The Figures

Figures 2–4 show graphs of measurements performed on a binary decomposition of the level set with escape time 7, which consists of 128 cells. Only 64 cells are depicted due to the mirror symmetry of M. The figures are centered on an axis as they look prettier that way, and the nature of the patterns seems more apparent. Thus, the measurements discussed below refer to the upper half of level set 7. In each case, the leftmost component of the graph represents the tine crossing level set 7 at the leftmost tip of M (almost touching the leftmost tip of M), the rightmost component represents the tine crossing the level set at the rightmost portion of the set (entering the main cardioid), and the longest measurement corresponds to the portion of the level set in the

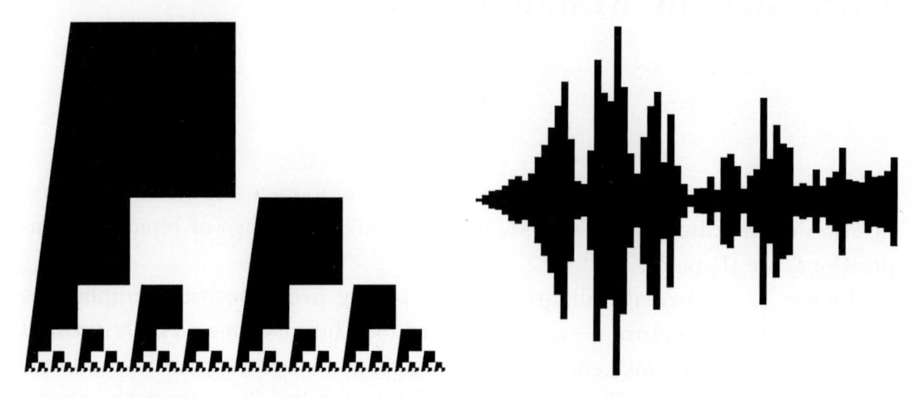

Figure 1. Schematic. **Figure 2.** Fork handles.

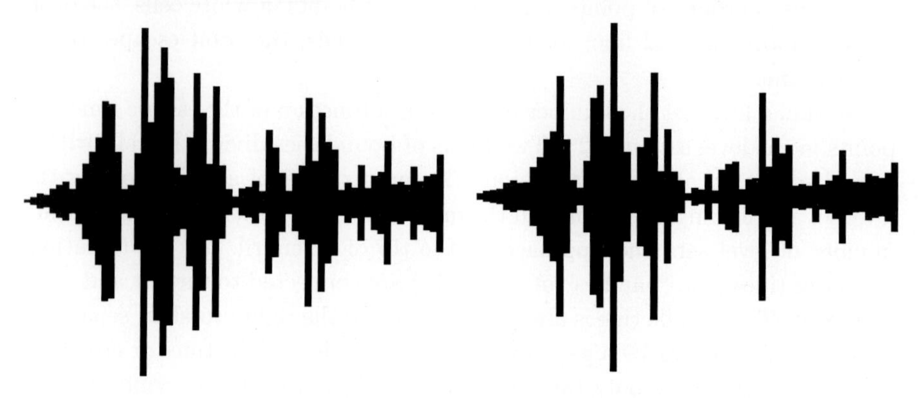

Figure 3. Left branches. **Figure 4.** Right branches.

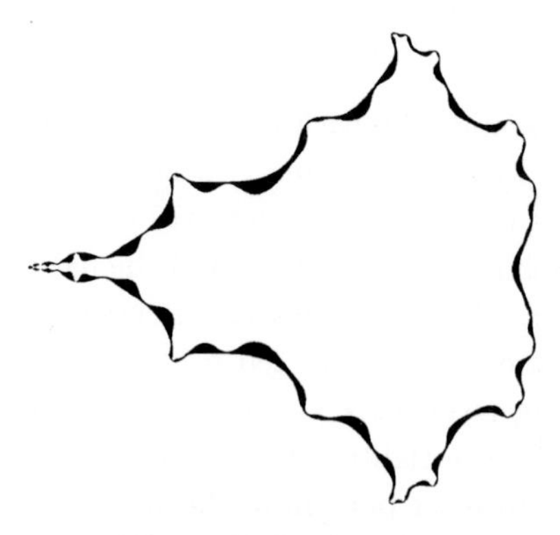

Figure 5. Level set 7.

midst of the Seahorse Valley.

Figure 2 depicts measurements performed on the outer tines of the forks crossing the level set. Odd numbered bars (counting from the left) correspond to the left-hand tine, even numbered bars correspond to the right-hand tine. Figure 3 depicts measurements on the left-hand crossbars of the bifurcations, on the boundary between level sets 6 and 7. Figure 4 depicts measurements of the right-hand crossbars of the bifurcations, also on the boundary between level sets 6 and 7. Figure 5 is a plot of the entire level set with an escape time of 7, and escape radius of 4. The measurements reflect the relative curve lengths, accurate to within a few percent. Graphs of the components of other level sets exhibit similar characteristics: tapering off to the left, almost constant towards the right, and multiply pinched and noisy in between.

Commentary

The magnitude of the difference between the longest and shortest components of bifurcations in the same level set is impressive. A visual inspection of pictures of binary decompositions suggests that the variance is smaller at level sets with escape times less than 7; increasing as escape time increases. A search for regularities in the patterns of local maxima and minima in these measurements should prove fruitful. Indeed, over some intervals in these measurements, local minima occur every fourth measurement. However, this relation does not appear to be true globally, nor does it seem to hold for local maxima.

Acknowledgments

The data for Figs. 2–4 was generated by "Bud-Oriented Zooming" software under development in the author's alleged spare time.

Reference

1. H.-O. Peitgen and P. H. Richter, *The Beauty of Fractals* (Springer-Verlag, 1986).

S. Dean Calahan

Parameter Space of Mandelbrot Sets from Variation of Coefficients

The images described here show how Mandelbrot sets of the form $z = cz^4 + bz^3 + az^2$ change with changes in the ratios $a:b$, $b:c$, and $a:c$. Assuming an appropriate escape radius, preserving the coefficient ratios preserves the shape of the set: the magnification required to resolve the main body to a particular size varies inversely with the length of the coefficient vector.

The Cartesian coordinates of vectors defining a spherical triangular grid on the unit sphere in the positive octant afford the parameters of the present maps. Figure 1 presents a broad overview of this parameter space; Fig. 2 magnifies the corner of image 11, the canonical Mandelbrot set.

The image parameters emphasize only the main body of each set, normalizing the height of each to approximately 225 pixels (.75 inch at 300 dpi). These maps ignore the seemingly disconnected islands emerging some distance from the main bodies of many of the sets. Attempting to encompass them reduces the image size unacceptably. Escape-time or potential mappings enhance the islands' appearance, but in many cases they appear to be isolated regions of high potential rather than attracting set material.

Figure 1

The images numbered 1, 11, and 15 represent Mandelbrot sets for the polynomials $z^4 + k$, $z^2 + k$, and $z^3 + k$ respectively. Along the lines between them only two powers of z contribute; all three powers of z contribute to the interior images. Notice that the corner at image 11 apparently depends more strongly on small parameter variations than do the other corners.

Figure 2

This map expands the corner at image 11, sharing with it images 7, 11, and 12. The extraordinary length of image 28 prevents the alignment of adjacent images to the triangular grid, dramatically emphasizing the apparent higher variability in this region.

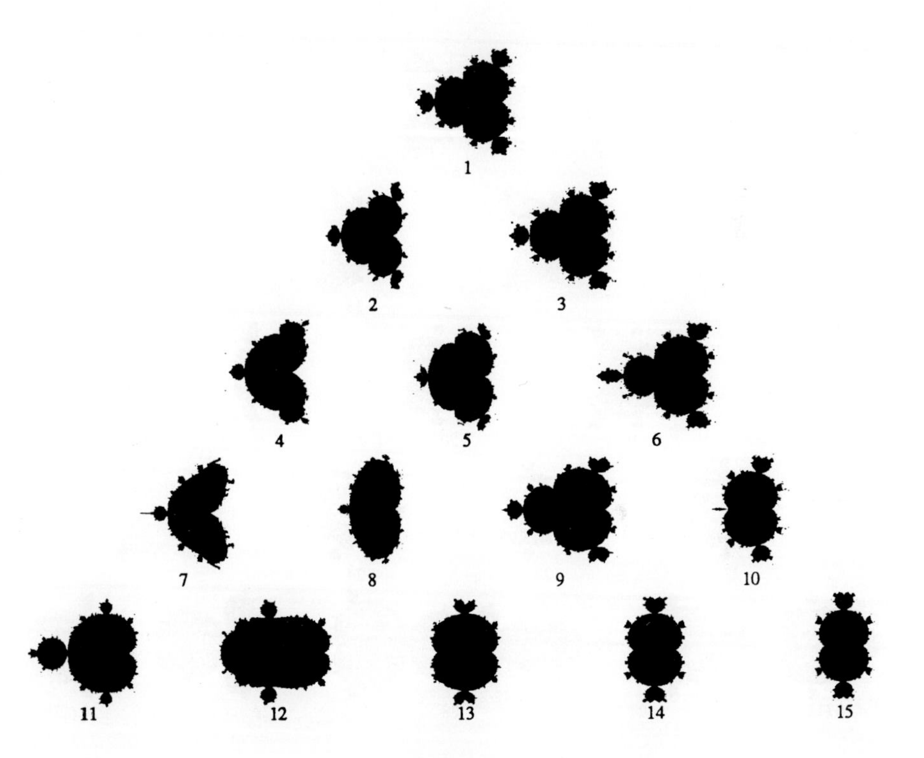

Figure 1.

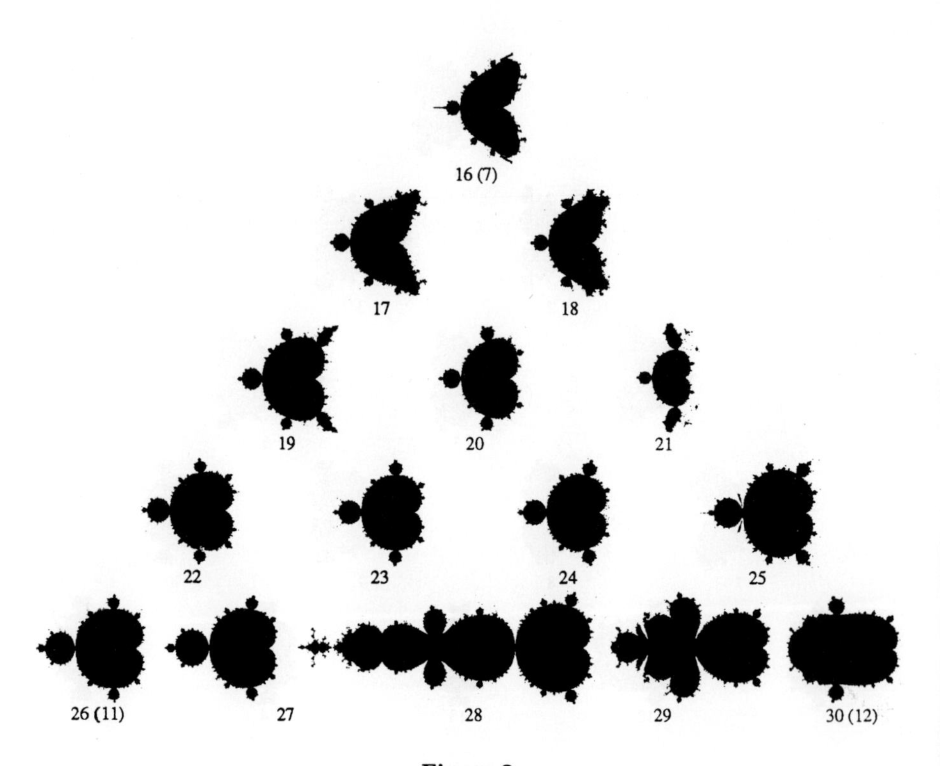

Figure 2.

Commentary

This mapping technique affords a large universe of possible explorations: Animation of the images by small changes in parameters between frames; exploration of the Mandelbrot sets at greater magnification; an orderly examination of their Julia sets; mapping other powers of z; incorporating more powers of z into similar parameter maps; allowing complex coefficients and exponents.

Any reader presently engaged in writing a fractal zooming program might consider whether to include a parameter mapping interface that generates figures of this kind. Other shapes also lend themselves as candidates for parameter maps. For example, the program that generated these images iterates polynomials up to degree five, so to accommodate a fourth power of z, add another vertex, forming a square with three edges identical to those here, a new fourth edge, and interior images from the contributions of four powers of z.

Acknowledgments

A Macintosh computer running MandleZot 3.0 (by Dave Platt) produced these images. Factors of Baloney Jim Flanagan, Bob Hagen, and Ed Osenbaugh invaluably critiqued and inspired the ideas behind this pattern. Thank you all.

Nicolas Chourot and Vedder Wright

Lizards

Described here is a pattern showing an approximate copy of a famous M. C. Escher drawing of lizards [1], using the pattern and tile-editing software Mosedit®. The program itself was jointly created by N. Chourot and J. Baracs of Montreal. Although the drawing could be constructed by hand as Escher did, its creation using the computer serves as a litmus test of the flexibility and effortlessness of the graphical interface of Mosedit®.

Every periodic plane-filling pattern or tiling must belong to one of 17 planar symmetry groups, which consist of certain combinations of basic maneuvers called symmetry operations. These operations are only four in number; translation (a rigid sliding motion), rotation, reflection, and glide (a combination of reflection and translation, like footprints in the sand). To construct his famous periodic drawings, M. C. Escher had to learn the fundamental laws of planar symmetry. The Montreal team had to learn similar things, combined with tiling theory, in order to develop the software.

The lizard pattern was initially created from a regular hexagonal cell whose alternate vertices contain three distinct centers of threefold rotational symmetry. A center of n-fold rotational symmetry in the plane is a pivot point around which the entire pattern (or portion of that pattern) can be rotated to coincide with itself n times in one complete revolution. The threefold centers of the lizard pattern are located at the creature's left cheek, at the right rear knee and at the inside of the left rear paw.

Reference

1. C. H. MacGillavry, *Fantasy and Symmetry: The Periodic Drawings of M. C. Escher* (Harry Abrams, 1976) pp. 76–77.

Koji Miyazaki and Manabu Shiozaki

Four-Dimensional Space Flowers

Described here are flower-like patterns derived from orthogonal projection into 3-space of typical 4-dimensional regular polytopes (120- and 600-cell) and semiregular polytopes which are derived from regular truncation around vertices of the 120- and 600-cell polytopes.

These semiregular polytopes have two kinds of regular and/or Archimedean-typed semiregular polyhedral cells, a fixed number of which fit together around every vertex in fixed order and any two cells have a face in common.

Each of the top row, from left to right, shows a portion of the cell-, vertex-, face-, and edge-center projection of the 120-cell polytope, and each of the bottom row, from left to right, a portion of the vertex-, cell-, edge-, and face-center projection of the 600-cell polytope. A vertex-center projection means a 3-dimensional solid model having a vertex at the body-center. Each of the edge-center, face-center, and cell-center projections is similarly defined.

The others are semiregular polytopes or their portions born between the above-mentioned projections of the 120- and 600-cell polytopes.

They have the following geometrical characteristics:

1. The outermost vertices lie on concentric spheres, rather than on a single sphere.
2. Their interiors are filled with a small kind of distinct cells, juxtaposed face-to-face.
3. The volume of a cell is reduced when it occurs in the outer layers of the polytope, and may even be reduced to zero in the outermost layer.
4. Each vertex has simple cartesian coordinates and the chord factors can easily be determined.
5. They have many planar polygonal sections determined by the choice of certain vertices, edges, and faces.
6. They can be stacked in periodic and in aperiodic arrays, meeting one another across planar polygonal boundaries, or sharing certain cells.
7. Various complicated designs can be continuously and systematically derived according to changes in the direction of the projection in 4-space.

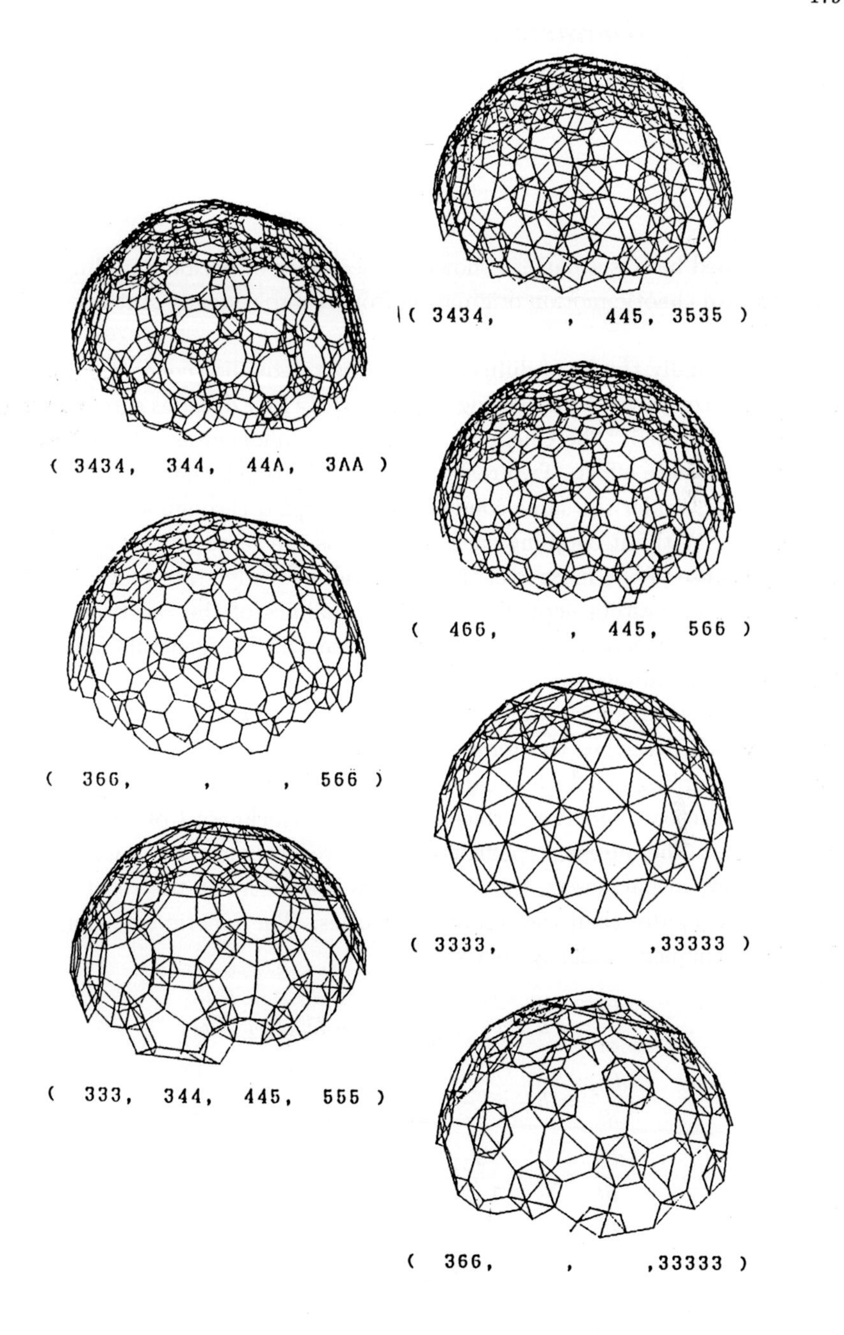

(3434, , 445, 3535)

(3434, 344, 44Λ, 3ΛΛ)

(466, , 445, 566)

(366, , , 566)

(3333, , ,33333)

(333, 344, 445, 555)

(366, , ,33333)

István Lengyel, Irving R. Epstein, and Helena Rubinstein

The Dolphin Head Bifurcation Diagram

Described here is a pattern showing a sequence of period doubling bifurcations toward chaotic motion originating from a stable steady state destabilized by diffusion.

One normally thinks of diffusion as acting to equalize concentration differences in space. However, as Turing showed nearly four decades ago in a remarkable paper entitled "The Chemical Basis for Morphogenesis" [1], diffusion can have the opposite effect. Two reactors or cells in which the same chemical reaction occurs under the same conditions are coupled through a semipermeable membrane. Without coupling each reactor has the same unique stable steady state. Because of the mass exchange this stable steady state may be unstable in the coupled system depending on the dynamics of the reaction, the ratio of diffusion coefficients, and the coupling strength. The general conditions for this type of instability are given in [2].

A model that shows diffusion-induced instability is the Degn-Harrison model, which describes the temporal behavior of a specific bacterial culture [3]. The coupling strength c can be expressed as a function of the volume V of the reactors, and the surface area A and thickness l of the membrane: $c = A/(V\,l)$. The diffusion coefficients of the reacting species must differ from one another; in particular, the components that activate the reaction should diffuse less rapidly than the inhibiting species. The kinetic equations in this 2-variable coupled model system are:

$$\frac{dx_1}{dt} = b - x_1 - \frac{x_1 y_1}{1+qx_1^2} + D_x c(x_2 - x_1)$$

$$\frac{dx_2}{dt} = b - x_2 - \frac{x_2 y_2}{1+qx_2^2} + D_x c(x_1 - x_2)$$

$$\frac{dy_1}{dt} = a - \frac{x_1 y_1}{1+qx_1^2} + D_y c(y_2 - y_1)$$

$$\frac{dy_2}{dt} = a - \frac{x_2 y_2}{1+qx_2^2} + D_y c(y_1 - y_2)$$

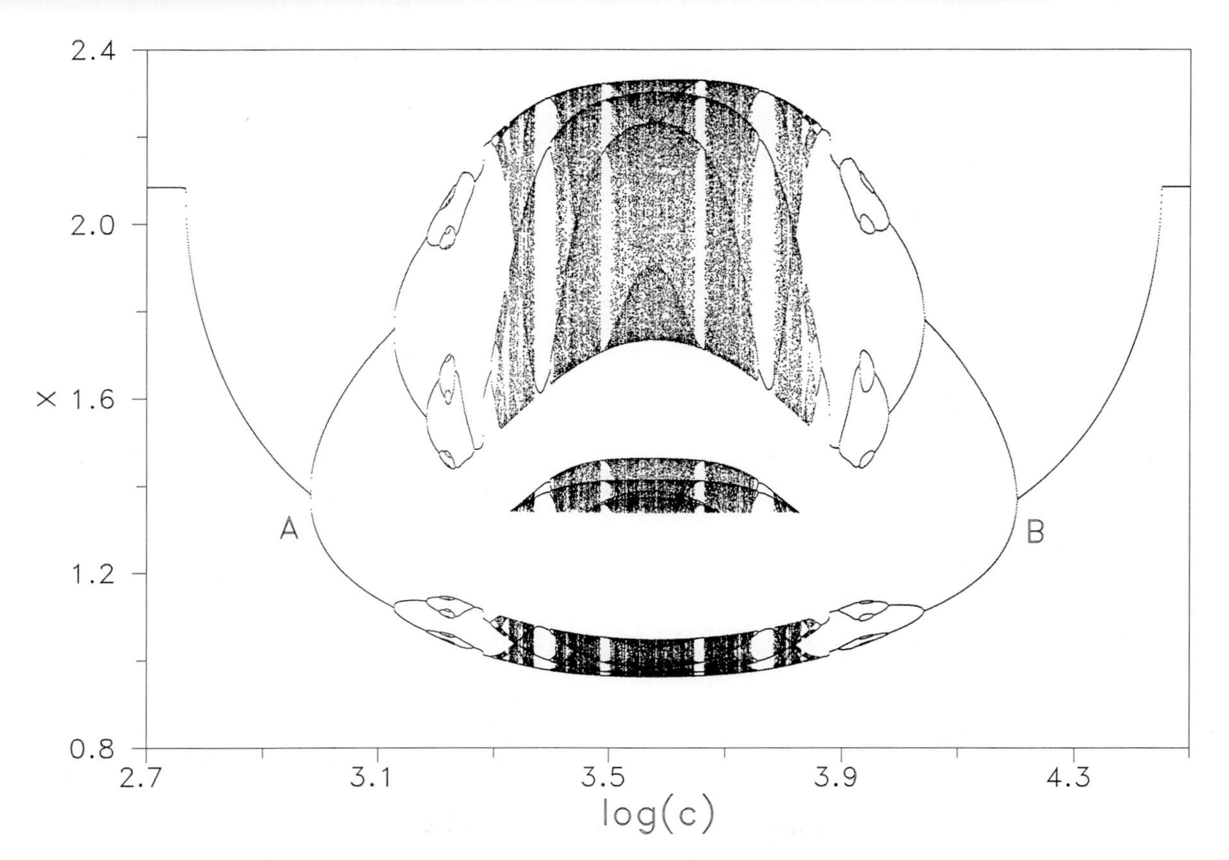

Figure 1. The dolphin head bifurcation diagram. Parameters of the differential equation system: $(a,\ b,\ q,\ D_x,\ D_y) =$ $(8.951,\ 11.0,\ 0.5,\ 10^{-5},\ 10^{-3})$.

where t is the time, x_1, y_1, x_2, y_2 are the reactants in the two reactors, and D_x, D_y are the corresponding diffusion coefficients. Without coupling the steady state of the system is $x_{1,\text{ss}} = x_{2,\text{ss}} = b - a$, $y_{1,\text{ss}} = y_{2,\text{ss}} = a[1 - q(b-a)]/(b-a)$. We have chosen parameters at which this steady state is stable and y diffuses faster than x.

To calculate the figure, we take the coupling strength as the bifurcation parameter. If c is too small or too large, the same steady state is established in both reactors. For small c, the reactors behave essentially independently, while for very large c they act as if we had only a single reactor of volume $2V$. Between these two limits the mass exchange destabilizes the steady state and new stable states appear in which the concentrations in the two reactors are different. For clarity we plot only one branch of the solutions. At points A and B oscillations appear and, after some period doubling and inverse period doubling sequences, choas is seen. The apparent symmetry is striking at first sight (especially in view of the logarithmic scale for c), but closer observation reveals that on the left-hand side there is an extra period doubling bifurcation.

Diffusion-induced instability requires relatively high coupling strength and a sizeable difference between the diffusivities of the reactants, conditions which probably cannot be attained under the standard experimental conditions applied in inorganic chemistry. In biological systems, however, the high surface/volume ratio and thin membrane walls of cells and the large differences between diffusion coefficients resulting from the wide range of molecular sizes and strengths of interactions with membranes make it far easier to realize the necessary conditions for diffusion-induced instability. It seems likely that nature has already discovered and made use of the dynamical possibilities of diffusion-induced instability to generate some of the wide variety of spatio-temporal behavior found in living systems.

References

1. A. M. Turing, "The chemical basis for morphogenesis", *Phil. Trans. Roy. Soc.* **B237** (1952) 37–72.
2. I. Lengyel and I. R. Epstein, "Diffusion induced instability in chemically reacting systems: Steady-state multiplicity, oscillation, and chaos", *Chaos* **1** (1991) 69–76.
3. V. Fairén and M. G. Velarde, "Dissipative structures in a nonlinear reaction diffusion model with inhibition forward", *Prog. Theor. Phys.* **61** (1978) 801–814.

Bob Brill

Embellished Lissajous Figures

Described here is an open-ended class of mathematically-generated line drawings based on Lissajous figures, but overlaid and sometimes entirely disguised with a variety of embellishments. Lissajous figures constitute a family of curves well known to scientists and engineers. They can be generated by the following C program:

```
for (t = 0.0; !closure(); t += Tstep) {
    x = Xamplitude * sin(Xfrequency * t + Xphase);
    y = Yamplitude * sin(Yfrequency * t + Yphase);
    plot (x, y);
}
```

To generate one of these curves, the user supplies values for the parameters (capitalized variables above). Lissajous figures are well behaved, being continuous at every point, forming elegant, sweeping curves that reveal their derivation from the sine function, closing seamlessly upon themselves, and fitting neatly into any desired rectangle (whose sides are $2 * X$amplitude by $2 * Y$amplitude). The closure function returns true when the current coordinates and heading are the same as the starting coordinates and heading.

Although such plots are handsome, they are much more interesting when embellished, as shown in the figures. Instead of plotting each point, as above, I have drawn a pair of arcs, each of which begins at the calculated point. Each arch is determined by three user-supplied parameters that specify the angle offset from the current heading, the degree of curvature (e.g., 360 is a circle, 90 a quarter circle, 1 a straight line), and the length of the straight line segments composing the arc. One arc is drawn counterclockwise and is offset to the left of the current heading, while the other is drawn clockwise and is offset to the right.

Many interesting figures can be drawn by varying the Lissajous parameters and the embellishment parameters, as described, but many other variations are also possible by altering the Lissajous equations or by changing the nature of the embellishment.

One fruitful idea was to change the heading before drawing the arcs. As each point is calculated, the heading is determined to lie in the direction of the undrawn line connecting the previously calculated point to the newly calculated point. Thus the embellishment follows the path of the curve. We can make the embellishment rotate in various ways by changing the heading after each recalculation as follows: heading $= 360 -$ heading or heading $= 360 * |\sin(t)| -$ heading or heading $= 360 * |\sin(t)|$ or even heading $= |\sin(t)|$. This by no means exhausts the possibilities. The embellishment parameters can also take $|\sin(t)|$ or $1 - |\sin(t)|$ as multipliers, causing the offset angles, curvature or line lengths to vary periodically between 0 and their specified values. Any of these modifications may be in effect separately or in combination.

Why stop at drawing two arcs? I have also drawn polygons, parameterized for a number of sides and side lengths. These also have been modified to cause the polygons to expand and contract in size. As for changing the Lissajous equations, there are many possibilities. I like the one where after the calculation of x, I add the line:

$$x * = \sin(t)\,.$$

When Yfrequency is 1 and Yphase is 0, this causes the value of x to be 0 whenever y is 0. The effect is like throwing a belt over the design and cinching it up tight across the waistline (i.e., the x axis).

Here is another such modification whose effect I invite you to discover. Initialize y to 0 outside the loop, then change the equations as follows:

$$x = X\text{amplitude} * \sin(X\text{frequency} * t + X\text{phase}) - y * \sin(t)\,;$$
$$y = Y\text{amplitude} * \sin(Y\text{frequency} * t + Y\text{phase}) + x * \sin(t)\,;$$

Space does not permit a complete description of everything I have tried. The interested reader is encouraged to explore new variations.

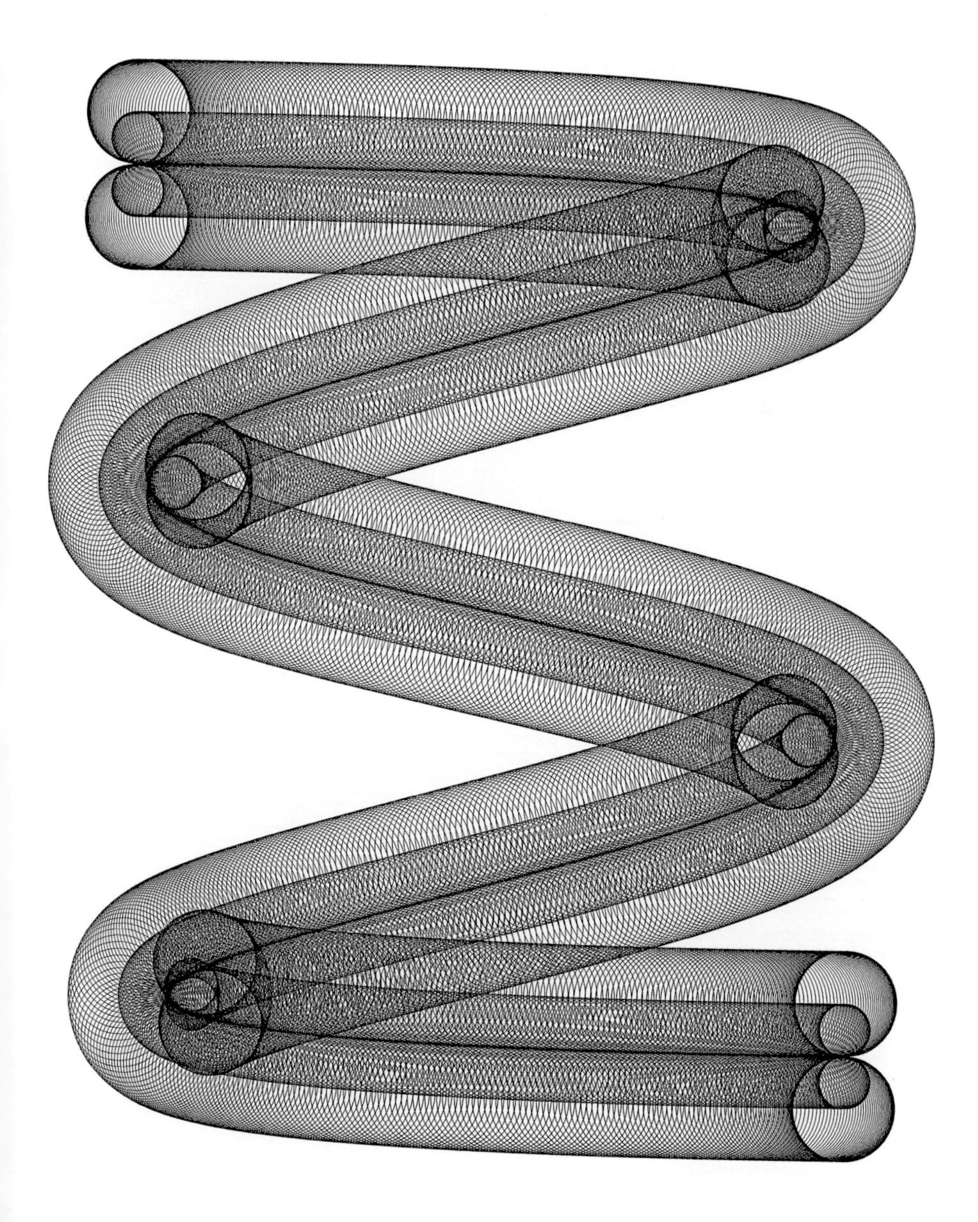

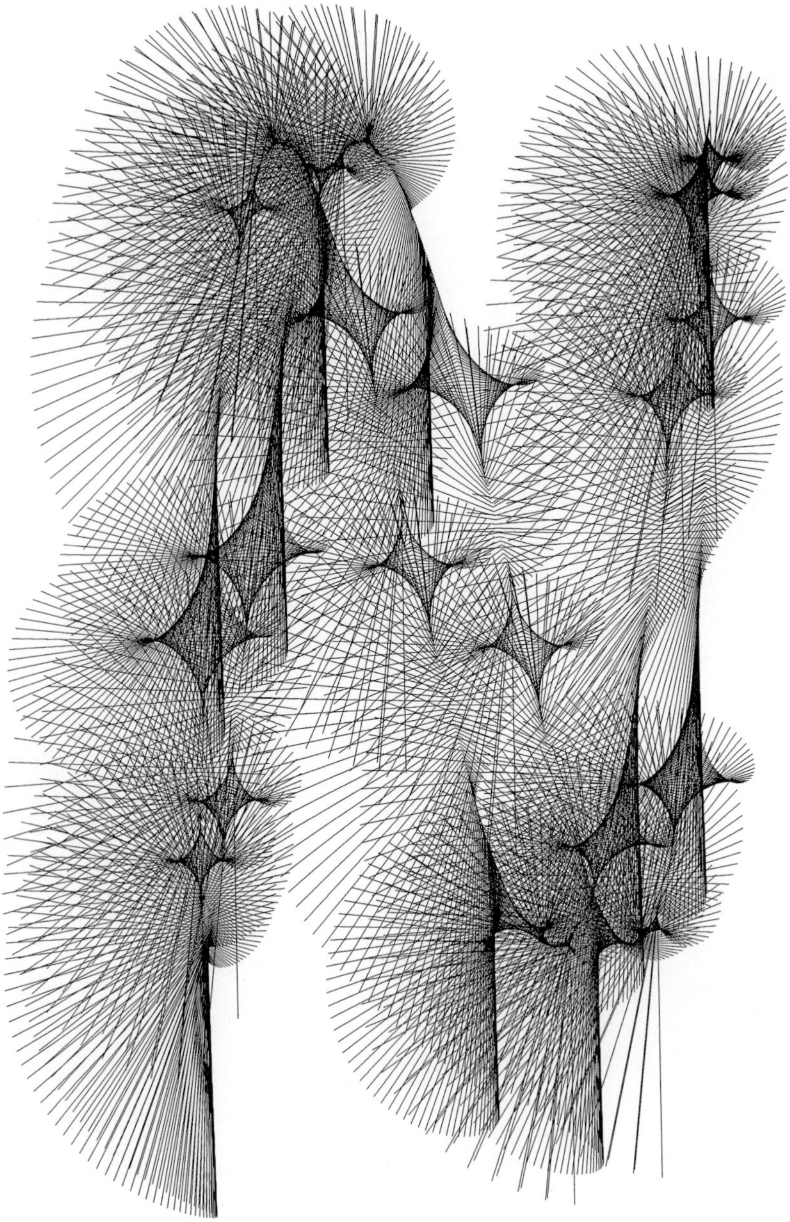

(Landscape)

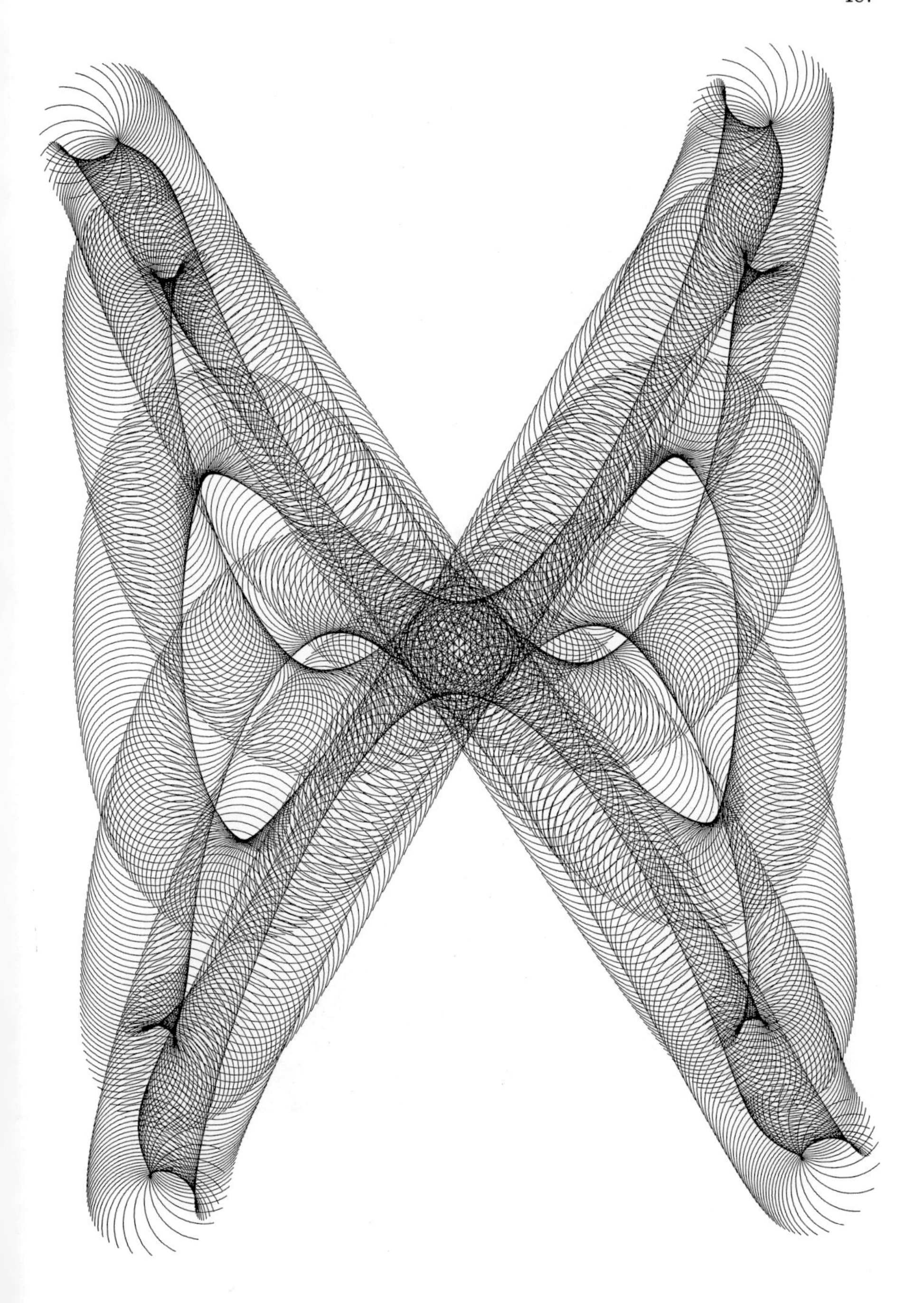

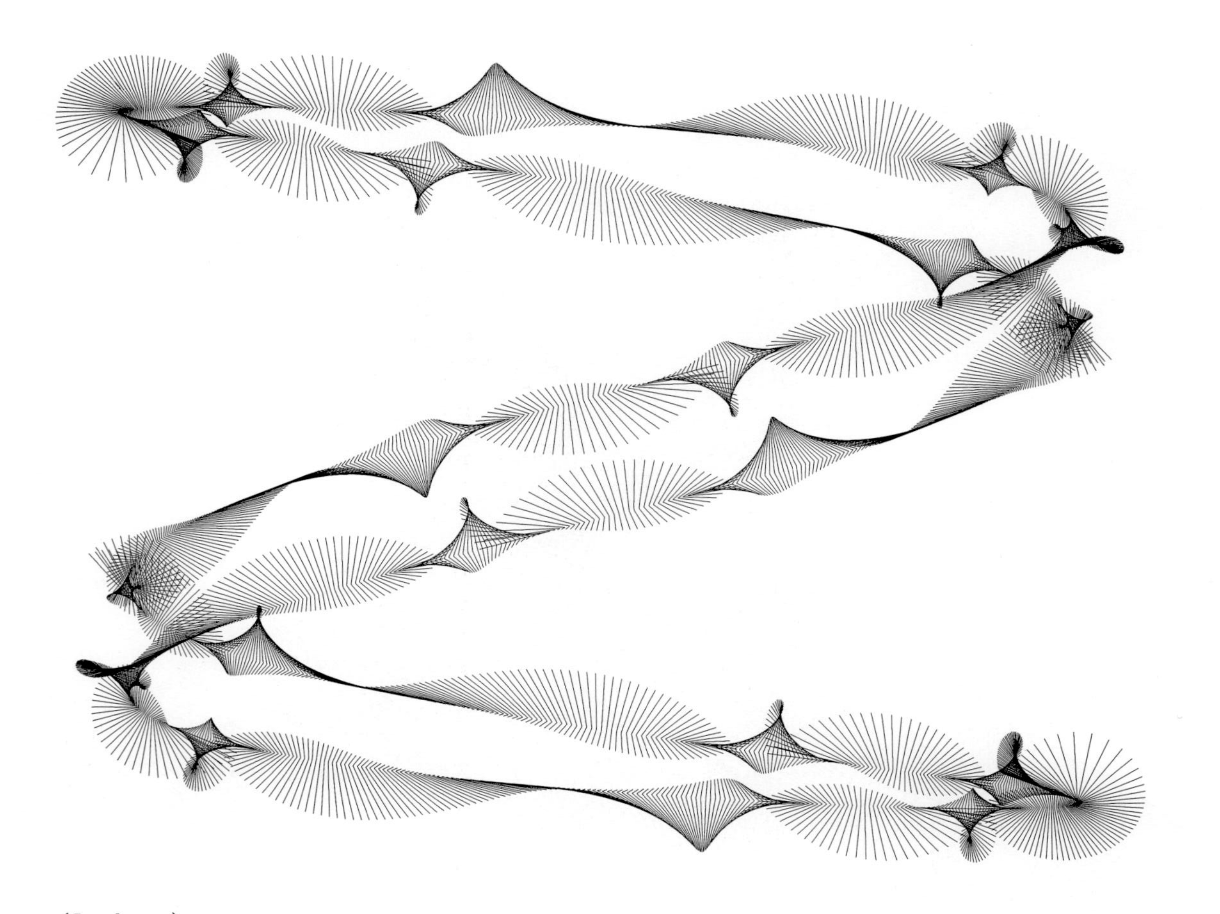

(Landscape)

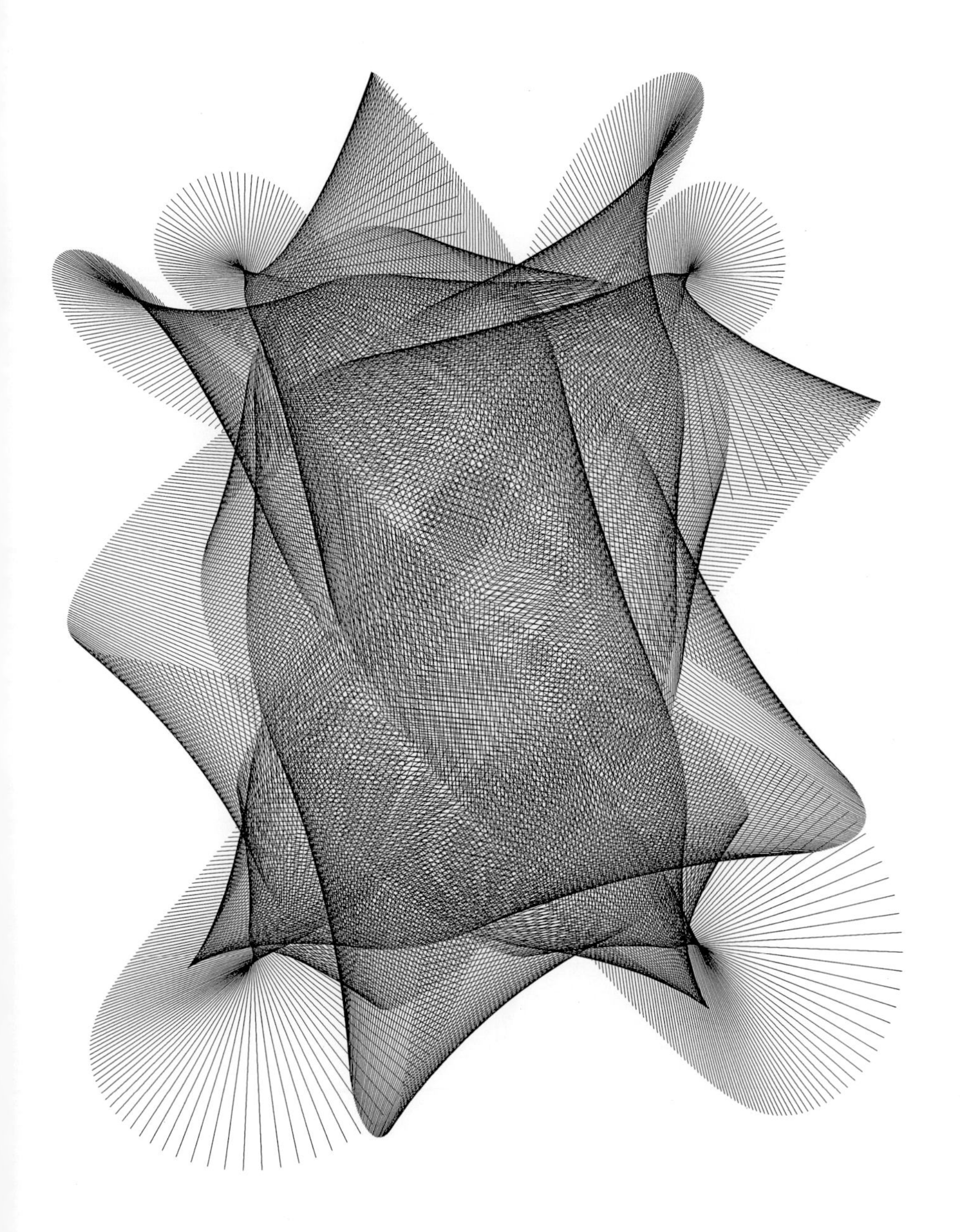

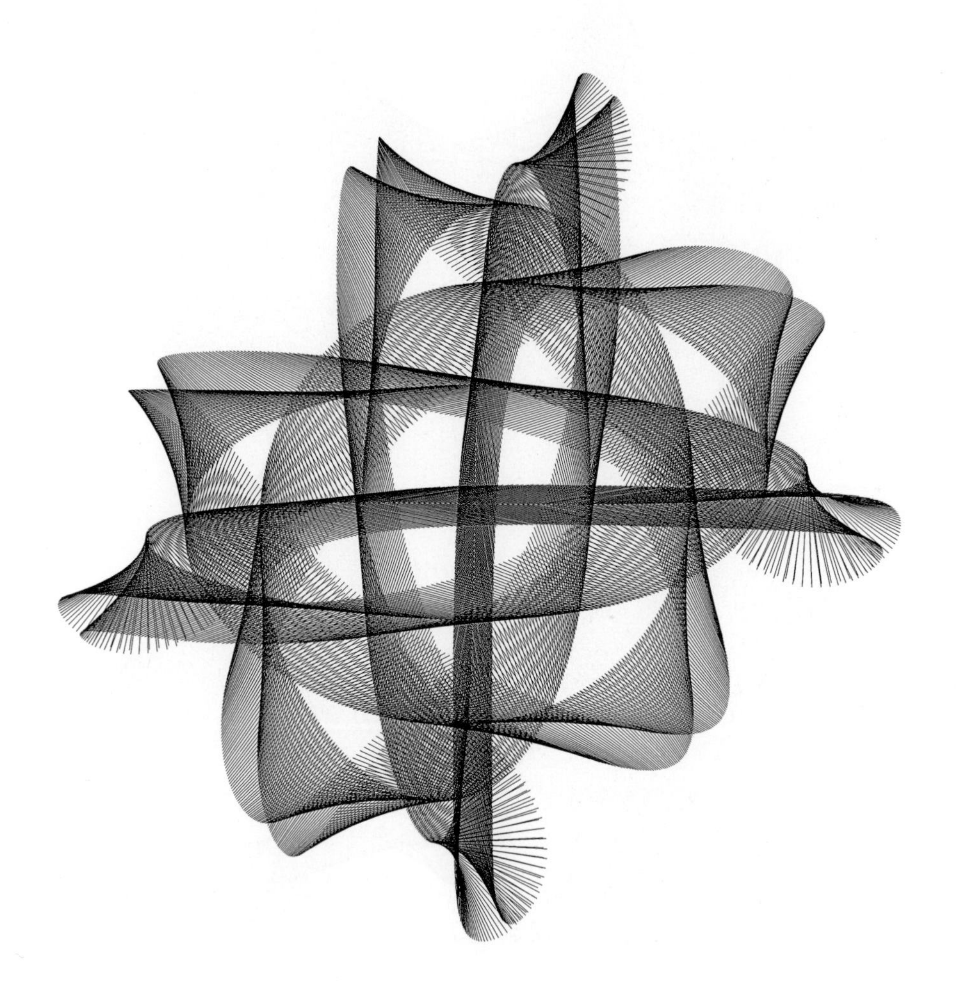

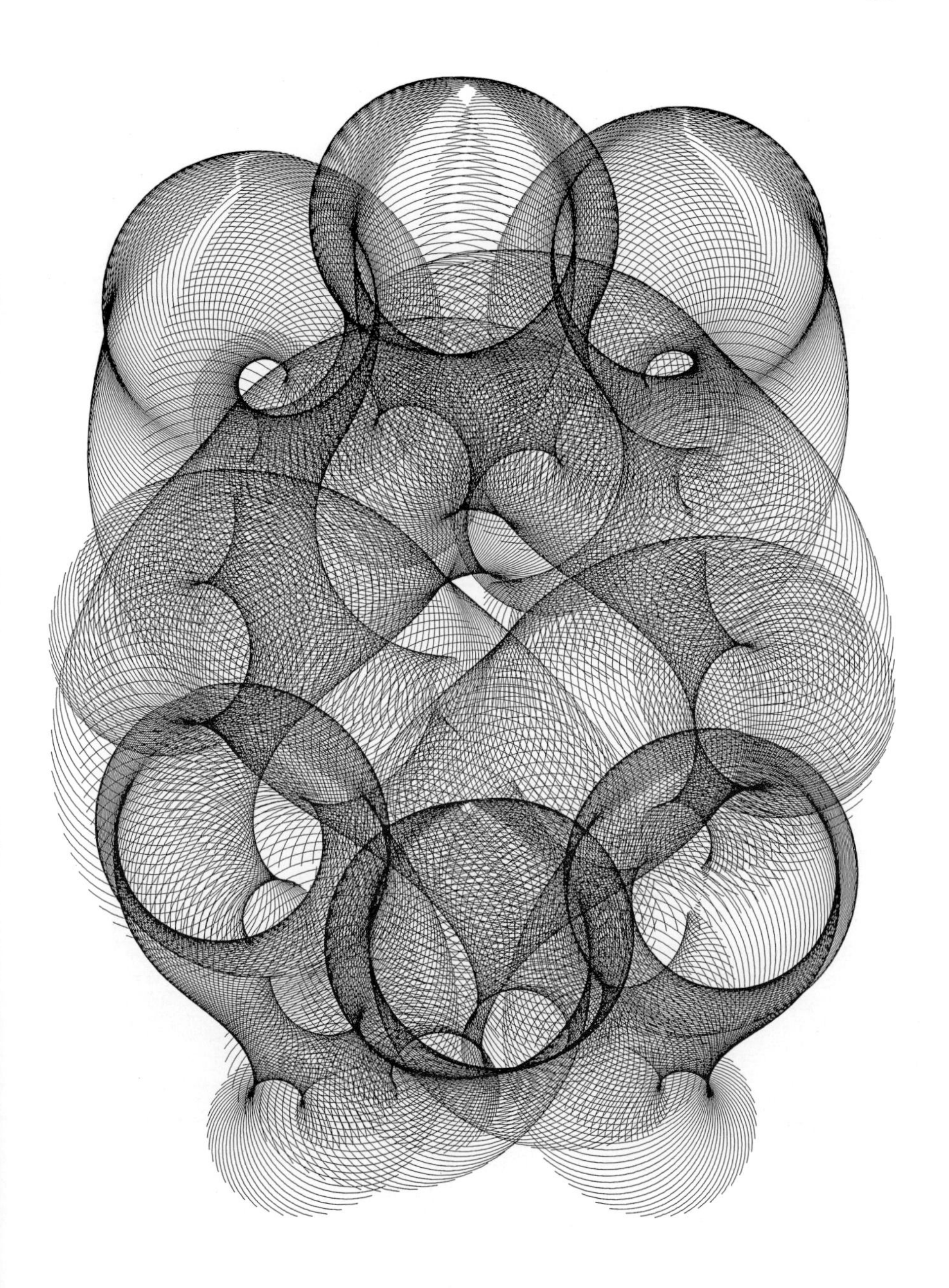

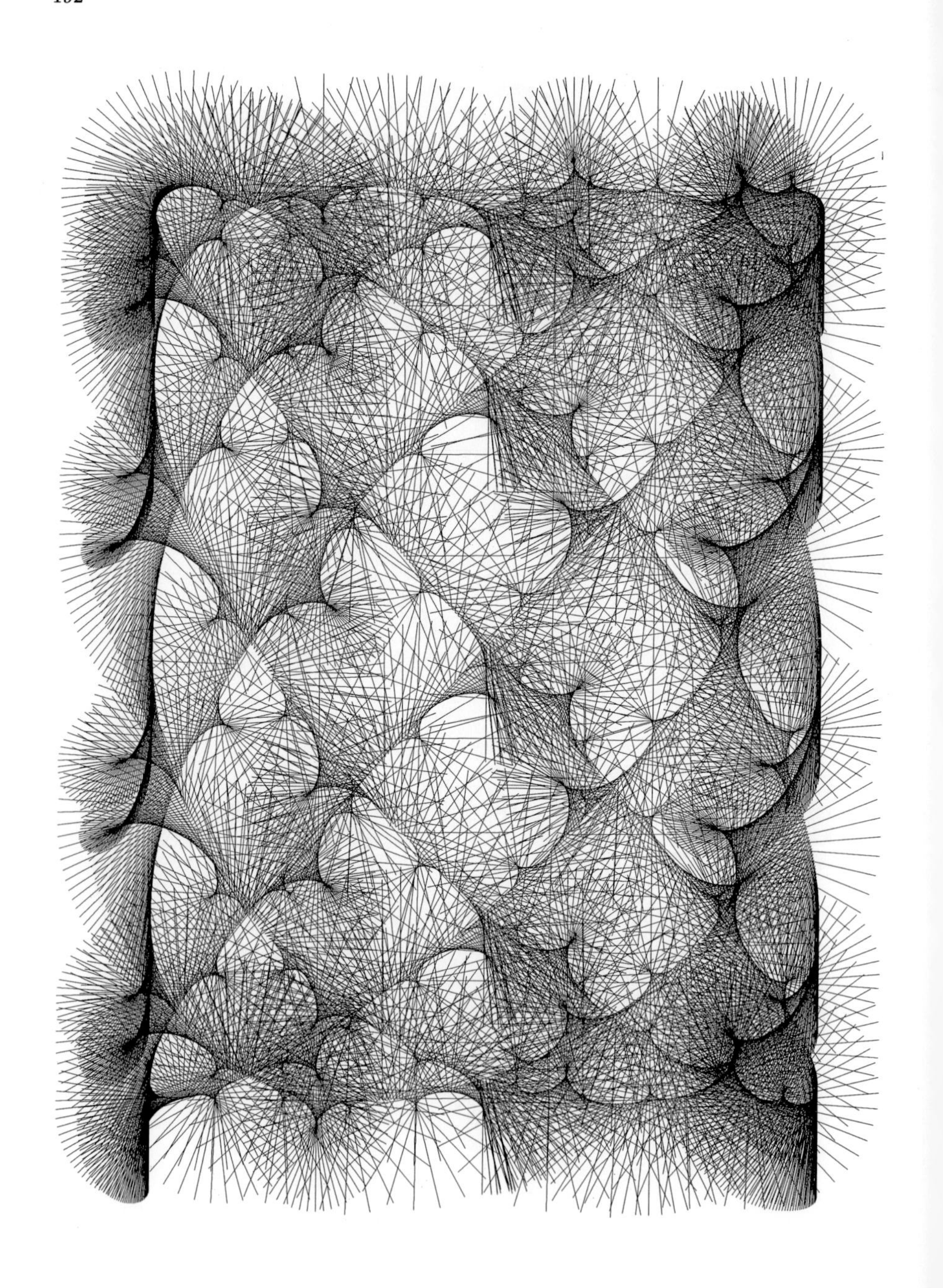

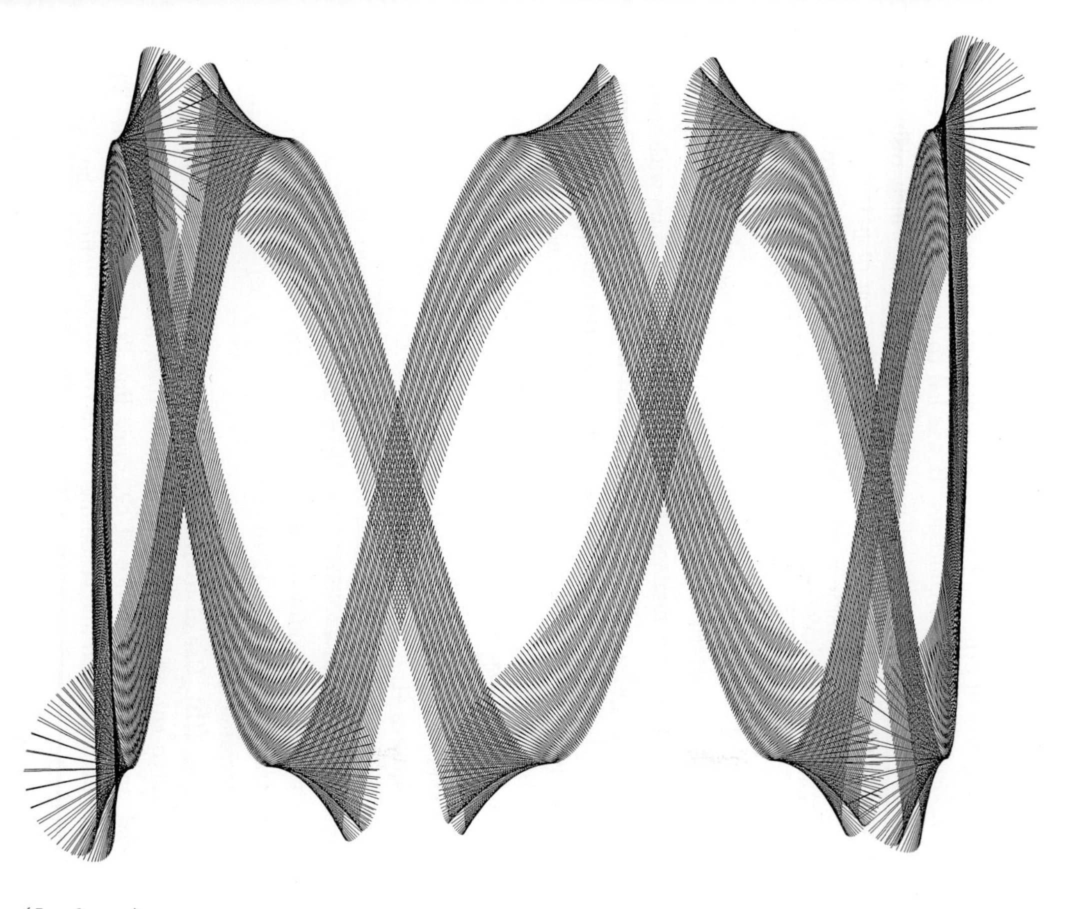

(Landscape)

S. J. Abas

Islamic Pattern

Unlike the arts of other cultures, Islamic art sets out deliberately to shun anthropomorphic forms and concepts. It was led to concentrate on the exploration of symmetry in geometrical patterns, an enterprise which resulted in an extraordinarily large, complex and elegant collection of patterns [1, 2, 3, 4]. Apart from their aesthetic merit, these patterns offer a treasure house of symmetry which make them of great interest to a large number of scientists and educators [5, 6].

The most sophisticated Islamic patterns rely on the use of concealed polygonal grids and the pattern depicted here is of this construction. It is rather an unusual example, for it manages to combine seven-rayed stars with squares and octagons. It is also a deceptive one, in that at a cursory glance it seems to contain tri-lobed symmetric regions (such as the two on the opposite sides of the center point) which on careful examination turn out not to be so.

We give below an algorithm for its construction which is simpler than the one published previously by Hankin [7].

1. Draw a grid of heptagons as shown in Fig. 1(a). This gives rise to a series of small squares s1.
2. Refer to Fig. 1(a) again. Centered on each of the squares s1, draw a circle c1 circumscribing s1 and a square s2 of side d, where d is the smallest distance between two of the nodes on the grid. From the nodes surrounding the squares, draw lines 11 and 12 to the vertices of the heptagon which lie on the edge forming the squares s1.
3. Refer to Fig. 1(b). Use c1 to draw the octagon shown then discard c1. Use 11 and 12 to cut off s2. Replace the two lines with the circle c2. From a point on the circumference of c2, draw line l3 to a vertex of s1. The intersection of l3, with line l4 joining two of the nodes, defines the radius of the circle c3.
4. Figure 1(c) shows how the pattern emerges by symmetrically performing the above steps in the region surrounding one of the squares s1.
5. Figure 1(d) shows the same when the same operations are continued on a larger region of the grid.

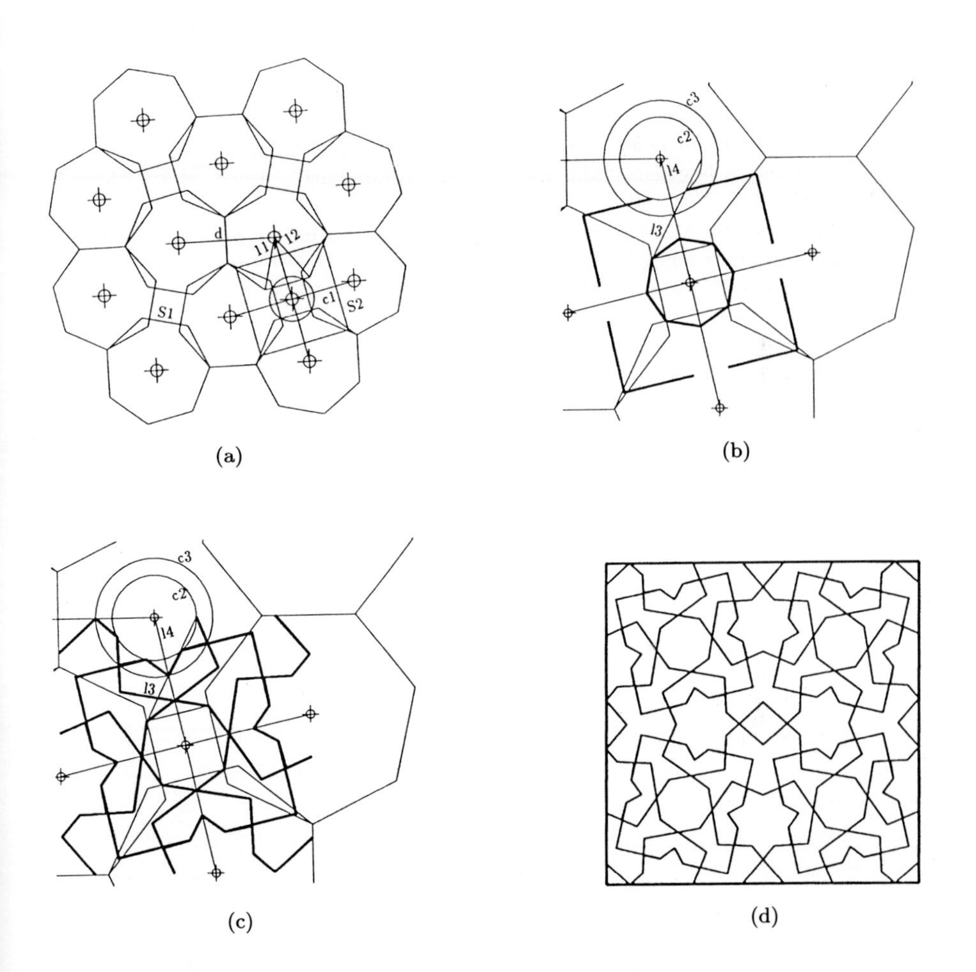

Figure 1. An unusual Islamic pattern containing seven-rayed stars, octagons and squares.

References

1. J. Burgoin, *Arabic Geometrical Pattern and Design* (Firmin-Didot, 1879 and Dover, 1973).
2. K. Critchlow, *Islamic Patterns: An Analytical and Cosmological Approach* (Thames and Hudson, 1976).
3. I. El-Said and A. Parman, *Geometrical Concepts in Islamic Art* (World of Islam Festival Publ. Co., 1976).
4. D. Wade, *Pattern in Islamic Art* (Cassell & Collier Macmillan, 1976).
5. J. Niman and J. Norman, "Mathematics and Islamic art", *Amer. Math. Monthly* **85** (1978).
6. E. Makovicky and M. Makovicky, "Arabic geometrical patterns — a treasury for crystallographic teaching", *Jahrbook fur Mineralogie Monatshefte* **2** (1977).
7. E. H. Hankin, "Some difficult saracenic designs", *Math. Gazette* **18** (1934) 165–168.

Paul D. Bourke
Swirl

The following is a means of drawing computer-generated swirling tendrils. The technique uses the following to generate a series of x, y points given any initial point x_0, y_0 and four constants a_{11}, a_{12}, a_{21}, a_{22}.

$$x_{n+1} = \sin(a_{11}y_n) - \cos(a_{12}x_n)$$
$$y_{n+1} = \sin(a_{21}x_n) - \cos(a_{22}y_n)$$

To create an image, each point x_i, y_i, after ignoring the first 10 terms in the series say, is drawn as an "infinitely small" point on the page. The resulting image shows all the possible coordinates the series can generate, that is, the attractor of a chaotic system. Any initial point x_0, y_0 (except for some rare special cases) gives the same set of coordinates and hence, the same image but in a different order.

The examples provided were created using the following parameters, the images are numbered left to right, top to bottom.

	a_{11}	a_{12}	a_{21}	a_{22}
1	−2.70	−0.08	−0.86	−2.20
2	−2.24	+0.43	−0.65	−2.43
3	+2.00	−1.00	−1.00	−2.00
4	+0.44	−1.22	+2.50	−1.50

It should be noted that the vast majority of values for a_{11}, a_{12}, a_{21}, to a_{22} do not yield interesting images, in such cases the attractor consists of only a few isolated areas.

Acknowledgments

Attributed to Peter de Jong, Léden, Holland by A. K. Kewdney in Computer Recreations, Scientific American.

Software called FRACTAL to generate these images and many other fractal and chaotic systems is available from the author for the Macintosh II family of computers.

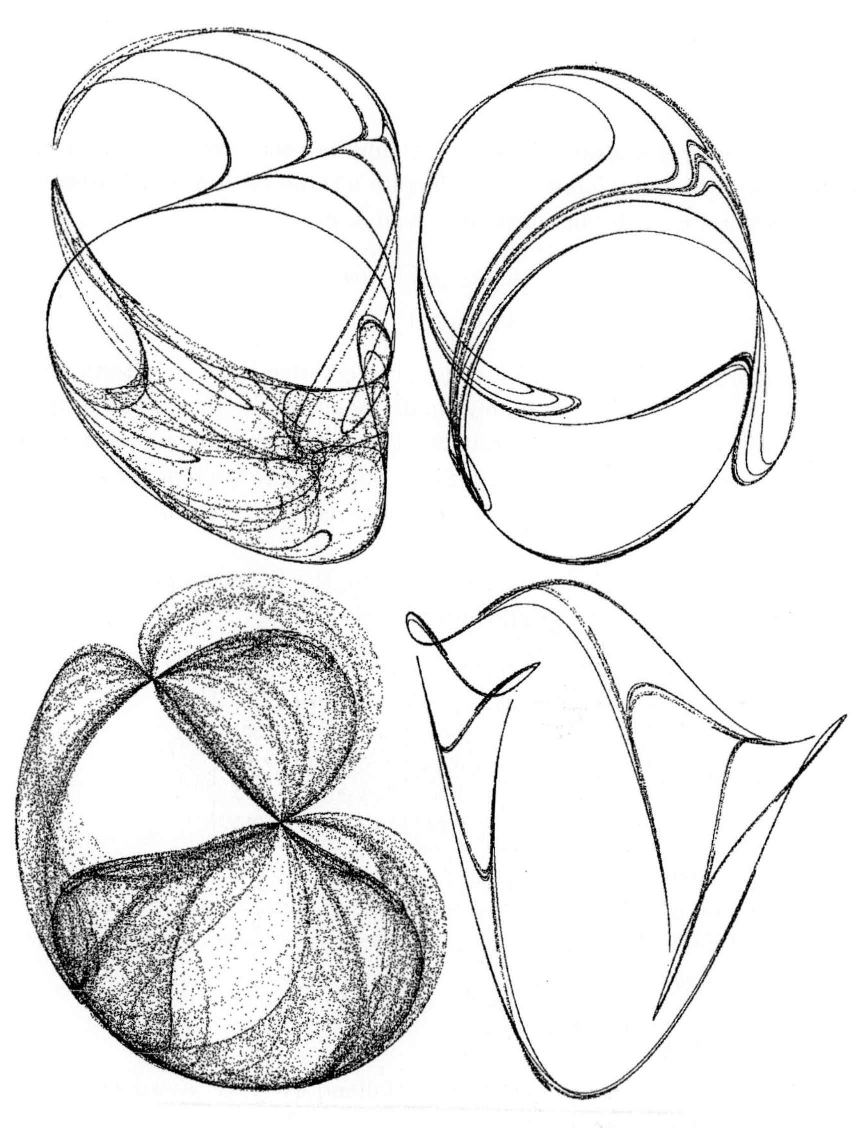

Swirl

Craig Cassin

Circlegraph 7.11

Described here is a pattern showing a combination of many circles. There are fourteen rows of circles which emanate from the center like petals in a flower. The twenty circles in each row are different in radius, line thickness, and position. This pattern was produced by a program, Circlegraph, written by the author and printed on an HP laserjet at 300 dots per inch.

series: 7.11
rows_of_circles: 14
circles_in_row: 20
inner_circle_line_width: 0.5
outer_circle_line_width: 2
circle_spacing: 70
inner_circle_radius: 1.8
outer_circle_radius: 1

Craig Cassin

Flow 2.8

Described here is a pattern showing a computer simulation of a flowing fluid. Thirteen points were selected, then the computer began to plot lines from left to right. The lines tend toward the closest point which was still to the right. Once that point is passed, the lines tend toward the next point. Moire patterns arise because the lines, which are close together, have ragged edges. This pattern was produced from a program, Flow, written by the author and printed on an HP laserjet at 300 dots per inch.

series: 2.8
limit: 1
damp: 0.5
conserve: 0.1
point: 0-1,7,0.5
point: 1-1.5,5,1
point: 2-2,3,1.5
point: 3-2.5,1,2
point: 4-3.5,0,1
point: 5-4,2,1.5
point: 6-4.5,4,2
point: 7-5,6,2.5
point: 8-5.5,8,3
point: 9-6.5,7,1.5
point: 10-7,5,2
point: 11-7.5,3,2.5
point: 12-8,1,3

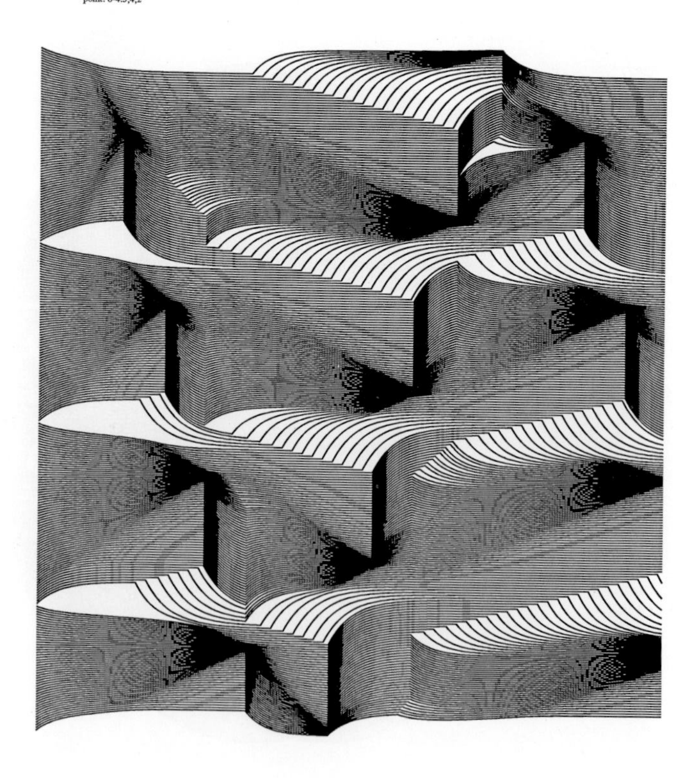

John D. Corbit

Botanical Biomorph Arising from the Iteration of $z \to z^3 + \mu$

The pattern described here is a biomorph image of the type first described by Pickover [1, 2, 3]. The pattern arises from the iteration of a very simple expression, namely, $z \to z^3 + \mu$, where μ and z are complex numbers. The image was rendered with the aid of a Julia set algorithm and the special convergence criterion required to reveal biomorphic forms [1]. According to this criterion, z is taken to be convergent and a point is plotted if either the real or the imaginary part of z is small after many iterations.

Figure 1 shows a map resulting from the iteration of $z \to z^3 + \mu$ in a region of the complex plane centered at real.z = 0.685426, imaginary.z = −0.175747, and with real and imaginary side lengths being real.side = imaginary.side = 0.04. The real and imaginary parts of μ were 0.5 and 0.0, respectively, and the number of iterations was 100.

Qualitatively, the pattern has a plant-like character. Overall, there is a central upheaval of frond-like foliage flanked by two botanic spirals of infinite depth. The leafy fronds are seen by perceiving the white areas as figure and the black areas as ground. Alternatively, spiky hair-like forms are prominent when one perceives black as figure and white as ground.

The computations were performed on a Macintosh II using Fractal 2.4 software written by Paul Bourke of Auckland University, Auckland, New Zealand. Four 450×450 pixel quadrants of the image were computed separately and then assembled in a graphics package giving a final image resolution of 900×900 pixels.

References

1. C. A. Pickover, "Biomorphs: Computer displays of biological forms generated from mathematical feedback loops", *Computer Graphics Forum* **5** (1986) 313–316.
2. A. K. Dewdney, "Computer recreations", *Scientific American* (July 1989) 110–113.
3. C. A. Pickover, *Computers, Pattern, Chaos, and Beauty — Graphics From an Unseen World* (St. Martin's Press, 1990).

Figure 1. A zoom on a region of the map for $Z \to Z^3 + \mu$. Note frond-like forms in the central area and the two spirals at the sides. See text for more information.

W. H. Cozad

Transient Microstructure

Described here is a pattern formed by transient values generated when the iterative logistic formula NEWX = K $*$ OLDX $*$ (1-OLDX) is repeatedly mapped over certain very small K intervals. The K interval for Transient Microstructure was 3.69925–3.7019. This interval is within the domain of the chaotic attractors, but I have found no clear relationship between chaos and the microstructure. The pattern varies with the value initially assigned to OLDX; here, it was 0.3.

The standard algorithm for the logistic map displays only attractor values. Screen output begins after 100 or 200 iterations when, in most cases, the production of transients has ceased and the process has fallen into the attractors' embrace. Mapping the transients' structure requires modifying the algorithm to display the output from each iteration. Interestingly, this modified algorithm will generate a macrostructure of transients if a larger K interval is used. Image A, below, shows the standard logistic map for the K interval 1–4, while image B includes an overlay of the transient macrostructure. The form of the macrostructure, like that of the microstructure, depends on the value initially assigned to OLDX.

The standard algorithm also must be modified by greatly reducing the total number of iterations. The transient microstructure's fragile filagrees are obscured if that number significantly exceeds the 25 iterations used for Transient Microstructure. I have also excluded the results of the first 13 of these iterations as they form relatively straight lines cutting horizontally across the filagrees.

Mapping transients to produce striking and significant patterns has been developed by C. A. Pickover and is discussed in several works by him. See, *Computers, Pattern, Chaos, and Beauty — Graphics From an Unseen World* (St. Martin's Press, 1990); "Personal programs: Close encounters with strange attractors", *Algorithm* **1**, 2 (1990); and "Graphics, bifurcation, order and chaos", *Computer Graphics Forum* **6**, 1 (1987).

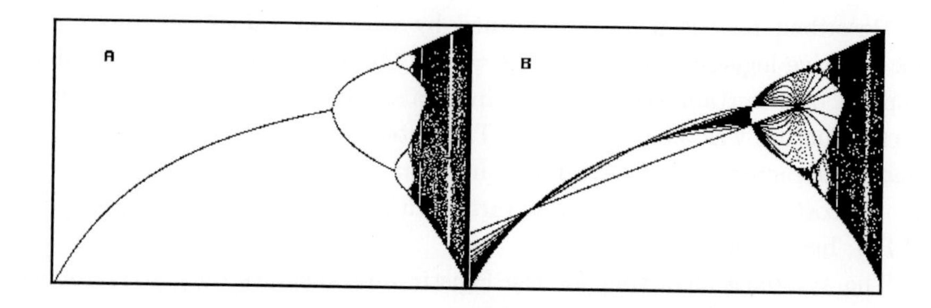

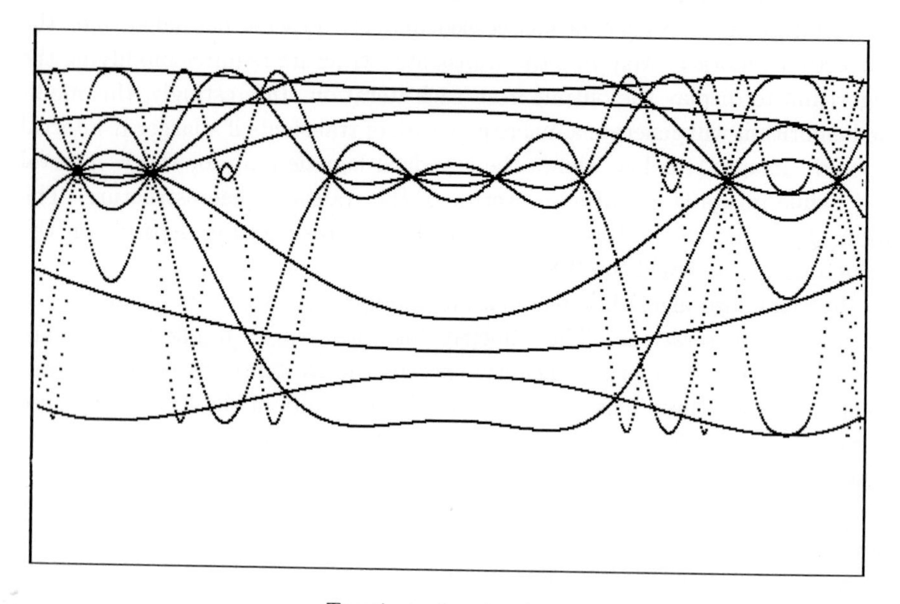

Transient microstructure

Douglas A. Engel

Engel Curves

The patterns described here show how Engel curves are generated and some interesting computer-generated Engel curve circuits. These curves were published in the February 1983 *American Mathematical Monthly.* In Fig. 1, a curve is generated by first creating a grid of m rows of n points, m and n coprime and only one of m and n even. Figure 1 has a 2×3 rectangular grid of 6 points, the grid then being divided into 3 pairs of points, each pair connected by a single line to form a net of valence 1. This pattern is then repeated 6 times in 3×2 array. The same array is then replicated, turned 90 degrees and superposed over itself to form a net of valence 2 composed of 1 or more closed curves. Patterns that form a hamilton cycle, a single closed curve using every point of the net like this one are not very common as the $m \times n$ pattern increases. Several investigators showed that only 42 of them exist for 3×4 nets, none symmetrical. In what follows, the $m \times n$ net is called the efactor and the final pattern the eproduct.

Engel curve sequences can now be described as shown in Fig. 2. The infinite sequences of efactors shown here produce only hamilton cycles. Two distinct curve networks can be achieved by multiplying an assymetrical efactor with itself in 2 ways. If one allows the efactors to be turned over then 2 different assymetrical efactors can be multiplied to get 4 different eproducts. Several investigators have attempted to describe these curves and elucidate their properties in terms of sequences, knots, topology, number of curves produced, and so on.

Properties that can be investigated include the simulation of Brownian motion as a random ecurve that is traced by a plotter, the circuit properties of intertwined ecurves, 3- and higher-dimensional Engel curve products where 3 or more edges meet at each vertex, the generation of fractal-like circuits as a sequence approaches infinity, and many other properties too numerous to mention here.

The final figures shows a 28×9 inverse frame curve (it is called an inverse frame curve if m is larger than n) surrounding a 10×11 perfect frame curve. This shows how two efactors made according to the same rule produce wildly different eproduct curves. The 10×11 frame curve is the first member of an infinite family that repeats at $10 \times (11 + 10n)$, $n = 0, 1, 2, \ldots$. The method of generating the efactors of frame curves is shown by the 4×5 pattern in Fig. 3. Only 3 other infinite families of perfect frame curves are known.

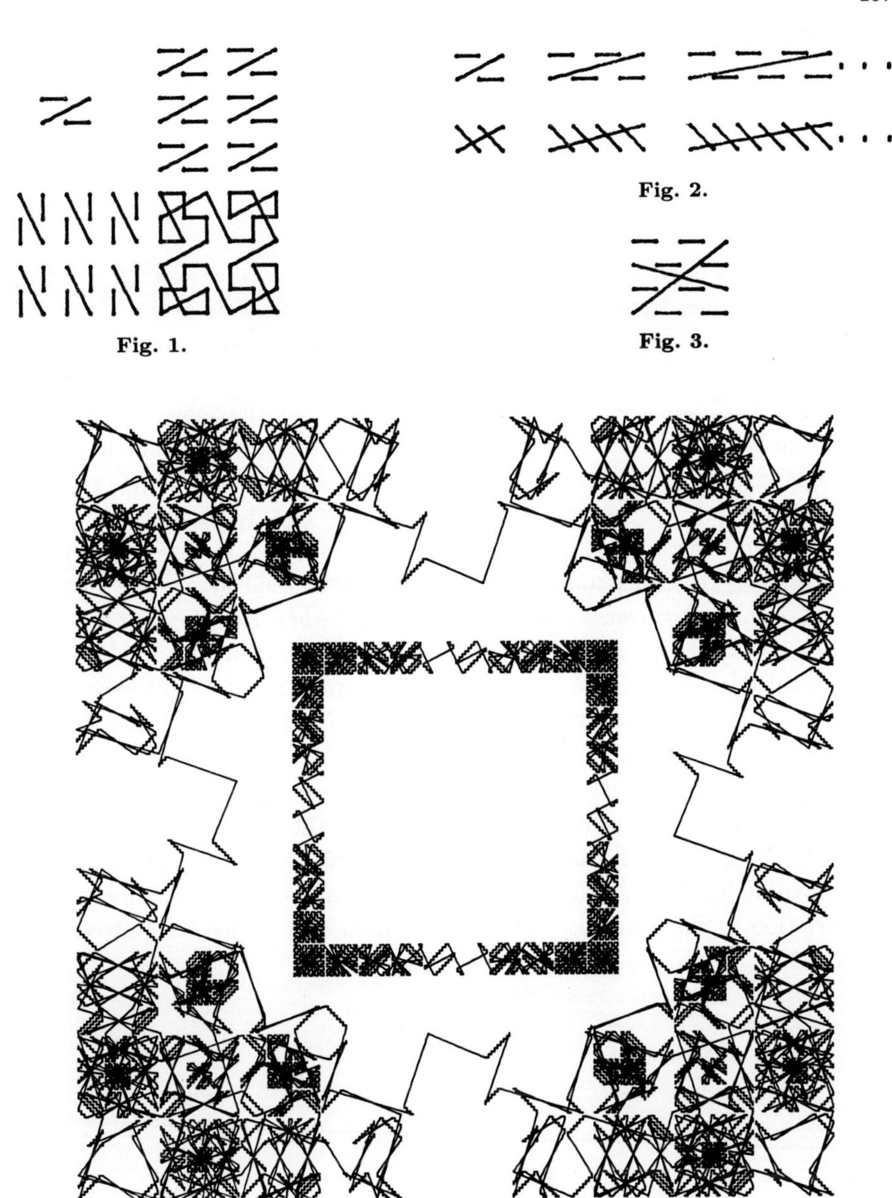

Fig. 1.

Fig. 2.

Fig. 3.

Fig. 4.

Figures 1–4 show the generation of Engel curves, curve sequences, and an inverse frame curve surrounding a perfect frame curve.

Ian D. Entwistle

"Spinning Chaos": An Inverse Mapping of the Mandelbrot Set

The pattern illustrates a mapping of the quadratic $fc(z) = z^2 + c$ in the complex $1/z$ plane. Iteration of $z^2 + c$ for the divergent points in the $1/z$ plane was controlled by the standard escape test $|z| < 2$. An alternative test was used to differentiate the points in the bounded set.

In order to obtain the inverse Mandelbrot(M) set, all the points in the plane need to be inverted. The inverted M set contains the those complex numbers such that the sequence $z = 0$, $z1 = z$, $z2 = z^2$, $z3 = (z^2 + z)^2$, $zn, \ldots,$ never satisfies $|z| > 2$ when the function $z^2(n-1) + 1/c$ is iterated. The position of an inverse point in the complex plane is given by the relationship $c *$ cinverse=1(1). Using complex algebra(2) substitution by $c = a + ib$ gives the relationship $c *$ cinverse $= a/(a^2+b^2) - ib/(a^2+b^2)$. Replacement of a and b in the LSM algorithm by the new values of parts of c then allows calculation of the inverse $|z|$. In order to create the artistic spinning effect in the pattern, the value of the real part of $|z|$ corresponding to the minimum value of the imaginary part of $|z|$ is used to control the print color. Note that the divergent point color is controlled by the iteration value at escape. Those points in the bounded set which are printed white because of the rounding down by the integer function are further controlled by the coloring routine for the divergent points. Adjusting the value of the Integer Factor can be used to further control the variation in the pattern. Values of the data to produce the pattern are appended.

Pseudo Code for Modified Level Set Method

Arguments: ny, nx, $x$min, $x$max, $y$min, $y$max, itermax
Variables: real x, y, $x2$, $y2$, cs, ci integer:sy, sx, iter

```
BEGIN
   FOR sy = 1 TO ny
         cy = ymin+sy * (ymax-ymin)/ny
   FOR sx = 1 TO nx
         cx = xmin+sx * (xmax-xmin)/nx
         bxy = 1/(cx^2 + cy^2) : x - 0 : y = 0 : xy = 0 : x2 = 0 : y2 = 0 : ci = 25
   FOR iter=1 TO itermax
         x = x2 - y2 + cx/bxy : y = xy + xy + cy/bxy
         x2 = x * x : y2 = y * y : cs = x2 + y2 : xy = x * y
         IF y2 < ci THEN ci=x2
         IF cs > 4 THEN exit LOOP and SELECT COLOR
   END LOOP
         ci=INT(ci *  INTEGER FACTOR)
         TEST ci AND PRINT
         SELECT COLOR ON BASIS OF iter VALUE
   END LOOP:END LOOP
```

Appendix

nx=1080, ny=2250, $x$min=5,6, $x$max=4.1, $y$min=3.6, $y$max=1.8, itermax=120, INTEGER FACTOR=100

References

1. R. Dixon, *Mathographics* (Basil Blackwell Ltd., 1987).
2. F. J. Flanigan, *Complex Variables: Harmonic and Analytical Functions* (Dover Publications Inc., 1972).
3. H.-O. Peitgen and D. Saupe, *The Science of Fractal Images* (Springer-Verlag, 1988).

Ian D. Entwistle

"Wings of Chaos": Mappings of the Function $\sinh(z) + c$ in the Complex Plane

The two figures illustrate patterns which are obtained when $f(z)$:– $\sinh(z)+$ is iterated in the z plane. They differ markedly from the full mapping of points which diverge and of those which belong to the bounded set (the Mandelbrot set) for the polynomial $z^2 + c$ derived patterns. Such patterns have been intensively studied and frequently generated [1]. The similar mappings for the transcendental function $\sinh(z) + c$ has received little attention. More recently, the strange behavior of mappings of the transcendental hyperbolic cosine function has been studied in both the z and c complex planes [2]. The dearth of reports on the z plane iterative mappings of the hyperbolic sine function prompted a study of this mapping which produces the two patterns, see Figs. 1 and 2. The Taylor expansion of $\sinh(z)$ contains only terms of uneven powers of z and so under recursion it was thought that the bounded set would not map to an outline shape similar to the topology of the Mandelbrot set for $z^2 + c$ even if an approximate expansion was used.

The outlines of the patterns in Figs. 1 and 2 approximate to the topology of the stable set for $\sinh(z) + c$. Unlike the outline geometry of the stable set for $\cosh(z) + c$, it is not sensitive to changes in the escape radius value. The circle-like leminiscates corresponding to increasing iteration values resemble more closely the shape of the bounded set and so the patterns generated on magnification are not so pleasing as for those of other functions on iteration in the complex plane. The behavior of the points "inside" the bounded set for $\sinh(z) + c$ can, however, be mapped utilizing various tests to give visually exciting patterns which are quite distinctive as illustrated by the two examples. The relationship $\sinh(z) + c = 0.5 * (e^z - e^{-z}) + c$ can be simplified by substitution, with $z = x + i * y$ using standard complex number algebra [3] to yield the two equations $z\text{real} = 0.5 * (e^{-x} - e^{x}) * \cos(y) + c\text{real}$ and $z\text{imaginary}=c\text{imaginary}-0.5 * (e^{x} + e^{-x}) * \sin(y)$. These equations simplify computation when used in the Level Set Method [4] if languages which support complex variables are not available. The test for boundedness $(|z|)^2 = 10$ was used in these mappings and all the divergent points were left unprinted. For Fig. 1, the minimum value of $(z\text{real})^2$ was used to determine whether the

Figure 1. Complex z plane map of $\sinh(z) + c$ using minimum values of $(z\text{real})^2$.

Figure 2. Complex z plane map of $\sinh(z) + c$ using minimum values of $|z|)^2$.

point was printed black or white. The minimum value of $(|z|)^2$ reached during iteration was similarly used to control the printing of the stable set of points in Fig. 2. Of particular note when computing the values of z is the symmetrical shape of the stable set. This allows the mapping to be achieved with calculation for only one quarter of the total pixel count. To allow classification of the minimum values on a size basis the real numbers were converted to integer equivalents by an integer factor (IFF). In both mappings, groups of $(z\text{real})^2$ or $(|z|)^2$ values were printed alternately in black and white. Data to allow generation of the patterns is appended.

Appendix

	IFF	Max. Iteration	Escape radius	Min$r$,	Max$r$	Min$i$,	Max$i$
Fig. 1.	80	200	10	−1.6	1.6	−3.2	3.2
Fig. 2.	50	200	10	−1.55	1.55	−3.2	3.2

References

1. H.-O. Peitgen and D. Richter, *The Beauty of Fractals* (Springer-Verlag, 1986).
2. C. A. Pickover, "Chaotic behaviour of the transcendental mapping($z - \cosh(z) + u$)", *The Visual Computer* **4** (1988) 243–246.
3. C. A. Pickover, *Computers, Pattern, Chaos, and Beauty — Graphics From an Unseen World* (St. Martins Press, 1990).
4. F. J. Flanagan, *Complex Variables: Harmonic and Analytical Functions* (Dover Publications Inc., 1972).
5. H.-O. Peitgen and D. Saupe, *The Science of Fractal Images* (Springer-Verlag, 1988).

Ian D. Entwistle

"Islands Among Chaos": Mappings of the Transcendental Function $\cosh(z) + c$ in the Complex Plane

The illustrated patterns result from the iteration of $f(z) :\rightarrow \cosh(z) + c$ for complex z and c planes. Previous studies of this function at iteration have led to important observations [1] about the morphology and behavior of the mapping. The bounded set is a single unconnected cardoid when the convergence test $|z| < 2$ is applied. For higher values of the escape radius this set is progressively distorted until it only resembles the $z^2 + c$ bounded set for points in the plane $< (-2, 0)$. The shape of the leminiscates are then no longer circle-like and the mapping becomes periodic $(2\pi i)$. The more complex dynamics of the $\cosh(z) + c$ iteration are therefore only realized by mapping with a large escape radius. Of particular interest in studies of this mapping is the possibility that the main cardiod centered at -0.14, 0i is connected to all the points in the bounded set as has been established [2] mathematically for the polynomials $z^n + c$ maps. In these mappings the morphology of the central cardoid is retained by the island miniatures, thus giving a high degree of self similarity to the fractal geometry of the maps. Greater variations in the patterns mapped to divergent points and the geometrical shapes of the islands which form the bounded set are observed for the function $\cosh(z) + c$. Figures 1–4 illustrate these variations. The central cardoid is the largest island cardiod on the 0 axis and has the geometric outline characteristic of the z^2 Mandelbrot set. It appears to be connected along the imaginary axis and along the spines emanating out to the pattern edge or ultimately to the periodic boundary. The other miniature cardoids, one of which is illustrated in Fig. 2 in the cell-like parts of the pattern are unconnected. This is similarly observed at the higher magnification of Fig. 2. Figures 3 and 4 illustrate the quite different behavior of the islands of the bounded set mapped in the area of the plane with positive imaginary coordinates belonging to the region where the main cardoid is extended. Figure 3 illustrates the non-quadratic behavior resulting in an unsymmetrical cardiod morphology. Connection to other parts of the bounded set is possible but awaits mathematical proof. The behavior of the divergent points in this region of the map is also strikingly different, producing the petal-like pattern in Fig. 3. Even more distinct is the chaotic morphology of larger islands in the bounded set. Figure 4 maps only the bounded set and shows the changes in morphology of the bounded set with considerable loss in self similarity. Where fractal geometry is evident, connection has not been established. The appearance of the miniature cardiod centered at 1.846, 1.60i in the lower left corner of the pattern suggests that some quadratic behavior still persists. The relationship $\cosh(z) = 0.5 * (e^z + e^{-z})$ can be simplified to two equations for computation by the substitution of $z = x + iy$. These equations $z\text{real} = 0.5 * (e^x + e^{-x}) * \cos(y)$, $z\text{imaginary} = 0.5 * (e^x - e^{-x}) * \sin(y)$ were iterated using published pseudo code [3] for the Level Set Method. Taylor's expansion of $\cosh(x)$ can be approximated to $1 + (x^2)/2$ in order to speedup the iteration. For Figs. 1–3, divergent points close to the boundary of the bounded set were left uncolored in order to outline more clearly the bounded set. In Fig. 4 only points in the bounded set were printed. Data to produce Figs. 1–4 is appended and all the patterns were produced using a grid 900×1680 points.

Appendix

Figure	Max. iteration	Escape radius	Min$i$	Max$i$	Min$r$	Max$r$
1	250	1000	−0.002	−0.002	−3.4005	−3.3985
2	300	1000	−1.187	−0.976	−3.9	−3.695
3	400	1000	1.325	1.475	0.704	0.854
4	200	1000	1.35	1.85	1.5	2.0

References

1. C. A. Pickover, "Chaotic behavior of the transcendental mapping ($z \to \cosh(z) + \mu$)", *The Visual Computer* **4** (1988) 243–246.
2. A. Douady and J. H. Hubbard, "Iteration des polynomes quadratique complexes", *CRAS Paris* **294** (1982) 123–126.
3. H.-O. Peitgen and D. Saupe, *The Science of Fractal Images* (Springer-Verlag, 1988).

Figure 1. Island Mandelbrot set.

Figure 2. Isolated bounded set island.

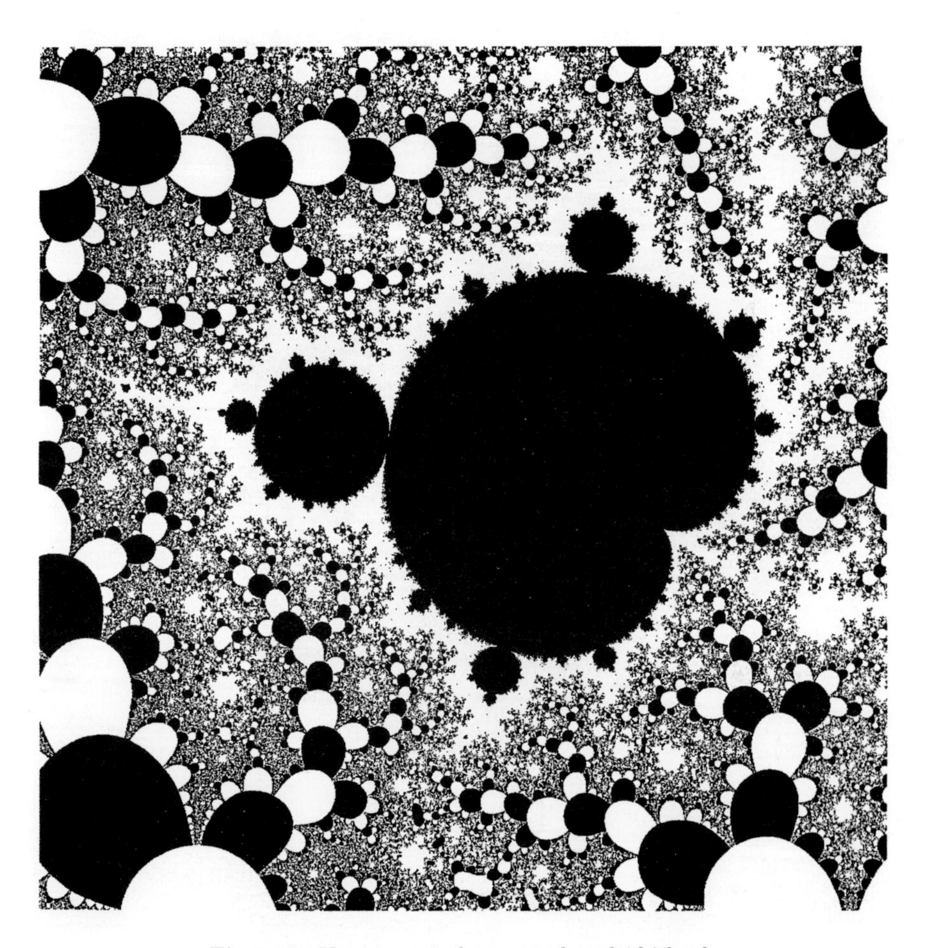

Figure 3. Unsymmetrical connected cardoid island.

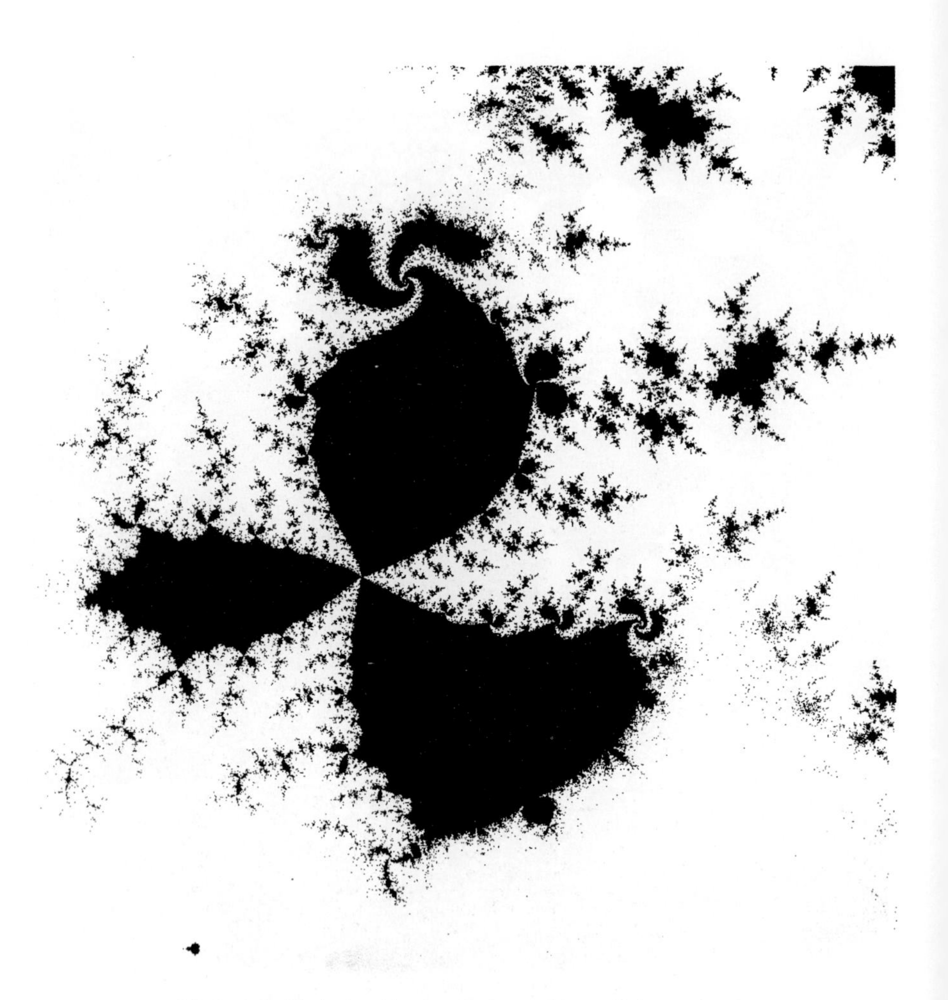

Figure 4. Variations in morphology of bounded set islands.

Ian D. Entwistle

Alternative Methods of Displaying the Behavior of the Mapping $z \to z^2 + c$

The patterns illustrate alternative behavior of the mapping $fc(z) = z^2 + c$ in the complex plane. Mandelbrot's study [1] of this function stimulated the publication of numerous mappings showing their beautiful fractal patterns [2]. Popularising the mapping algorithm [3] has resulted in a plethora of studies of the iterative properties of other mathematical functions [4]. Alternative tests for divergence have been used to produce many new pattern variations [5]. Commonly, the divergence of the modulus of z was used to control the pattern colors but more recently, the minimum size of real or imaginary values of z have been used to obtain novel mappings of various functions [6]. Other factors can influence the output of a complex function iteration. Both the choice of algorithm and computational language can affect a variation in a mapping. In many studies, avoidance of the effects of computation requires extended precision arithmetic and in others the use of scanning algorithms such as the Limited Set Method [7] or Escape Time Algorithm [8] since precision mappings need iteration for every point in the plane. For low power functions of z being mapped over relatively large areas of the plane, algorithms such as the Distance Estimator Method [9] can be used. Faster but less precise mappings can be achieved using modified image processing algorithms such as Mariani's [9]. Implementation of the algorithm in the native CPU code is favored for fast iteration. Where speed rather than accuracy is required, integer arithmetic can be helpful since it requires fewer CPU operations. Figures 1 and 2 were mapped using integer arithmetic and illustrated an effect caused by slight miscoding. Since the modification can be controlled, it represents an alternative mapping of the function $z^2 + c$. In the integer method, the floating point values are constantly adjusted to integer values which are within the normal single precision limits for integers. This process can produce rounding errors which result in the Figs. 1 and 2 differing from standard mappings and they may not be similarly reproduced on other CPUs. Utilizing the assembler code on the ARM2 32 bit RISC processor is particularly advantageous for integer maths since this CPU can perform large arithmetic shifts in one operation. Although careful control of these shifts was maintained, the escape radius test introduced an unforseen problem. In order to affect the divergence test the value of the escape radius 4 was multiplied using a single arithmetic shift 2^{27} before comparison with the value of $z$imaginary^2 obtained directly from the integer iteration. This approach appears to result in anomalous escape rates. When the assembler code is altered so that the integer value of the imaginary part of $|z|^2$ is divided by 2^{27}, using a logical right shift before a comparison with the escape value 4, the mapping resembles that produced using conventional coding. The petal shaping [6] of the divergent-mapped curves in the patterns results from the use of only the imaginary part of $|z|^2$ and not from the inaccuracy of the arithmetic. Data used to create the patterns is appended.

Appendix

	Min$r$, Max$r$	Min$i$, Max$i$	Max. iterations	Pixels x, y
Fig. 1.	−2.0, 1.0	−1.5, 1.5	1024	1280, 1960
Fig. 2.	−1.786623, −1.785227	−0.00095, 0.00095	1024	1280, 2600

References

1. B. B. Mandelbrot, *The Fractal Geometry of Nature* (W. H. Freeman, 1983).
2. H.-O. Peitgen and D. Richter, *The Beauty of Fractals* (Springer-Verlag, 1986).
3. A. K. Dewdney, "A computer microscope zooms in for a look at the most complex object in mathematics", *Scientific American* **255**, 8 (1985) 8–12.
4. C. A. Pickover, *Computers, Pattern, Chaos, and Beauty — Graphics From an Unseen World* (St. Martin's Press, 1990).
5. I. D. Entwistle, "Julia set art and fractals in the complex plane", *Computers and Graphics* **13**, 3 (1989) 389–392.
6. I. D. Entwistle, "Methods of displaying the behaviour of the mapping $z \to z^2 + u$", *Computers and Graphics* **13**, 4 (1989) 549–551.
7. H.-O. Peitgen and D. Saupe, *The Science of Fractal Images* (Springer-Verlag, 1988).
8. M. Barnsley, *Fractals Everywhere* (Academic Press, 1988).
9. R. Silver, "Mariani's algorithm", *Amygdala* **4** (1987) 3–5.

Figure 1. Mapping of the full Mandelbrot(M) set using integer arithmetic.

Figure 2. A magnified mapping of an "island" M set using integer arithmetic.

Ian D. Entwistle

"Serpents and Dragons": A Mapping of $f(z) \to \sinh(z) + c$ in the Complex Plane

The pattern shows the fractal features of the function $z \to \sinh(z)+c$ when it is mapped in the complex plane by the iteration of z at a constant c.

For real values of z, the curve for $f(z) = \sinh(z) = 0.5*(e^z - e^z)$ is a double nonperiodic exponential curve without a turning point. If it is iterated in the c plane, the resulting map is also nonperiodic unlike the similar $\cosh(z) + c$ map [1]. For $zn = f(zn-1, c)$, $n = 1, 2, 3, \ldots\infty$ the lemniscates $n = 1, 2, \ldots$ are circle-like but morphologically unlike the familiar Mandelbrot set shapes of the polynomials $z^n + c$ or that of the bounded set of $\cosh(z) + c$. The maps of $\sinh(z) + c$ in the z plane are also quite dissimilar to the much studied Julia sets of $z^n + c$ and have not been studied in much detail. The expansion of $\sinh(z)$ contains only $z^{2n} + 1$ terms and so cannot under iteration become approximated to z^2 as $\cosh(z)$ appears to. The greatly different behavior on iteration may result from this difference. For computation with languages which do not support complex variables $\sinh(z)$ can be divided into real(x) and imaginary(y) parts by substitution with $z = x + iy$ using complex algebra [3] for use in the Pseudo Code.

The pattern illustrates several aspects of $\sinh(z)+c$ behavior. The Julia set points are mapped using the minimum value of $|z|$ reached during iteration to control the printing color [4]. The divergent points which map the "serpents" have an escape value for $|z| > 1000$. An additional test was then applied to separate the points with absolute values of real z or imaginary $z < 100$. The other divergent points were then mapped with the opposite color. The pattern contains three distinct types of fractal geometry, namely, the "serpents" formed from the divergent points, the "dragons" mapped by the bounded points and the interior points of the "dragons". The periodic nature of the mapping of $\sinh(z)+c$ in the complex z plane is not evident from the pattern but mapping with a larger window reveals that the periodicity is 2pi and the pattern is formed from one 2pi unit along the imaginary axis.

A modified version of the Level Set Method [5] was used to generate the pattern from the appended data. In order to clarify the division between the bounded and divergent points, divergent lemniscates $n > 15$ were not printed.

Figure 1. "Serpents and Dragons": A Julia set map of the function $z \to \sinh(z) + c$.

Pseudo Code: Modified Level Set Method

```
Arguments: nx, ny, xmin, xmax, ymin, ymax, itermax, cx, cy.
Variables: real x, y, cr, cs, css, csi integer sy, sx.
BEGIN    FOR sy = 1 TO ny
         FOR sx = 1 TO nx
              y = ymin+sy * (ymax−ymin)/ny   :x = xmin+sx * (xmax−xmin)/nx :
                   csi = 25
         FOR iter=1 to itermax
              x = 0.5 * (e^x − e^−x) * cos(y) + cx   :y = −0.5 * (e^x − e^−x) * sin(y) + c
              cr = x * x; ci = y * y : css = cr + ci
              IF css < csi THEN csi = css
              IF css > 1000 THEN EXIT LOOP TO TEST
         END LOOP
              csi=INT(csi * 110):REM ASSIGN COLOR VALUES TO csi
              IF ABS(x) < 100 OR ABS(y) < 100 THEN ASSIGN COLOR VALUE
         END LOOP:END LOOP
```

Appendix

ny=2420, *nx*=1320, *x*min=−1.5, *x*max=2.1, *y*min=−2.1, *y*max=1.6, *cx*=1.1666, *cy*=−.8084, itermax=150.

References

1. C. A. Pickover, "Chaotic behavior of the transcendental mapping $(z-\cosh(z)+u$", *The Visual Computer* **4** (1988) 243–246.
2. C. A. Pickover, *Computers, Pattern, Chaos, and Beauty — Graphics From an Unseen World* (St. Martins Press, 1990).
3. F. J. Flanagan, *Complex Variables: Harmonic and Analytical Functions* (Dover Publications Inc., 1972).
4. I. D. Entwistle, "Julia set art and fractals in the complex plane", *Computers and Graphics* **13**, 3 (1989) 389–392.
5. H.-O. Peitgen and D. Saupe, *The Science of Fractal Images* (Springer-Verlag, 1988).

Ian D. Entwistle

"Star of Chaos": A Multiple Decomposition Mapping of $f(z) : z \to -z^7 + c$ in the Complex Plane

The pattern illustrates an elaborated Julia set fractal obtained from iteration of the function $z^7 + c$ in the complex plane.

Iterative complex plane maps of the function $z^2 + c$ have been widely explored by graphical methods because of the fascinating fractals that can be generated [1]. The striking beauty of these maps and the relationship of their geometry to the corresponding z plane mappings has prompted the study of the similar behavior of higher polynomial functions. On iteration, such functions generate $z(n)$ values approaching infinity within a few iterations (n) and very few of the points in the complex plane are part of the "bounded set" if the value of c is in the main cardoid of the corresponding Mandelbrot set. The resulting Julia sets fractals have outlines which are quasi-circles, i.e., they are homomorphic to a circle. The maps are not visually so appealing as those obtained for $z^2 + c$ since there is no complex fractal behavior close to the boundary between the divergent and bounded points. Most points are mapped to only a few distinct quasi-circles. Variations in the escape values employed or control of the point mapping positions on the basis of the $z(n)$ value have been utilized to produce visually exciting maps for the functions $z^p + c$ ($p = 3, 4, 5$) [4]. Other variations related to the pattern have also been studied [5]. In order to generate the complex pattern evident in the figure, secondary data is generated from the values of $z(n)$ after completion of iteration for each point. This new value controls the printing color of the point and results in the transformation to tree-like patterns which radiate to the pattern edges. This is achieved using a modification of the Binary Decomposition Method for Julia Sets (BDM/J) [6] derived from the mathematics of binary decomposition [7]. A simplified description has been reported which allows multiple decomposition maps to be generated [8]. Application of BDM/J to the function $z^7 + c$ can be simplified for computation using languages which do not support complex variables. Substitution of $z^7 + c$ by $z = x + i * y$ using standard complex algebra [9] leads to the two equations (1 and 2)

utilized for iterations 1) $z(\text{real})=x^7-7*x*y^6-21*x^5*y^2+35*x^3*y^4$ and 2) $z(\text{imaginary})=7*x^6*y+21*y^5*x^2-35*x^4*y^3-Y^7$. Since most of the points are iterated rapidly to divergence, the computation load is relatively low for Julia sets of the type illustrated. Most of the points are defined with less than fifteen iterations. The data for generation of the pattern is appended.

Appendix

Iterations	Min(r),	Max(r)	Min(i),	Max(i),	$c(r)$,	$c(i)$
30	−1.5,	1.5	−1.5,	1.5	−0.3,	0.7
Pixelation	1280 × 1780					

(r = real, r = imaginary)

References

1. H.-O. Peitgen and D. Richter, *The Beauty of Fractals* (Springer-Verlag, 1986).
2. C. A. Pickover, *Computers, Pattern, Chaos, and Beauty — Graphics From an Unseen World* (St. Martins Press Inc., 1990).
3. I. D. Entwistle, "Julia set art and fractals in the complex plane", *Computers and Graphics* **13**, 3 (1989) 389–392.
4. C. A. Pickover and E. Khorasani, "Computer graphics generated from the iteration of algebraic transformations in the complex plane", *Computers and Graphics* **9** (1985) 147–151.
5. I. D. Entwistle, "'Fractal Turtle' and 'Elephant Star': Multiple decomposition mappings of $f(z) : z \to z^5 + c$ in the complex plane", in *The Pattern Book: Fractals, Art, and Nature*, ed. C. A. Pickover (World Scientific, 1995) p. 155.
6. H.-O. Peitgen and D. Saupe, *The Science of Fractal Images* (Springer-Verlag, 1988).
7. [1], pp. 40–44, 64–76.
8. J. D. Jones, "Decomposition: They went that away — Another way of viewing the object", *Amygdala* **13** (1988) 2–3.
9. F. J. Flanagan, *Complex Variables: Harmonic and Analytical Functions* (Dover Publications Inc., 1972).

Michael Frame and Adam Robucci

Complex Branchings

Described here are patterns showing branching structures generated by several iteration schemes. Figure 1 shows a magnification of the Mandelbrot set; the picture is centered at $c_0 = -1.74995643$ and is in the main cardioid cusp of the 3-cycle midget. The height of the picture is 1.4×10^{-7}. Figure 2 shows a magnification of the Julia set for $z^2 + c_0$. This picture is centered at $z = 0$ and has height 0.282. Both figures were obtained by iterating $f(z) = z^2 + c$ for the Mandelbrot set, scanning across c values and starting each iteration with $z = 0$, while for the Julia set c is fixed at c_0 and the scan is across the z plane. Similarities between magnifications of the Mandelbrot set and the corresponding Julia sets are quite common. Indeed, a theorem of Tan Lei [2] guarantees that at any Misiurewicz point c, the Mandelbrot set magnified around c converges to an appropriately rotated and magnified portion of the Julia set of $z^2 + c$. Some of the power of Tan Lei's theorem lies in the fact that Misiurewicz points lie arbitrarily close to every point of the boundary of the Mandelbrot set [2]. Note, however, that c_0 is not a Misiurewicz point.

Figure 3 shows the Julia set of the complex cosine function $g(z) = d_0 \cos(z)$ for $d_0 = 2.967$. The branching and spiral patterns are quite similar to those of Fig. 2. We suspect these like features are due to the cosines being dominated by its quadratic term in this region, though to be sure, the similarity of the pictures is not "exact" and the relation between the additive constant in $f(z)$ and the multiplicative constant of $g(z)$ is not transparent.

When we first saw Fig. 1, the branching pattern reminded us of the usual IFS synthesis of a tree (see [1], p. 95). To this end, Fig. 4 modifies the IFS tree to produce a "linear approximation" of the pattern: this picture is generated by a five function piecewise-affine iterated function system. For example, the upper left branch is covered by the function $T(x, y) = (.3x - .52y + .45, -.52x + .3y + .7)$ if $y \geq 0$, and $T(x, y) = (.0005x - .26y + .45, -.0087x + .15y + .7)$ if $y < 0$.

References

1. M. Barnsely, *Fractals Everywhere* (Academic Press, 1988).
2. B. Branner, "The Mandelbrot set", in *Chaos and Fractals: The Mathematics Behind the Computer Graphics*, eds. R. Devaney and L. Keen (American Mathematical Society: Providence, 1989).

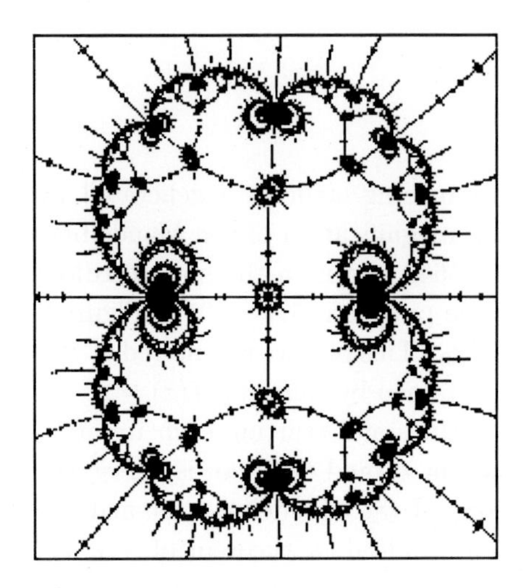

Figure 1. Part of the Mandelbrot set, centered at c_0.

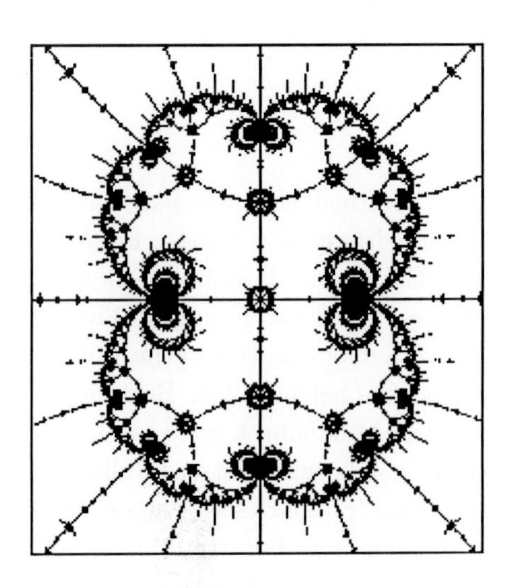

Figure 2. Part of a Julia set for $z^2 + c_0$.

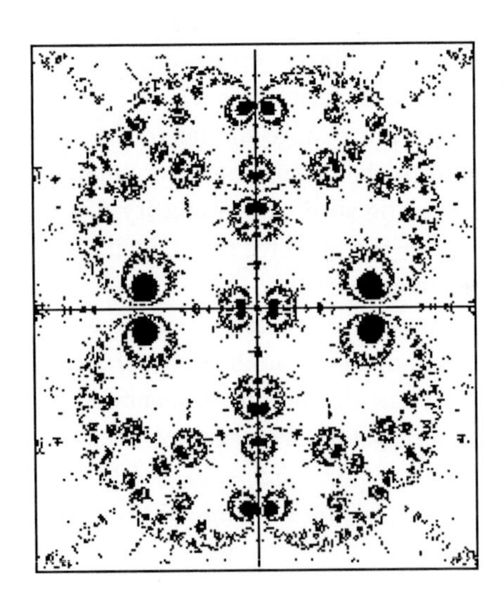

Figure 3. A Julia set for $d_0 \cos(z)$.

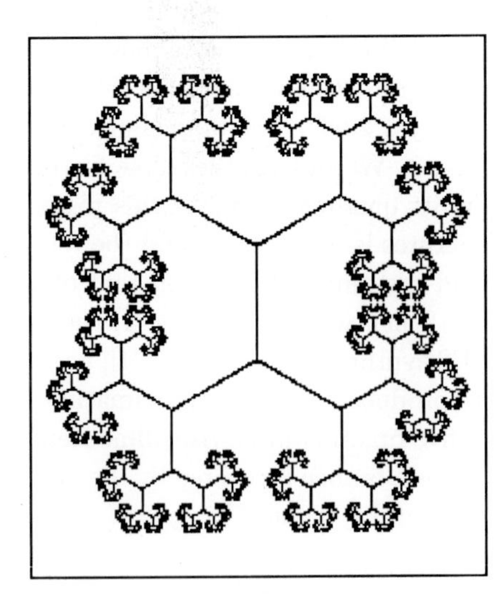

Figure 4. An IFS "linearization".

*Philippe Gibone**

Nevada Sets

Shown here are two images showing fractal features. In other words, the patterns are quite intricate and show self-similar edges. I call these patterns Nevada sets. You may write to me for more details on how I created these pictures.

Figure 1.

*13, Rue Corot, 77330 Ozoir La Ferriere, France.

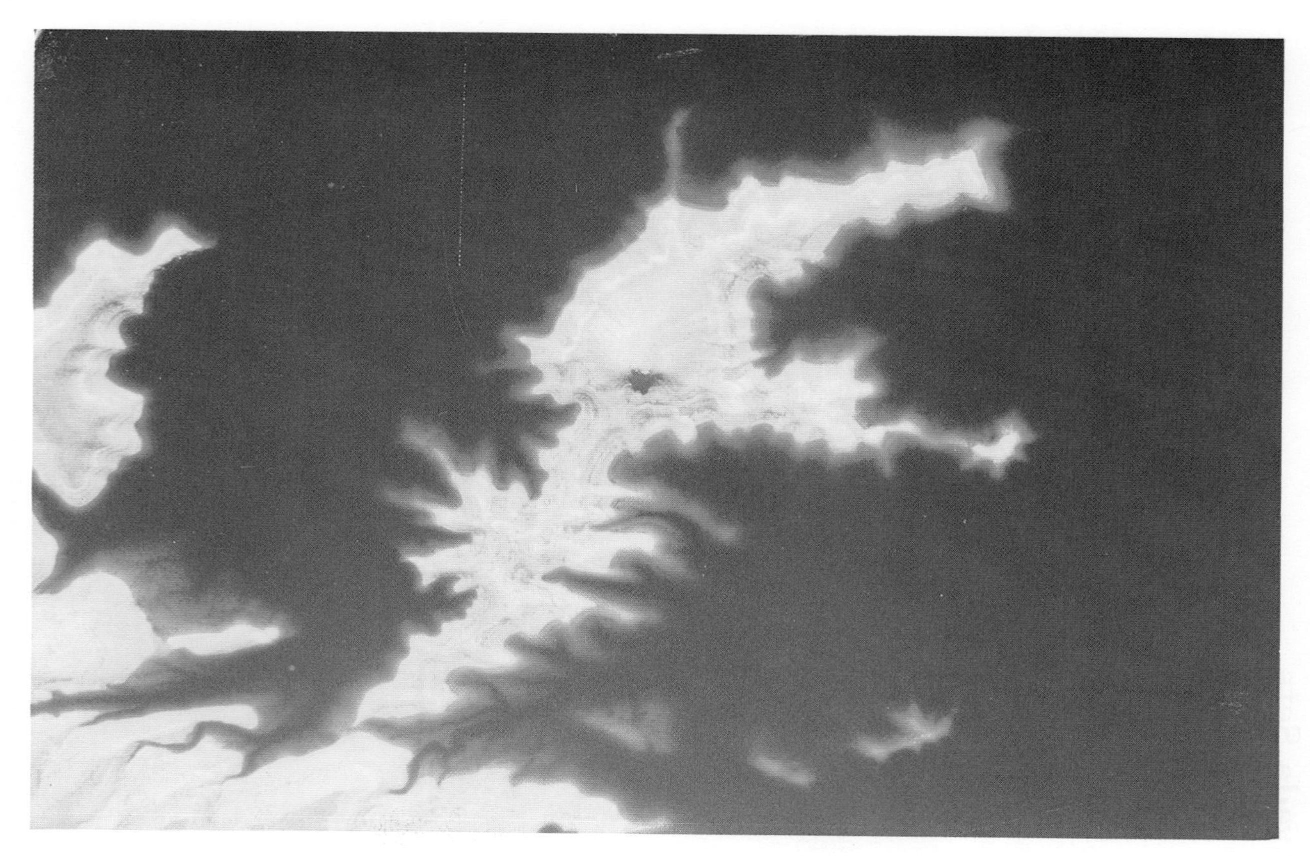

Figure 2.

Ira Glickstein

"Handmade" Patch Quilt Generated by L-System

Described here is a pattern showing a "handmade" quilt. The very idea of computer-generated handmade patterns seems an oxymoron! The precise regularity characteristic of a machine would seem to work against the artistic quality of the product and deprive it of whatever it is that makes handmade patterns special.

Yet, as the machine-made patchwork quilt in Fig. 1 and the handmade version in Fig. 2 demonstrate, both can be created using a combination of "L-systems" and "turtle graphics". L-systems, introduced by Aristid Lindenmayer, is a method of parallel string rewriting that makes use of an initial set of short strings to iteratively generate a much longer string. Turtle graphics is a method of interpreting a string to generate multi-dimensional images of varying complexity.

Both patterns were generated using a program written by Ivan Rozehnal that will accompany a forthcoming book, *Explorations with L-systems*, by Narendar S. Goel and Ivan Rozehnal, Department of Systems Science, T. J. Watson School, State University of New York, Binghamton, NY, USA.

Without resort to "random" variation, L-systems can generate remarkably natural-looking patterns. The instructions for the patch quilts are quite simple:

A		
B → AB+B+B +		
B → Bx		
x → >{F+F}>{+F}>{+F+f}		
iterations:	20	
	Machine-made	Handmade
theta angle:	90.0°	90.002°
initial angle:	0.0°	4.000°

Without becoming mired in the details of L-systems, here is an explanation of the above recipe:

1. Begin with the symbol **A**.
2. Substitute the symbol string **AB+B+B+** wherever **A** appears.
3. Substitute the symbol string **Bx** wherever **B** appears.
4. Substitute the symbol string >{**F+F**}>{**+F**}>{**+F+f**} wherever **x** appears.
5. Repeat steps 2 through 4 a number of times (in this case 20 iterations, resulting in a long string of 9,412 characters).
6. Interpret the string as instructions for a "turtle" that moves and turns, leaving behind a trail of "ink". (**F** means move forward one increment. + means turn right by the angle theta. The other symbols control color changes and shading).

The first pattern, for a machine-made quilt, uses an exact value, 90.0° for the turn angle theta. The "trick" for making the second pattern look handmade is setting the theta to an inexact value, in this case 90.002°.

Figure 1. Machine-made.

Figure 2. Handmade.

Earl F. Glynn

Spiraling Tree/Biomorphic Cells

Pictured here are different magnifications for a Julia-like set for the iteration $z \to z^{1.5} - 0.2$. Depending on one's point of view, Fig. 1 resembles either a tree with two infinite spiraling branches, or biomorphic cells about to escape through an orifice. Figure 2 shows the tip of the right spiraling branch.

The quantity z^x is calculated either through DeMoivre's Theorem with real parameter x instead of the usual integer parameter n,

$$z^x = r^x(\cos x\theta + i \sin x\theta)\,,$$

or, by using an equivalent formula with complex exponential and logarithm

$$z^x = e^{x \ln z}\,.$$

For more information on Julia sets, see the references below.

References

1. H.-O. Peitgen and P. H. Richter, *The Beauty of Fractals* (Springer-Verlag, 1986)
2. H.-O. Peitgen and D. Saupe, eds. *The Science of Fractals Images* (Springer-Verlag, 1988).

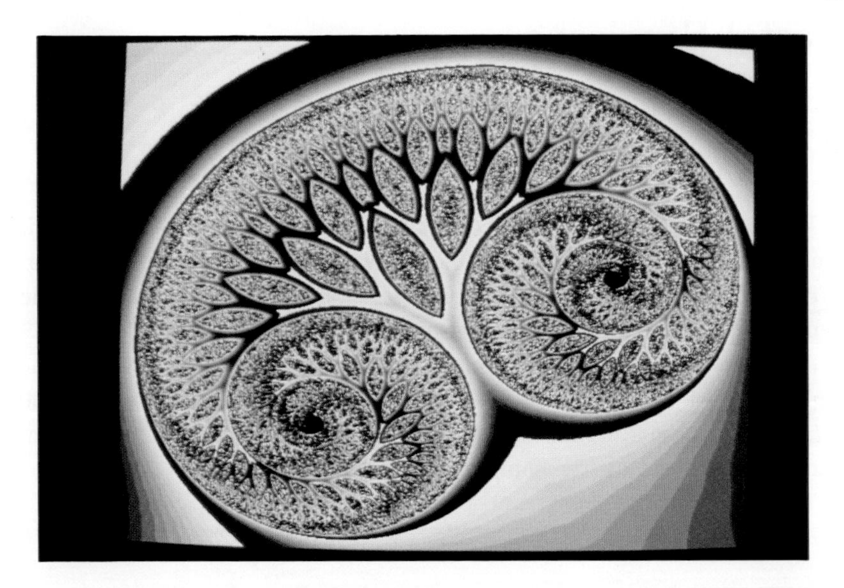

Figure 1. Spiraling tree/biomorphic cells. (3X Magnification)

Figure 2. Tip of spiraling tree branch. (25X Magnification)

Earl F. Glynn

Fractal Hurricane

Described here are two views of a Julia-like set computed using the complex sine function. Figure 1 shows a broad view with many "storms". Figure 2 is an enlargement of one of the "hurricanes". Note that this fractal hurricane even has miniature tornadoes along many of its arms.

Julia sets with parameters from near the boundaries of their corresponding Mandelbrot set often result in interesting images. The parameter, $c = 2.505 + 0.6i$, used in these Julia-like set images is from near the filaments seen in the Mandelbrot-like set for the complex sine function.

Figure 1. Stormy areas from Julia-like iteration $z \to \sin(z) + (2.505 + 0.6i)$.

Figure 2. Fractal hurricane.

Earl F. Glynn

Intertwined

Pictured here are two views of a Julia-like set computed using the complex sine function.

Figure 1 shows "arms" from the top and bottom reaching out for each other but never touching. Figure 2 shows an enlargement of the tips of the arms. While the arms from the $\sin(z)$ Julia-like sets are "shy", the arms from the $\cos(z)$ Julia-like sets are very direct and unite quickly as the Julia set parameter is varied in a similar way.

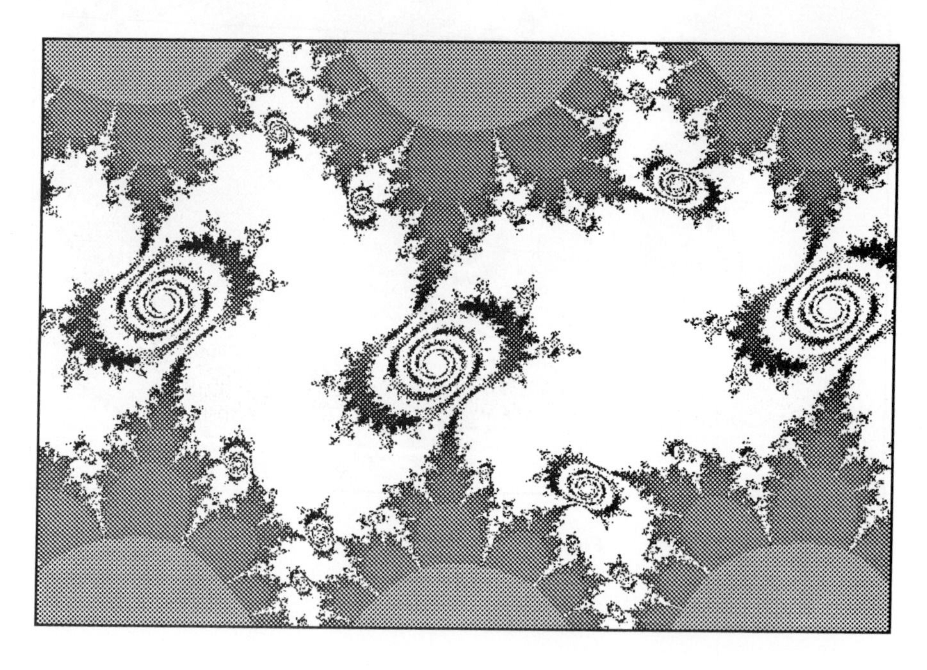

Figure 1. Macroscopic view of Julia-like set for iteration $z \to \sin(z) + (3.70 + 0.6i)$.

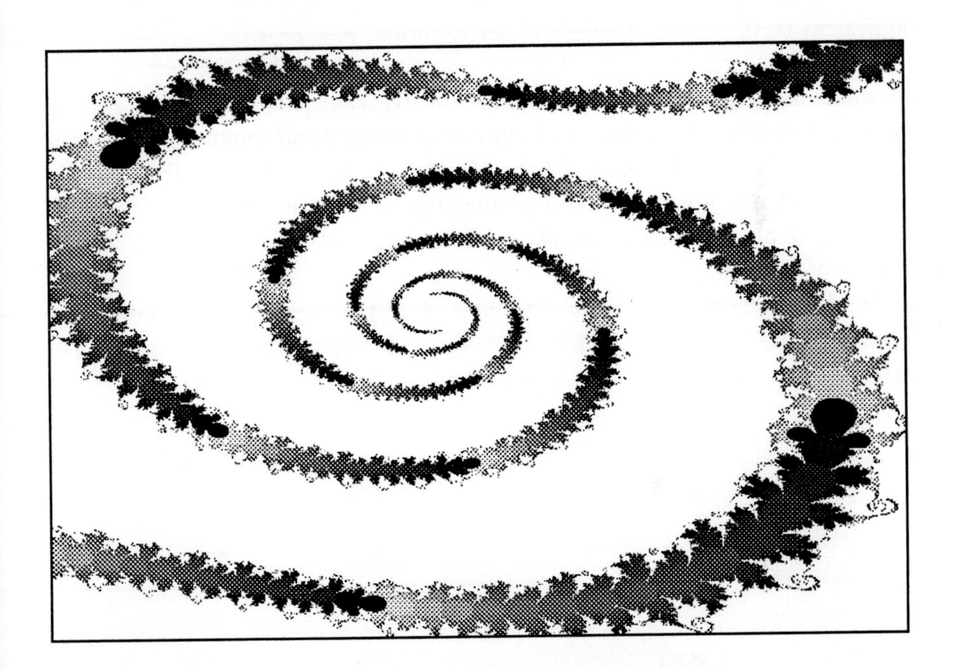

Figure 2. 2.5X magnification of Julia-like set for iteration $z \to \sin(z) + (3.65 + 0.6i)$.

Earl F. Glynn

Mandelbrot Iteration $z \to z^{-2} + c$

Pictured here are two different views of the Mandelbrot iteration $z \to z^{-2} + c$. The initial condition for this iteration was changed from the usual $z_0 = 0$ to $z_1 = c$ to avoid division by zero. This provided an initial condition equivalent to the usual Mandelbrot set iteration, $z \to z^2 + c$.

Figure 1 was created with iteration parameter ModulusMax = 2, which is the value usually used with the Mandelbrot iteration. Given the nature of the iteration function, this value of ModulusMax seemed too small. By increasing the value of ModulusMax slightly the Mandelbrot image changes dramatically.

Figure 2 was created with ModulusMax = 10. Increasing the value of ModulusMax to as much as 200 results in an image like Fig. 2 but with a lower density of "circles".

With an exponent of -2 in the iteration function, the Mandelbrot image is a figure with 3 vertices. This generalizes for other negative integers. The figure for the iteration with an exponent of $-n$ has $n+1$ vertices.

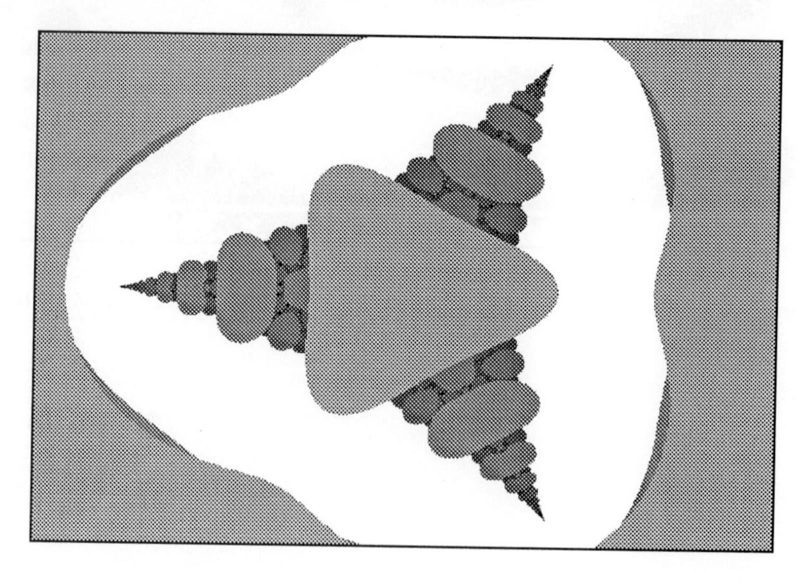

Figure 1. Mandelbrot iteration $z \to z^{-2} + c$; ModulusMax = 2.

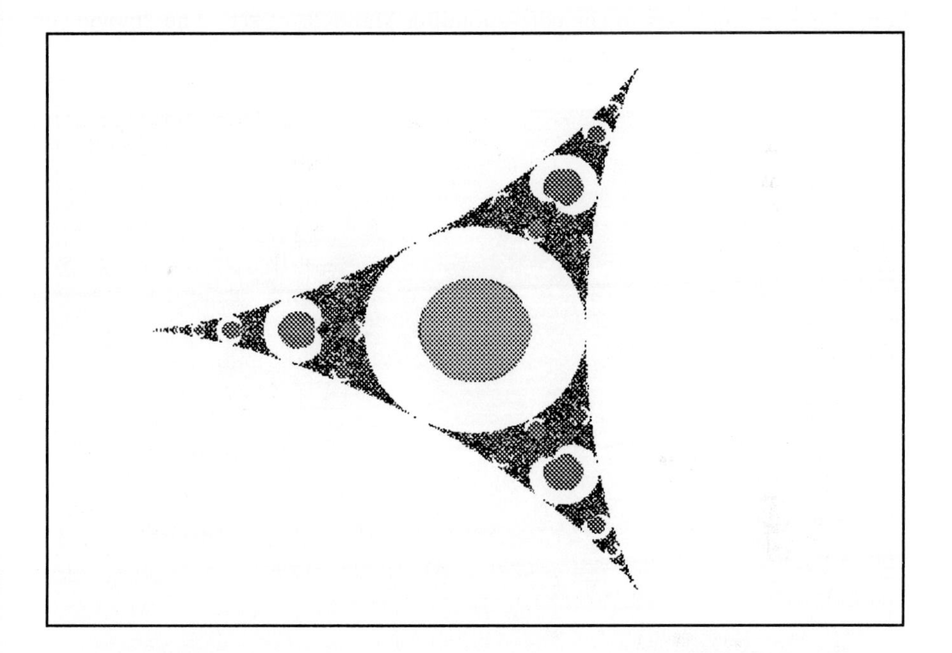

Figure 2. Mandelbrot iteration $z \to z^{-2} + c$; ModulusMax = 10.

Earl F. Glynn

Julia Iteration $z \to z^{-2} + c$

Pictured here are two different views of the Julia set iteration $z \to z^{-2} + c$ (related to the previous article, *Mandelbrot Iteration* $z \to z^{-2} + c$, Fig. 2). Both of the parameters c for the Julia set calculations shown here were chosen from along the real axis in the corresponding Mandelbrot set. The "movie" of the transformation of Fig. 1 into Fig. 2 is quite interesting.

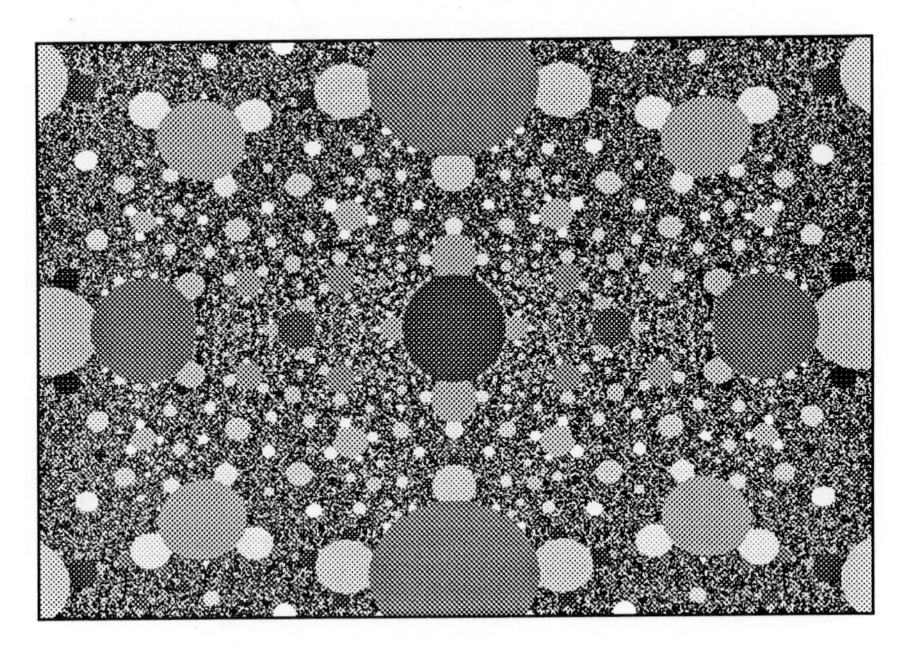

Figure 1. Julia iteration $z \to z^{-2} - 1.25$.

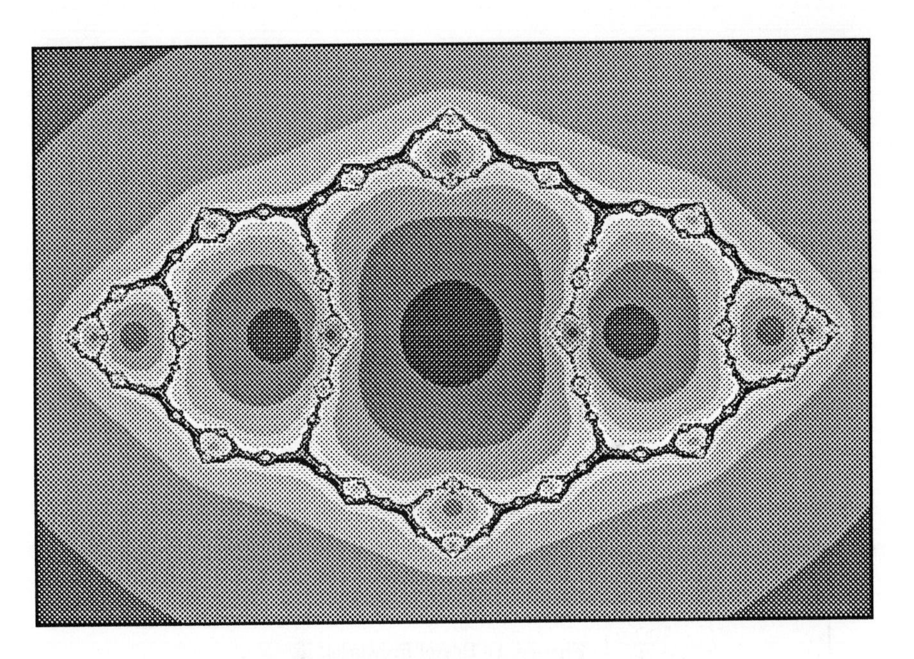

Figure 2. Julia iteration $z \to z^{-2} - 1.00$.

Earl F. Glynn

Bessel Fireworks and Bessel Parallelogram

Shown here are two views of Julia sets formed from the J_0 Bessel function. Figure 1 shows "fireworks", while Fig. 2 shows an interesting fractal "parallelogram".

Figure 1. Bessel fireworks.

Figure 2. Bessel parallelogram.

Earl F. Glynn

Fractal Moon

Pictured here is a view of the mandelbrot set,

$$z \to (\bar{z})^{-2} + c,$$

where,

$$\bar{z}$$

is the complex conjugate of z.

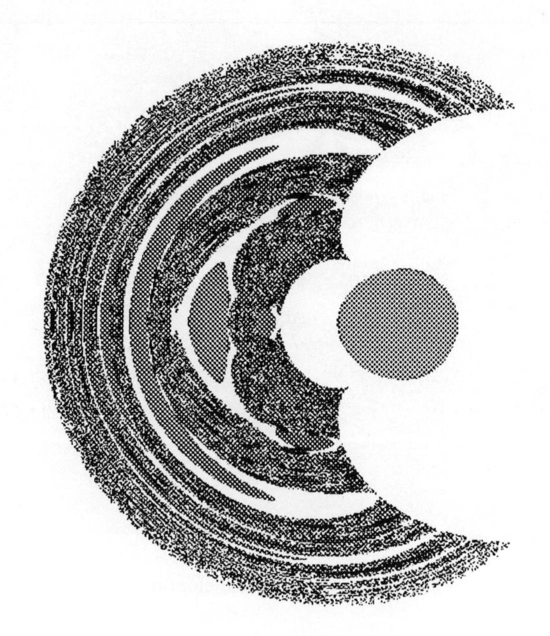

Figure 1. Fractal "Moon".

Branko Grünbaum

Isogonal Decagons

The pattern illustrates the variability of shape among isogonal polygons, that is, polygons whose vertices are all equivalent under symmetries of the polygon. Shown is the case of decagons, which is representative of a great wealth of forms, but still fits reasonably on a single page.

A (planar) polygon is called regular if all its vertices, as well as all its edges, are equivalent under symmetries of the polygon. (Other definitions are possible, but all reasonable ones are equivalent to this.) It is easy to show that if n is odd, each isogonal n-gon is necessarily regular. The situation for even n is more interesting. If $n = 4m + 2$ for a positive integer m, then each 2-dimensional isogonal n-gon can be determined by two parameters b and c, where b is a positive integer, $b \leq m$, and c is real-valued with $0 \leq c \leq n/4$. The vertices can be specified by

$$V_k = \left(\cos \frac{2\pi(bk + (-1)^k c)}{n}, \sin \frac{2\pi(bk + (-1)^k c)}{n}\right),$$

where $k = 1, 2, \ldots, n$, and edges are the segments $[V_j, V_{j+1}]$ for $j = 1, 2, \ldots, n-1$, and $[V_n, V_1]$. In the illustration $n = 10$, to avoid clutter, the vertex V_k is denoted simply by k. Shown are two sequences of isogonal decagons, corresponding to $b = 1$ and $b = 2$, respectively. Each sequence starts and ends with a regular polygon, which is indicated by a variant of the usual symbol. If n is a multiple of 4, the situation is quite similar but somewhat more complicated. The reader may enjoy investigating the cases $n = 12$ and $n = 14$.

Besides the attractiveness of their shapes, these sequences bring up several aspects of elementary geometry. First, they illustrate the difficulty of devising meaningful classification schemes for polygons. Second, although most texts shy away from admitting polygons which exhibit various coincidences between vertices, it is clear from the examples shown that exclusion of such coincidences would make it very cumbersome to describe the different possibilities. The labels in the diagram are meant to indicate that each of the polygons shown is a decagon — even though some seem to have fewer vertices. Edmund Hess is the only author who investigated isogonal polygons in any detail (in a very long paper published in 1876); he failed to reach any meanitngful conclusions, in

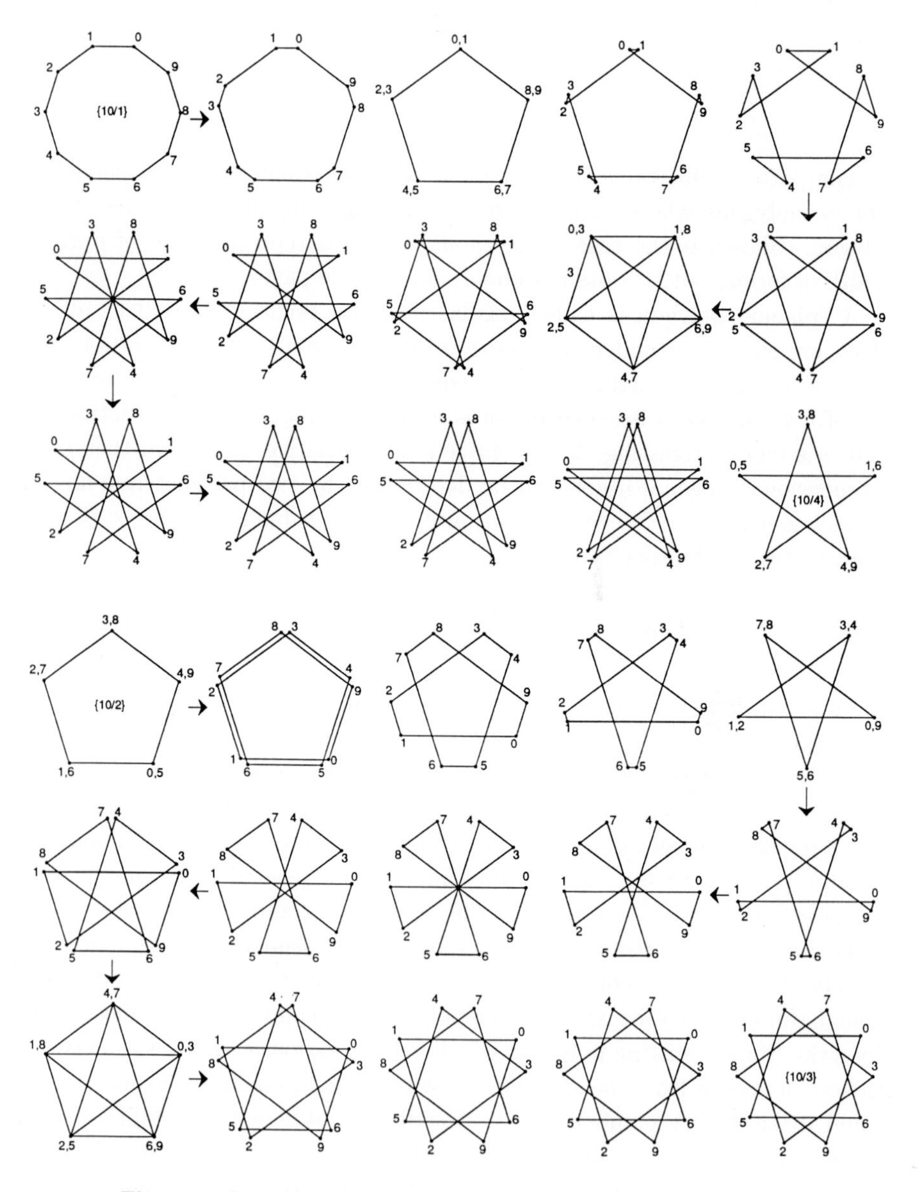

Figure 1. Representatives of the two sequences of isogonal decagons.

part due to his inconsistent treatment of coincidences. An exposition of Hess' paper is given in Max Brückner's well-known book, *Vielecke und Vielflache* (Teubner, 1900), but this is even less helpful than the original. There seems to have been no more recent work on this topic.

Third, the possibility of coincident vertices leads to regular polygons that are omitted from the enumerations given by various authors. This error started with Poinsot's famous paper of 1810. Although he never forbids coincident vertices, Poinsot ignores this possibility. Instead, he calls attention to "disconnected" polygons which, however, he correctly excludes from consideration. The problem is that he should not have brought them up at all, since they do not satisfy his own definition of regular polygons. All later authors (the present one included) blindly followed Poinsot down the garden path. It is ironic that in the only serious treatment of polygons which preceded Poinsot's, in a 1769 paper by A. L. F. Meister, the definitions of polygons and regular polygons are given correctly and applied consistently. However, Meister's work seems not to have been read by anybody except the mathematical historian Siegmund Günther, and in his 1876 book, Günther misquoted Meister at the crucial place, making it appear that Meister is saying the same thing as Poinsot. As already mentioned, all later authors accepted the inconsistency without protest.

The enrichment of the family of regular polygons leads naturally to additional regular polyhedra; hence, Cauchy's famous theorem, to the effect that the five Platonic and four Kepler-Poinsot polyhedra are the only regular ones, loses its validity. An account of all these developments is in preparation.

Joe Jacobson

Analytic Computer Art

Described here is "analytic computer art" consisting of geometrical designs based on explicit mathematical functions. A common motif in analytical computer art is the polar coordinate curve. This has the form $R = f(A)$ where R is the radius, f is a mathematical function, and A is the angle. The angle parameter A is swept through some range of values, the radius R is calculated, and the computer polar coordinate points (R, A) are converted to rectangular coordinates and plotted. The resulting curves frequently (but not always) exhibit angular symmetry, that is, they look the same after being rotated through suitable angles. What is special here is that simple polar coordinate curves are swept through the interval 0 to 360 degrees and incremented by a fixed amount between sweeps. The pictures were generated on a Tektronix 4052 intelligent terminal.

Reference

1. J. Jacobson, "Analytic computer art", *Proc. 2nd Symp. Small Computers in the Arts* (1982) pp. 47–60.

```
89 REM FLUTED SCALLOPS
100 PAGE
109 SET DEGREES
110 WINDOW −501, 501, −501, 501
111 VIEWPORT 15, 115, 0, 100
112 PAGE
113 PRINT "ENTER L"
114 INPUT L
115 PAGE
120 FOR B=100 TO 400 STEP 10
130 FOR A=0 TO 360 STEP 5
140 GOSUB 180
145 IF A>0 THEN 150
146 MOVE X, Y
147 GO TO 160
150 DRAW X, Y
160 NEXT A
170 NEXT B
175 GO TO 211
180 R=B*(1+0.25*ABS(SIN(L*A)))
190 X=R*COS(A)
200 Y=R*SIN(A)
210 RETURN
211 FOR N=0 TO L−1 STEP 1
212 R=100
213 T=N*(180/L)
214 X=R*COS(T)
215 Y=R*SIN(T)
216 MOVE X, Y
217 X=R*COS(T+180)
218 Y=R*SIN(T+180)
219 DRAW X, Y
220 NEXT N
221 END
```

Figure 1.

William J. Jones

Euler's Triangle

Euler's Triangle is, essentially, a graph of *Euler's Phi-function*, which for a positive integer N is defined as the number of positive integers less than N which are relatively prime to N. For $N = 2, 3, 4, \ldots$, the successive rows of the triangle display a circle where N and each of the numbers $1, 2, 3, \ldots, N-1$ are relatively prime; no circle where the two numbers share a common factor greater than 1. Thus, the circle in the top row of Fig. 1 stands for (2, 1), the two in the second row for (3, 1), (3, 2), and so on. When N is prime, there is a solid row of $N - 1$ circles.

The overall pattern of Euler's Triangle has a uniform look to it, but because of the unpredictable behavior of primes, we know the pattern never repeats. The density of circles within the triangle is the same as the probability that two integers chosen at random are relatively prime, which has been proven to be $6/\pi^2$. The lateral symmetry of the triangle reflects the theorem $gcd(n, r) = gcd(n, n - r)$.

The size and shape of the white spaces between the circles is interesting. They "almost always" cover an odd number of positions, but there are some even ones, as the two $1 - 2 - 3$ triangles sitting over the row for $N = 37$ in Fig. 1.

Since the triangle is infinite and never repeats, it represents a unique and sizeable piece of abstract real estate. Figure 2 shows a hexagonal tract centered 6002 rows down and 206 positions out from the left side of the triangle. Using, say, a person's social security number as the row and date of birth as the indentation, a unique hexagonal region could be staked out and displayed for everyone on earth claiming two such numbers.

Figure 1. The top 89 rows of Euler's triangle.

Figure 2. Hexagon-shaped region centered at (6002, 206).

William J. Jones

Euler's Crossing I and II

These patterns were created by a computer program called *Carpet*, which is based on one of the three basic concepts described in the article, "Wallpaper for the mind", in the September 1986 *Scientific American*. The program selects representatives in a uniformly random way from a domain of hundreds of millions of patterns. The following is a slightly simplified description of the defining process for the domain.

One of fifteen functions (e.g., cosine, exponential, etc.) is selected for half the X range of pixels, another for half the Y range. The functions are calculated and sorted in arrays; call them $f(x)$ and $g(y)$.

A two-term polynomial in x and y is selected with coefficients and exponents taken from random domains. Care is taken such that exponents of x and y are not both 0, so the polynomial is not constant in x or y.

The polynomial is evaluated for each pixel x, y with x and y replaced by $f(x)$ and $g(y)$, respectively, from the calculated arrays.

The result is truncated to an integer, and this number modulo the number of colors is the color index of the color of the pixel and its reflections in the remaining three quadrants of the pattern. When rendered in color, the RGB (red–green–blue) components for each of the color indices are determined using another random process for each pattern, and new color assignments may be made while the pattern is being displayed.

For the patterns "Euler's Crossing I" and "Euler's Crossing II", the program chose Euler's Phi-function for both $f(x)$ and $g(y)$. The difference between the two patterns is in the bit selected from the four-bit color value for plotting in black and white.

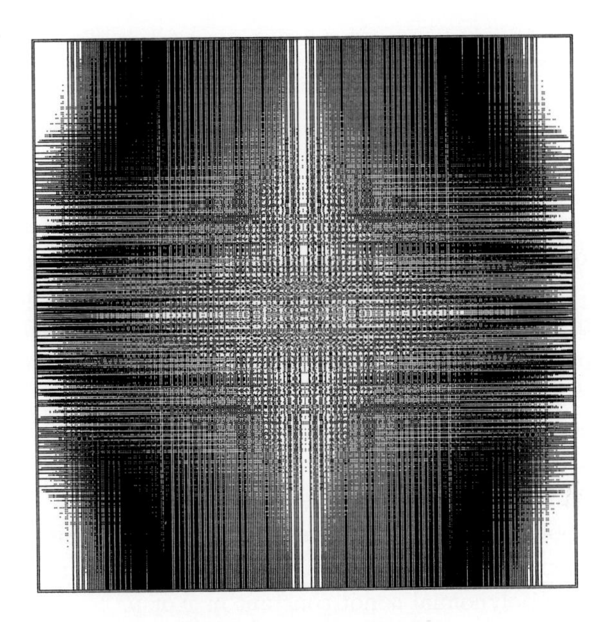

Figure 1. Euler's Crossing I.

Figure 2. Euler's Crossing II.

William J. Jones

Stripes

This pattern was created by a computer program called *Carpet*, which is based on one of the three basic concepts described in the article, "Wallpaper for the mind", in the September 1986 *Scientific American*. The program selects representatives in a uniformly random way from a domain of hundreds of millions of patterns. The following is a slightly simplified description of the defining process for the domain.

One of fifteen functions (e.g., cosine, exponential, etc.) is selected for half the X range of pixels, another for half the Y range. The functions are calculated and stored in arrays; call them $f(x)$ and $g(y)$.

A two-term polynomial in x and y is selected with coefficients and exponents taken from random domains. Care is taken such that exponents of x and y are not both 0, so the polynomial is not constant in x or y.

The polynomial is evaluated for each pixel x, y with x and y replaced by $f(x)$ and $g(y)$, respectively, from the calculated arrays.

The result is truncated to an integer, and this number modulo the number of colors is the color index of the color of the pixel and its reflections in the remaining three quadrants of the pattern. When rendered in color, the RGB (red–green–blue) components for each of the color indices are determined using another random process for each pattern, and new color assignments may be made while the pattern is being displayed.

"Stripes" employs an exponential function for both $f(x)$ and $g(y)$, yielding an op-art effect; it is not exactly easy on the eyes.

William J. Jones

Outer Space

This pattern was created by a computer program called *Carpet*, which is based on one of the three basic concepts described in the article, "Wallpaper for the mind", in the September 1986 *Scientific American.* The program selects representatives in a uniformly random way from a domain of hundreds of millions of patterns. The following is a slightly simplified description of the defining process for the domain.

One of fifteen functions (e.g., cosine, exponential, etc.) is selected for half the X range of pixels, another for half the Y range. The functions are calculated and stored in arrays; call them $f(x)$ and $g(y)$.

A two-term polynomial in x and y is selected with coefficients and exponents taken from random domains. Care is taken such that exponents of x and y are not both 0, so the polynomial is not constant in x or y.

The polynomial is evaluated for each pixel x, y with x and y replaced by $f(x)$ and $g(y)$, respectively, from the calculated arrays.

The result is truncated to an integer, and this number modulo the number of colors is the color index of the color of the pixel and its reflections in the remaining three quadrants of the pattern. When rendered in color, the RGB (red–green–blue) components for each of the color indices are determined using another random process for each pattern, and new color assignments may be made while the pattern is being displayed.

"Outer Space" uses sin(pi* cos(x)), essentially, for $f(x)$, and a fraction with (2+cos(y)) in the denominator for $f(y)$.

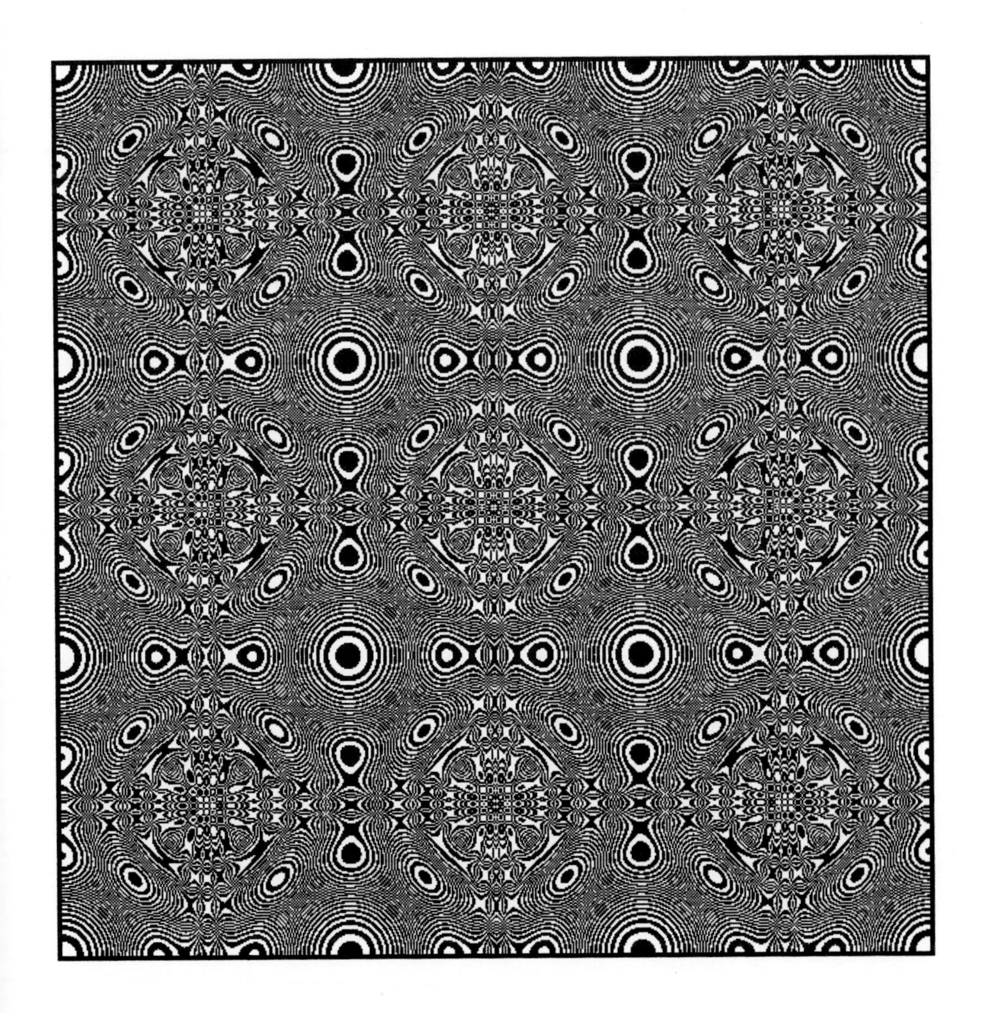

William J. Jones

Parquet

This pattern was created by a computer program called *Carpet*, which is based on one of the three basic concepts described in the article, "Wallpaper for the mind", in the September 1986 *Scientific American*. The program selects representatives in a uniformly random way from a domain of hundreds of millions of patterns. The following is a slightly simplified description of the defining process for the domain.

One of fifteen functions (e.g., cosine, exponential, etc.) is selected for half the X range of pixels, another for half the Y range. The functions are calculated and stored in arrays; call them $f(x)$ and $g(y)$.

A two-term polynomial in x and y is selected with coefficients and exponents taken from random domains. Care is taken such that exponents of x and y are not both 0, so the polynomial is not constant in x or y.

The polynomial is evaluated for each pixel x, y with x and y replaced by $f(x)$ and $g(y)$, respectively, from the calculated arrays.

The result is truncated to an integer, and this number modulo the number of colors is the color index of the color of the pixel and its reflections in the remaining three quadrants of the pattern. When rendered in color, the RGB (red–green–blue) components for each of the color indices are determined using another random process for each pattern, and new color assignments may be made while the pattern is being displayed.

"Parquet" uses a logarithm function for $f(x)$ and an exponential for $f(y)$. The parquet effect appears to come from discontinuities at discrete values of x and y, creating discernable rectangular regions. The discontinuities arise, not from the functions, but from regions of dominance of the two terms in the polynomial and from truncation to produce color index values.

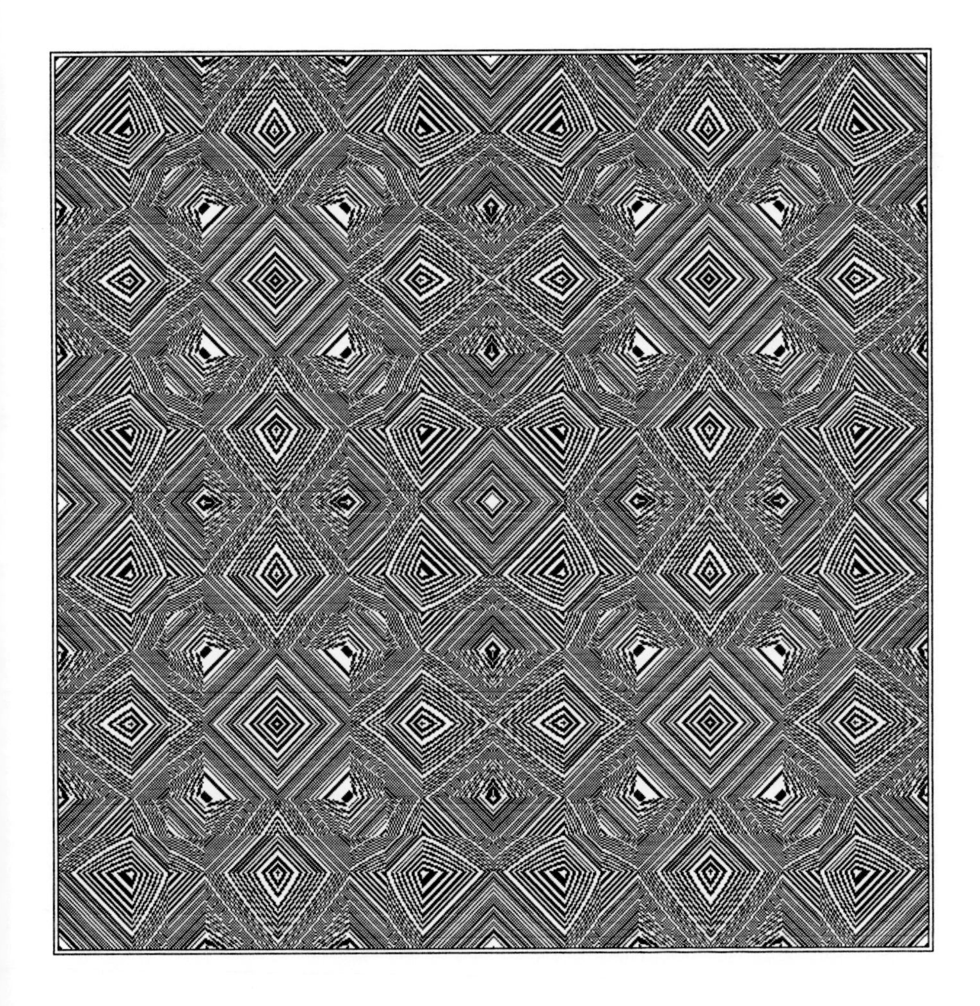

William J. Jones

TESS Patterns

TESS is a program which generates repeating surface pattern designs spontaneously. The program first selects one of eighteen *schemes*, or sets of rules, for creating a small square pattern in the upper left corner of the computer screen. Then it displays the pattern and propagates it in both directions so as to fill the screen. As it displays or prints the rows, the program may sometimes indulge in some *row transformations*, such as reversing and sliding, so as to add some more variety.

Figures 1 and 2 display the same pattern but in different scales. The scheme was "Stripes", which yielded a simple base pattern with horizontal and vertical stripes, but the program decided to *quadruplex* it, that is, form a new base pattern twice as large in both dimensions with the original base pattern subjected to various transformations in each of the quadrants.

Continuing clockwise, Fig. 3 is also a product of the "Stripes" scheme, quadruplexed with a color permutation as well as rotation. In Fig. 4, the base pattern was from a scheme called "8-way", with symmetry about NS axis, EW axis, and both diagonals. Various transformations involving color permutations destroyed some of that symmetry. Color renderings of this pattern are especially pleasing.

Figure 1.

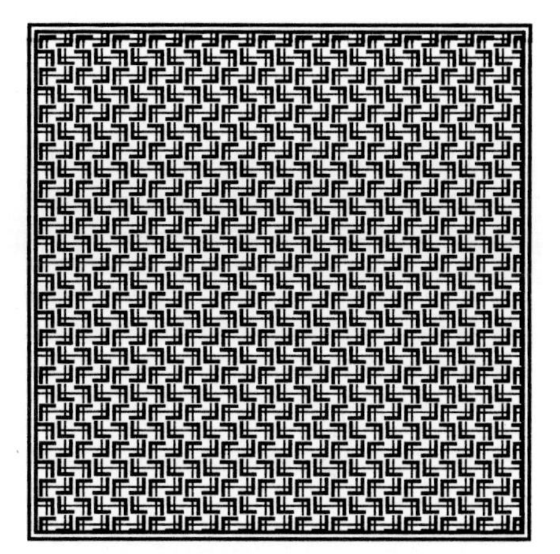

Figure 2.

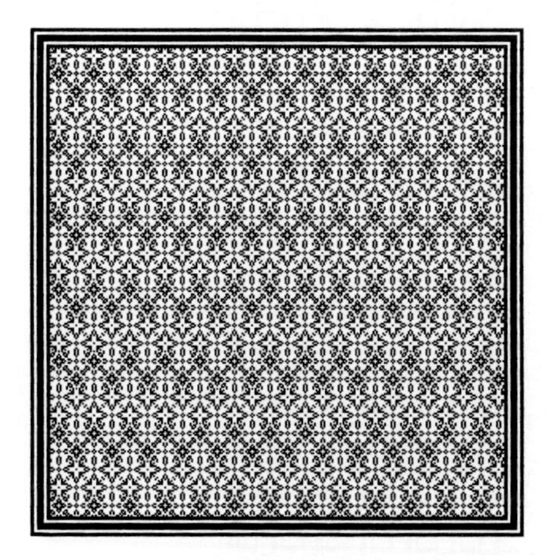

Figure 3.

Figure 4.

William J. Jones
Skew Squares

Skew Squares is a product of a program called TESS which creates huge domains of surface pattern designs by applying rules defining eighteen *schemes*, or classes, of designs to create a *base pattern*, then replicating that pattern to fill the computer screen. The parameters for this particular pattern are now lost in antiquity, since the file from which it was printed has been erased, and TESS was deliberately designed to sample a different part of its pattern domain each time it runs. The pattern was probably based on either a scheme called "Stripes" or one called "Circ/Sq", producing an off-center square, then subjected to a transformation that reproduced the base square into 2×2 array while rotating it.

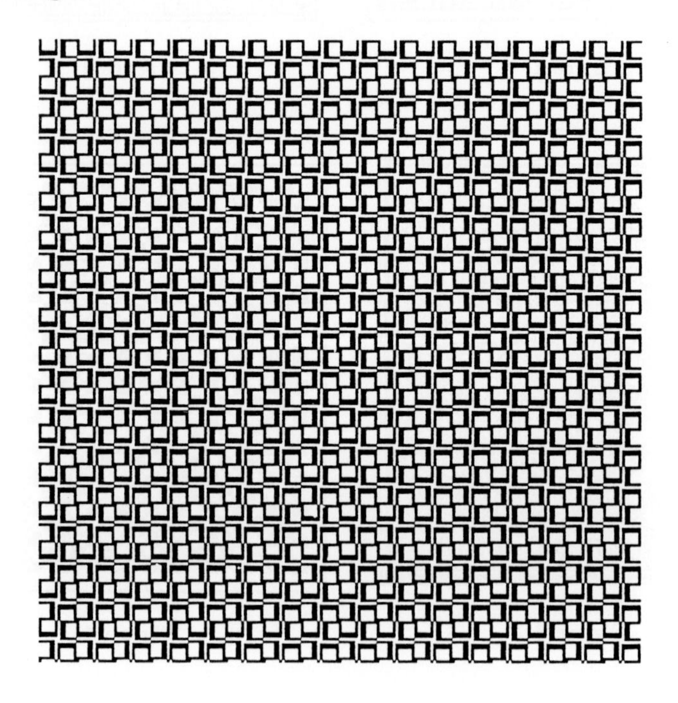

Huw Jones

Fractal Gaskets

The patterns introduced here comprise a family generated by an extension of methods used in creating the well-known Sierpinski triangle or gasket (Fig. 1). These deterministic fractals can be created either by a standard recursive method [1] or by an adaptation of Barnsley's chaos game [2], which produces a randomly generated sequence of points having the required fractal shape as attractor.

Suppose we start with vertices at the corners of any regular polygon with n sides, say. Imagine that the polygon has very small copies of itself placed within it at each of its vertices. Now make the small copies grow, still placed at the vertices and still remaining equal to each other, until they just touch. To produce a fractal gasket [3], the contents of the original polygon need to be shrunk to fit inside the sub polygons, so the size of these copies when compared with the size of the original polygon is important in defining the production rule. In evaluating this ratio of sizes, the exterior angle of the polygon, $360°/n$, is needed to decide the length of the portion to be cut out of the side of the original polygon. If $360°/n$ is not less than $90°$, that is, n is 4 or less, the edges of the smaller polygons meet exactly at the edge of the original polygon (triangle or square in these cases). When n is 5, 6, 7, or 8, the shape cut out at an edge is a triangle; for larger values of n, more complicated shapes are cut out. The value of

$$k = \text{trunc}(\{90\}/\{360/[n-1]\}) = \text{trunc}([n-1]/4)\,,$$

where $\text{trunc}(x)$ is the truncated integer part of the real value x, determines how many sides of a small polygon are projected onto the portion cut out of the side of the larger polygon. Let us say that the cut out length is given by

$$2s\sum_{i=1}^{k}\cos(360i°/n) = 2sc\,.$$

Thus, if S is the side length of the original polygon, we have

$$S = 2s + 2sc = 2s(1+c)\,.$$

The required shrink factor is

$$f = s/S = 1/\{2(1+c)\}\,.$$

A fractal gasket can then be constructed by recursively drawing n scaled down polygons within each larger polygon, each smaller polygon having one vertex in common with its "parent".

Alternatively, using an adaptation of Barnsley's chaos game, the gasket can be constructed from an original point P by selecting one of the vertices V of the original polygon at random and then plotting a new point P' at a fraction f of the distance from V to P. For the triangular gasket, f will be $\frac{1}{2}$ as specified in Barnsley's method and the familiar Sierpinski triangle will be created. Calculation of the position of P' is simple. Merely apply the formula

$$P' = (1-f)V + fP$$

to the x and y coordinates of V and P in turn. Some results of this process are shown in Figs. 2 to 5. The pentagonal-based process (Fig. 3) gives a result similar to "rosettes" used in tiling patterns produced by the Dutch artist Albrecht Dürer in 1525 [4]. The hexagonal form (Fig. 4) has internal and external bounding curves which take a familiar form — the Koch curve. For large numbers of sides in the original polygon, the pattern produced becomes narrower, resembling a laurel wreath (Fig. 5). If this procedure is attempted with a square, the resulting image is not fractal. A set of random points evenly distributed within the square is generated, as the four half size sub-squares completely fill the original square.

The fractal dimension [5] for each such gasket can easily be found using the formula

$$D = \log(n)/\log(1/f)\,,$$

as each gasket is exactly self-replicating, containing n copies of itself at scale factor f. Thus, the fractal dimensions for the gaskets illustrated with $n =$ 3, 5, 6, and 9 are 1.585, 1.672, 1.631, and 1.621 respectively. When $n = 4$, $f = \frac{1}{2}$, so $D = \log(4)/\log(2) = 2$, verifying that the algorithm does not produce a fractal object when applied to a square.

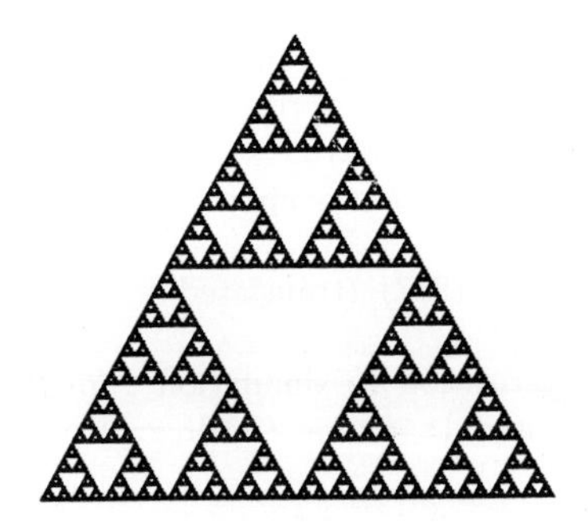

Figure 1. A Sierpinski triangle generated recursively.

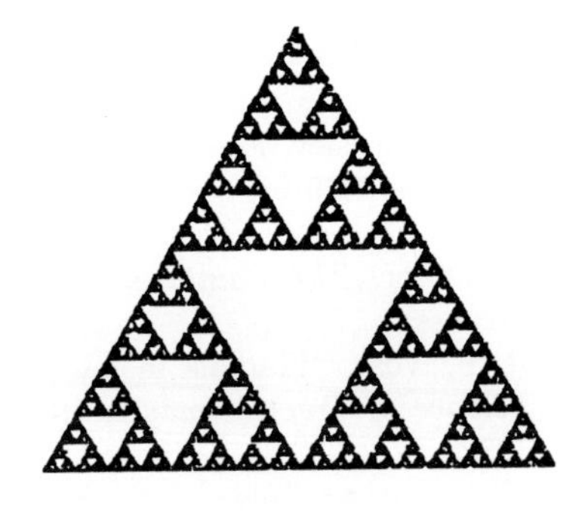

Figure 2. A Sierpinski triangle generated using the chaos game.

Figure 3. A pentagonal gasket.

Figure 4. A hexagonal gasket.

Figure 5. A nonagonal gasket.

References

1. D. Saupe, "A unified approach to fractal curves and plants", in *The Science of Fractal Images*, eds. H.-O. Peitgen and D. Saupe (Springer-Verlag, 1988).
2. M. F. Barnsley, *Fractals Everywhere* (Academic Press, 1988).
3. H. Jones, "Dürer, gaskets and the chaos game", *Computer Graphics Forum* **9**, 4 (1990).
4. A. Dürer, *The Painter's Manual* (Abaris Books, 1977) (translated by W. L. Strauss).
5. R. F. Voss, "Fractals in nature: From characterisation to simulation", in *The Science of Fractal Images*, eds. H.-O. Peitgen and D. Saupe (Springer-Verlag, 1988).

Huw Jones

An Octahedral Fractal

The pattern illustrated is an octahedral fractal object generated in a way analogous to the technique used to produce a Sierpinski tetrahedron (Fig. 1). The tetrahedron is the simplest form of polyhedron, having four triangular faces and four vertices. In its regular form, it is one of the five Platonic solids, these being the only absolutely regular tetrahedral solids. To construct a Sierpinski tetrahedron, the original shape is replaced by four similar half linear scale tetrahedra, at the same orientation as the original, each being placed within the original tetrahedron adjacent to a vertex. This procedure is recursively continued. If repeated indefinitely, this produces an exactly self-replicating fractal object with fractal dimension [1]

$$D = \log(4)/\log(2) = 2\,,$$

as the object contains four copies of itself at half scale. This is an example of a fractal object which has an integer dimension, although the dimension is lower than that of the least dimensioned Cartesian space within which the object can exist. At each stage of recursion, one half of the existing volume of the object is eliminated. This generates a geometric progression which sums to unity, showing that the object is eventually reduced to zero volume. In producing images of the Sierpinski tetrahedron, recursion is stopped after a set number of stages, giving an approximate depiction, as in Fig. 1.

An octahedral fractal can be created from a regular octahedron in a similar way. The regular octahedron is another of the Platonic solids, having eight triangular faces and six vertices (it is the "dual" of the cube or hexahedron, which has six faces and eight vertices). A canonical form of the octahedron can be drawn with vertices at $(-1,\ 0,\ 0)$, $(1,\ 0,\ 0)$, $(0,\ -1,\ 0)$, $(0,\ 1,\ 0)$, $(0,\ 0,\ -1)$, and $(0,\ 0,\ 1)$ in 3D Cartesian space. The first stage in creating an octahedral fractal involves replacing the original octahedron by six half scale copies of itself, each sitting adjacent to a different vertex of the original and lying within the space occupied by the original (Fig. 2). This is already a nonmanifold object, having some edges from which four faces emanate (these lie in the coordinate planes at $45°$ to the axes), but has no through holes as there are squares in diamond orientation lying in the coordinate planes which are not

Figure 1. A Sierpinski tetrahedron after 3 subdivisions.

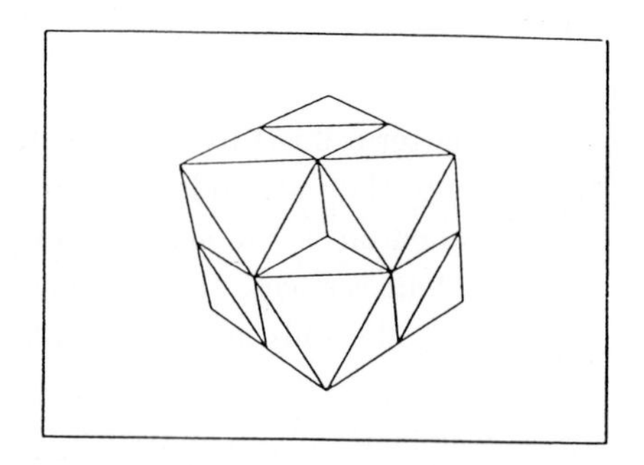

Figure 2. An octahedral fractal after 1 subdivision.

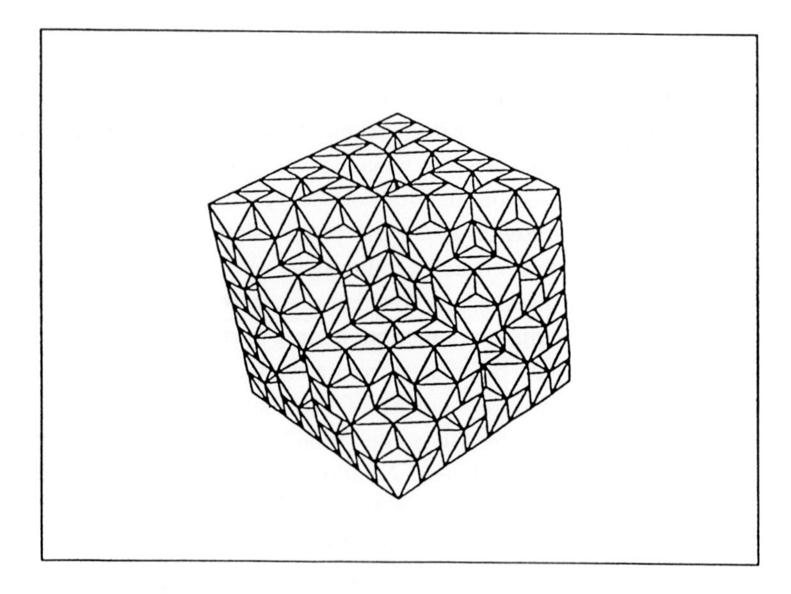

Figure 3. An octahedral fractal after 3 subdivision.

penetrated. These smaller octahedra can now be recursively replaced in the same way to produce an octahedral fractal (Fig. 3). Even at higher levels of subdivision, this continues to have nonpenetrated planes, so it has no through holes. The object is exactly self-replicating and has fractal dimension

$$D = \log(6)/\log(2) = 2.585\,,$$

as it contains six half scale copies of itself. A geometric progression, eliminated after each stage, can be simply generated as in the case of the Sierpinski tetrahedron. This again sums to unity, showing that the octahedral fractal has zero volume. It is interesting to note that the faces of the original octahedron are reduced to Sierpinski triangles or gaskets, as are the faces of the original tetrahedron in the creation of a Sierpinski tetrahedron.

A coauthored paper (with A. Campa), describing the properties of this and similar objects, giving different methods of depiction and construction of the object, is currently being refereed for publication.

Reference

1. R. F. Voss, "Fractals in nature: From characterisation to simulation", in *The Science of Fractal Images*, eds. H.-O. Peitgen and D. Saupe (Springer-Verlag, 1988).

Haresh Lalvani and Neil Katz

A Packing with Icosahedral Symmetry 1

Described here is a pattern showing a three-dimensional packing of rhombicosi-dodecahedra, viewed along its fivefold axis of symmetry. The polyhedra are packed in such a way that the center of each polyhedron lies at the vertex of a larger rhombicosi-dodecahedron. The model is only one of a family of sixty-four packings [1], where each of seven fivefold polyhedra (along with a point) are placed at the vertices of seven fivefold polyhedra (and a point) so that the polyhedra meet each other at a face, an edge, or a vertex. In the example, each polyhedron shares a square with its immediate neighbor.

The process may be recursive. The act of replacing each vertex with a polyhedron may be continued, recursively, replacing each vertex of the packing with a polyhedron. Each polyhedron, as well as the entire model, has three axes of symmetry — a twofold axis, a threefold axis, and a fivefold axis. Polygons lying on planes perpendicular to these axes of symmetry may be color-coded with three primary colors: twofold is yellow, threefold is blue, and fivefold is red.

The image was created at Skidmore, Owings & Merrill, using the following:

Computer:	IBM RS/6000
Software:	IBM AES (Architecture and Engineering Series)
Plotter:	Versatek 8900 Electrostatic Color Plotter

Reference

1. H. Lalvani, *Patterns in Hyper-Spaces* (Lalvani, 1982) p. 24.

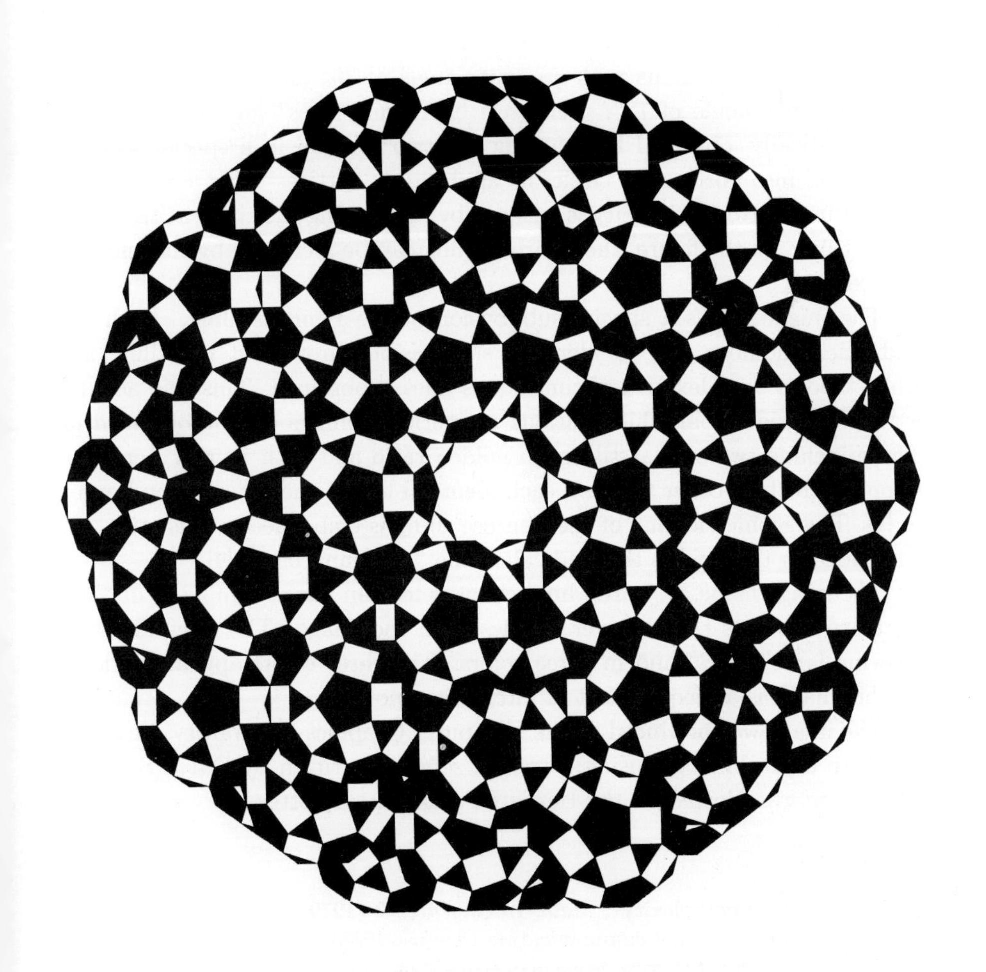

Haresh Lalvani

Hard Lines, Soft Circles

This image is based on the well-known Penrose tiling which uses two rhombii [1], a "thin" rhombus with an acute angle of 36° and a "thick" rhombus with an acute angle of 72°. Penrose's "matching rules" for tiling the plane nonperiodically, obtained by marking the two rhombii in a special way, are abstracted into portions of circles at appropriate vertices using a notation by De Bruijn [2]. Nonperiodicity is ensured by completing the circle at the vertex of the tiling. In the figure, the Penrose tiling can be identified by the encircled vertices.

The image shows a special subdivision of the Penrose tiling. Each thin and thick rhombus has an identical subdivision throughout the tiling and, when tiled, a complex grid of continuous lines is obtained. This grid acts like a "master grid" from which various nonperiodic tilings can be obtained [3]. These include two similar tilings in golden ratio and with regular pentagons (along with three other shapes), each identical to the one reported by Penrose in [1]. The geometric dual of the Penrose tiling is embedded in this grid. Also embedded in this grid is the grid called the "Ammann bars" [4].

The image made of "hard" lines changes into an image composed of "soft" circular shapes as the observer recedes from the image. The original plotter drawing is $21'' \times 28''$ and in three colors. The three colors appear randomly on the image as an experiment in interactive plotting.

The image was produced at the Computer Graphics Laboratory, New York Institute of Technology, using a special software written by David Sturman. The image was plotted on the Hewlett-Packard pen-plotter.

References

1. R. Penrose, "Pentaplexity", *Math. Intelligencer* **2** (1979) 32–37.
2. N. De Bruijn, Personal communication (August 1985).
3. H. Lalvani, "Morphological aspects of space structures", in *Studies of Space Structures*, ed. H. Nooshin (Multi-Science Publishers) (in press) Figs. 240–248.
4. D. Levine and P. Steinhardt, "Quasicrystals 1: Definition and structure", Department of Physics, University of Pennsylvania, USA (August 1985) (preprint).

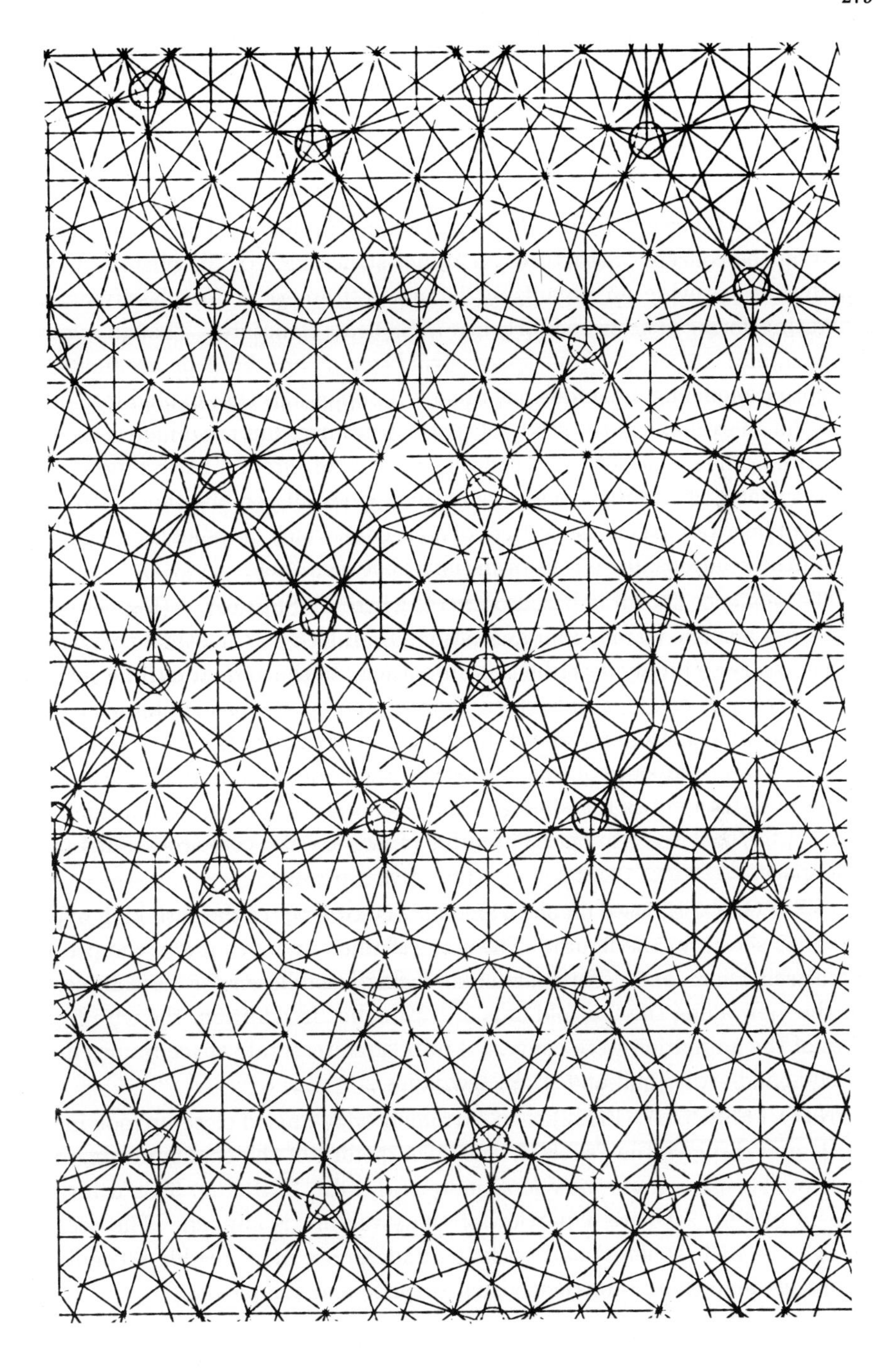

Haresh Lalvani

Tesspoly 3, into Tesspoly and Sectus

The three images in Figs. 1–3 are related in a sequence. The broad study is based on the concept "Structures on Hyper-structures" [1, 2], where n-dimensional cubes provide an organizing lattice or framework for classifying and generating form. The framework has a Boolean structure. Though typical, the examples are taken from space structures of one kind or another, this conceptual modeling technique has broader implications that extend into other fields. As an example in the study of space structures, this technique is used for subdividing the surface of a polyhedron, say, a cube. From n basic subdivisions of the surface, 2^n subdivisions are obtained from all combinations of n, and these can be arranged on the vertices of an n-cube which maps all the structures and all the relationships (transformations) between them.

The sequence shown is based on n=4, where TESSPOLY 3 (Fig. 1) is derived from a single frame of an animation. In order to understand this figure, it is necessary to briefly describe this animation. In the animation, the vertices of a hyper-cube "grow" into sixteen "wire-frame" cubes of increasing size till they reach a fixed size, and where the surface of each cube is subdivided differently. In a later sequence of this animation, the underlying hyper-cube shrinks while the sixteen cubes remain the same size, and the shrinking continues till all the cubes coalesce into one. The subdivisions are also color-coded in 4-cube color space.

TESSPOLY 3 is based on one frame in this sequence of the shrinking hypercube (not shown in the figure). The sixteen different cubes are in the process of coalescing into one. This particular frame was chosen for its complexity and the view along the fourfold axis of each cube. In addition, the wire-frame cubes of the animation were converted into "architectural" space frames using the node-and-strut (ball-and-stick) technique.

INTO TESSPOLY (Fig. 2) is an extremely wide-angled close-up look inside Fig. 1, and displays the "space frame"-like quality of the configuration. The struts of the space frame now look more real. SECTUS (Fig. 3) is a further close-up of Fig. 2 where a portion of the structure is clipped. The section through the ball (spheres) and sticks (cylinders) are more readily recognized. All images were chosen for their visual surprise and complexity.

The three images were done at the Computer Graphics Laboratory, New York Institute of Technology, using their modeling and rendering software and special software written by Patrick Hanrahan.

References

1. H. Lalvani, "Multi-dimensional periodic arrangements of transforming space structures", Ph.D. Thesis, University of Pennsylvania (University Microfilms, Ann Arbor, 1982).
2. H. Lalvani, *Structures on Hyper-Structures* (Lalvani, 1982).

Figure 1. Tesspoly 3

Figure 2. Into tesspoly

Figure 3. Sectus

Wentian Li

Tetrahedron

Described here is a pattern showing the spatial-temporal configuration generated by applying a next-nearest-neighbor two-state cellular automaton rule to a random initial configuration. Cellular automata are discrete dynamical systems which update configurations by local coupling [1]. For example, the next-nearest-neighbor two-state cellular automata are rules of the following form [2]:

$$a_i^{t+1} = f(a_{i-2}^t,\ a_{i-1}^t,\ a_i^t,\ a_{i+1}^t,\ a_{i+2}^t)$$

where $a_i^t \in (0, 1)$ is the variable value at site i and time t, and $f()$ is the function that represents the rule.

This pattern is generated by the cellular automaton rule which is encoded in the following binary string (*rule table*):

11000011101111001110001110010000,

in which each bit represents the function $f()$ with a particular input. For example, the first bit (on the left) represents $f(11111)$, the second represents $f(11110), \ldots$, and the last bit represents $f(00000)$.

The spatial-temporal pattern produced by this rule (x axis is space, y axis is time with the time arrow pointing down) is packed with seemingly three-dimensional tetrahedrons with different sizes, arranged in a fractal-like hierarchical structure. What is interesting about this rule is not only its highly symmetrical spatial-temporal pattern, but also the fact that this pattern is the *attractor* of the dynamics. In other words, if one starts from a different random initial condition, the pattern produced is still the same! What else can we ask for from a dynamical system which naturally produces a pretty picture?

References

1. S. Wolfram, *Theory and Application of Cellular Automata* (World Scientific, 1986).
2. W.-T. Li, "Complex patterns generated by next nearest neighbors cellular automata", *Computers and Graphics* **13**,4 (1989) 531–537.

Tetrahedron

Siebe de Vos and Raymond Guido Lauzzana

Binary Matrix Symmetry

The pattern described here illustrates binary matrix symmetry. Black signifies $+1$ and white -1, while gray indicates that a cell may be either $+1$ or -1. The upper pattern contains all of the possible 3×3 matrices of this type. The lower pattern illustrates the symmetry groups, S-groups, for the upper pattern. The members of an S-group are equivalent under $k\pi/2$ rotation: τ, reflection: ρ, and complement: χ. Our interest in this problem developed when we discovered that there are fifty-one unique TriHadamards as described in the illustration entitled *TriHadamards.* Since then, we have been looking for an analytic solution to the number of N-Hadamards.

A 3×3 matrix can be partitioned into three parts, the center: □, the corners: ⊞, and the side-middles: ✢. The upper pattern illustrates that the set of 3×3 binary matrices is the Cartesian product of this partition:

⊞ × ✢ × □.

The vertical axis contains all of the possible corner configurations, the horizontal axis contains all of the side-middle variations, and the two center possibilities are indicated using the color gray. The ordering along the axes was chosen to highlight the S-groups. The origin is considered to be in the upper left-hand corner and the population of black cells increases from left to right and top to bottom.

One important characteristic of this partitioning is that the cells contained in each partition are equidistant from the center. All three of the transformations, τ, ρ, χ, preserve this distance, and therefore, the sets defined by the partitioning are closed under these transformations. This makes it possible to analyze each of the sets independent of their combination. This partitioning defines four and only four possible types of subsets. The three types described above plus a fourth type, ⧆, only occur in matrices of size greater than three.

We have also found a calculus for analyzing S-groups of the four types and their relationship to the S-groups of an $N \times N$ matrix. An example of an expression using this calculus is:

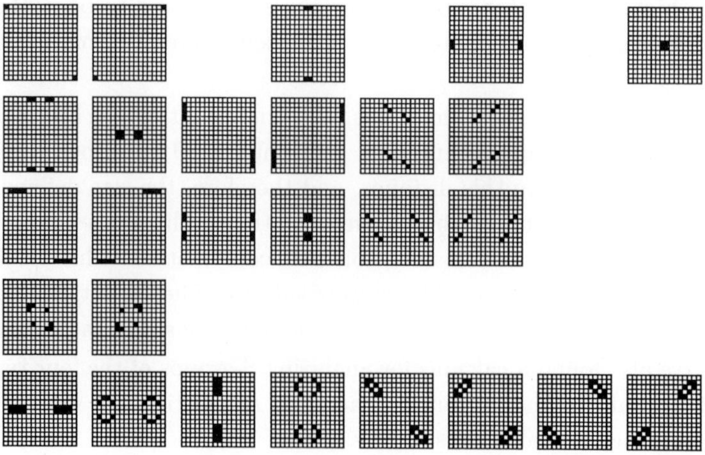

We are in the process of formalizing this calculus and identifying its operations, rules, and theorems. We have identified several basic equivalencies that can be expressed using this calculus, such as:

As yet we have not found an analytic solution to the problem:

How many S-Groups are in an $N \times N$ matrix?

Raymond Guido Lauzzana and Siebe de Vos

TriHadamards

This pattern is based on matrices derived from the work of the French mathematician, Jacques Hadamard. He defined a class of orthogonal two-valued matrices with unusual symmetry properties [1]. These matrices are now known as the Hadamard matrices and are commonly used in optics, acoustics, signal processing, image analysis, and pattern recognition [2]. The matrices have many properties in common with the Fourier Series. Hence, the coefficients of the Hadamard matrix are called the sequencies, corresponding to the Fourier frequencies. A transformation, known as the Hadamard transform, is homomorphically invertable, such that $H(H(I)) = 1$. This property, which the Fourier transform also shares, has made the Hadamard transformation popular for data compression, and many high-speed algorithms have been developed to transform signals into the Hadamard domain.

The Hadamard matrix is a matrix composed of only $+1$ and -1. The elements of the matrix are often illustrated using black and white to indicate $+1$ and -1, respectively. It is defined for $(n > 0)$, recursively on a 2×2 kernel, in the following manner.

$$H^n = \begin{bmatrix} +H^{n-1} & -H^{n-1} \\ -H^{n-1} & -H^{n-1} \end{bmatrix}, \quad \text{where} \quad H^0 = \begin{bmatrix} +1 & -1 \\ -1 & -1 \end{bmatrix}.$$

We have extended this recursive definition for Hadamard matrices to 3×3 kernels, calling them TriHadamards with the following definition:

$$T^n = \begin{bmatrix} T^0_{[0,0]} \cdot T^{n-1} & T^0_{[0,1]} \cdot T^{n-1} & T^0_{[0,2]} \cdot T^{n-1} \\ T^0_{[1,0]} \cdot T^{n-1} & T^0_{[1,1]} \cdot T^{n-1} & T^0_{[1,2]} \cdot T^{n-1} \\ T^0_{[2,0]} \cdot T^{n-1} & T^0_{[2,1]} \cdot T^{n-1} & T^0_{[2,2]} \cdot T^{n-1} \end{bmatrix}.$$

The kernel, T^0, may be any 3×3 matrix containing only $+1$ and -1. Since there are only two possible values for the elements of the kernel, there are 512 kernel matrices. Each of these kernels may be referenced by an index similar to binary counting. The pattern depicted here is the TriHadamard generated from the 299th kernel. Any of the kernels may be used to generate a TriHadamard, but many of them are equivalent under complement, rotation, and reflection. There are fifty-one unique kernels. We have only recently discovered TriHadamards and have begun to investigate their properties. Some of the symmetries associated with the kernels are described with the illustration entitled *Binary Matrix Symmetry.*

References

1. J. Hadamard "Resolution d'une question relative aux determinants", *Bull. Sci. Math.* **17**, I (1993).
2. S. Agaian, *Hadamard Matrices and Their Applications* (Springer-Verlag, 1980).

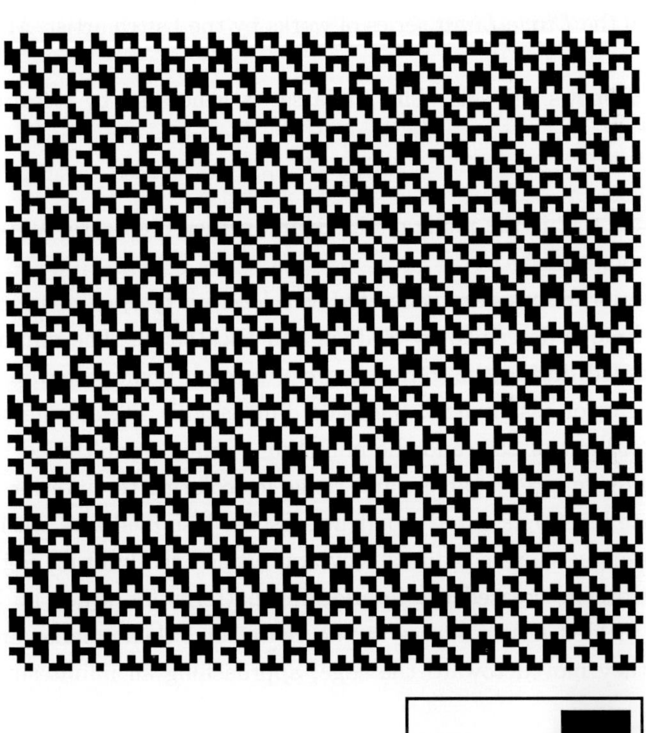

James Elliott Loyless

Fractal Limit — A Fractal Pattern Emulating M. C. Escher

Described here is a pattern showing a simple Julia set seeded near zero with the interior features revealed as a level set. It demonstrates a haunting similarity to the *Circle Limit* series of works by the Dutch artist, M. C. Escher [1, 2].

Figure 1 shows the set. This Julia set is generated by iterations of the formula

$$z \to z^2 - c$$

where $z_0 = x + yi$ is the plane of the figure, centered on zero, and $c = a + bi$ is the seed point 0.05.

The interior features are generated as a *level set of equal dynamic distance* using a technique described by Peitgen and Richter [3]. The minimum value of z is recorded during the iteration cycle for each point tested, and used to describe the color assignment. The side dimension is 1.0, alternation goes to the iteration limit of 250, and the minimum z value is multiplied by a factor of 250 to provide a reasonable count range for coloring the interior.

To create his *Circle Limit* woodcuts, Escher inscribed arcs within a circle with increasingly shorter radii as they approached the limiting edge. Within this framework he created the interlocking patterns for which he is famous. The effect yielded larger figures at the center and then smaller and smaller figures as they moved toward the edge, approaching an infinitesimally small size [1].

It should not be surprising to find similarities between Escher and fractal geometry, since Escher was a student of higher geometry and mathematics. Through illustrations by the geometer, H. S. M. Coxeter, he discovered the hyperbolic model which Jules Henri Poincaré used to show the whole of an infinite flat plane inside a finite circle [2]. This Coxeter illustration bears a striking resemblance to the Julia set shown in Fig. 1.

"Deep, deep infinity! ... to sail over a calm sea at the prow of a ship, toward a horizon that always recedes ..."

(M. C. Escher [1])

References

1. J. L. Locher, *The World of M. C. Escher* (Abrams, 1972) pp. 15–16, 142–143.
2. B. Ernst, *The Magic Mirror of M. C. Escher* (Random House, 1976) pp. 102–111.
3. H.-O. Peitgen and P. H. Richter, *The Beauty of Fractals* (Springer-Verlag, 1986) pp. 60–62.

Figure 1. "Fractal Limit", a Julia set.

James Elliott Loyless

Fractal Limit — A Perspective View

Described here is a pattern showing a Julia set with interior features revealed that demonstrate a striking similarity to the *Circle Limit* series by the Dutch artist, M. C. Escher. The set is projected in a vanishing point perspective to achieve both the interesting effect of hovering over a fractal landscape and a new way of looking at fractals. The detail in the foreground combined with a broader view in the distance gives the impression of an *in situ* zoom, as the eye tracks the shrinking horizon from top to bottom of the page.

Figure 1 shows the work by M. C. Escher entitled *Circle Limit IV.* To create this woodcut, Escher inscribed arcs within a circle with increasingly shorter radii as they approached the limiting edge, based on the hyperbolic model of Poincaré, which can show an infinite flat plane inside a finite circle [1]. The effect yields larger figures at the center and smaller figures toward the edge, approaching infinitesimal size.

Figure 2 shows the Julia set created by the writer and entitled *Fractal Limit.* This set is generated by iterations of the formula

$$z \to z^2 - c$$

where $z_0 = x + yi$ is the plane of the figure, centered on zero, and $c = a + bi$ is the seed point 0.05.

The interior features are generated as a *level set of equal dynamic distance* using a technique described by Peitgen and Richter [2]. The minimum value of Z is recorded during the iteration cycle for each point tested and used to describe the color assignment. The side dimension is 1.0, alternation goes to the iteration limit of 250, and the minimum z value is multiplied by a factor of 250.

To create the projected view, shown in Fig. 3, the writer derived an algorithm through trial and error. Rather than project a previously computed set onto a vanishing point plane, the algorithm reverses this process by computing the virtual x-y coordinates for each point before the iterations are started. This approach allows greater detail and efficiency in calculation, and fits well into the conventional iteration procedure.

Figure 1. "Circle Limit IV".

Figure 2. "Fractal Limit".

Figure 3. "Fractal Limit" in perspective view.

The algorithm assumes a printer dot matrix medium consisting of 1440 dots across and 1200 dots down (180 dpi). The constants can be scaled for any other format:

wide = dot number across the page (from 1 to 1440)
high = dot number down the page (1200 at top; 1 at bottom)
trak = (40370 − 32 * high + 1.66667 * wide) / (1300 − high)
horn = 2400 / (1300 − high)
bittrak = (trak − 29) * 240
bithorn = 1200 − (horn − 2) * 300
x coord. = left real no. + (bittrak − 1) * pixelgap
y coord. = top imaginary no. − (bithorn − 1) * pixelgap

References

1. B. Ernst, *The Magic Mirror of M. C. Escher* (Random House, 1976) pp. 102–111.
2. H.-O. Peitgen and P. H. Richter, *The Beauty of Fractals* (Springer-Verlag, 1986) pp. 60–62.

John MacManus

Jungle Canopy

The jungle canopy shown here (Fig. 1) lies within the Mandelbrot set which is generated by the iteration of $z = z * z + c$ (to a maximum of 50 iterations with an escape radius of 2) where $z = x + yi$ [1–3]. The jungle canopy image is located between $x = -0.3984473$ and -0.3742513, and between $y = 0.1272181$ and 0.1452246 in the complex plane. This gives the sky (white) and the foliage (light gray).

Simultaneous with the plotting of this set boundary, the values of x and y are separately monitored. When below a threshold value (here 0.01 for both x and y) either the initial value of x or y is plotted. The black is the path of x, and the dark gray that of y. This technique is a development of ideas outlined in [4] of how to obtain details inside the M-set. In the jungle canopy, x and y are plotted separately, both inside and outside the set, and across the boundary. It is intriguing to trace how the paths of x and y interweave.

The image is of visual interest because it has a sense of depth and a painter's quality of movement.

References

1. B. B. Mandelbrot, *The Fractal Geometry of Nature* (W. H. Freeman, 1982).
2. A. K. Dewdney "Computer recreations", *Scientific American* **255**, 9 (1986) 14–23 and **255**, 12 (1986) 14–18.
3. K. H. Becker and M. Dorfler, *Dynamical Systems and Fractals* (Cambridge University Press, 1989).
4. C. A. Pickover, "Inside the Mandelbrot set", *Algorithm* **1**, 1 (1989) 9–12.

Figure 1. Looking up into a jungle canopy.

Tom Marlow

A "Y" Hexomino Tiling

Described here is a pattern showing a tiling of a 23×24 rectangle by the "Y" hexomino.

Hexominoes are formed by joining six equal squares edge to edge, and 35 different shapes are possible [1]. According to Gardner [2], the shape used here is the only hexomino for which it is undetermined whether it can tile a rectangle. The figure shown was obtained by exhaustive computer search which tested all relevant rectangles which would fit within a 23×29 area and also a few larger. No other solution was found.

The method was basically an exhaustive backtracking search by the computer of all possible arrangements of the hexomino within the rectangle under consideration. The eight possible orientations of the hexomino were used in turn each time that there was a backtracking return to a square after it had become impossible to continue. For rectangles of the size of the solution, this proved very demanding in time for a micro computer. The speed was improved by methods that are difficult to describe in detail and are quite arbitrary. Each time there is an attempt to place a hexomino on the next vacant square, it is necessary to check if a further five squares are empty. The technique is to check the five squares for vacancies in the order that is most advantageous. This order first tests those that are most likely to have been occupied by previous placings and/or, if occupied, will authorize skipping over the largest number of subsequent orientations. Secondly, the eight orientations are used in an order that maximizes the chance of such skipping. The programme took many hours of running time for the larger rectangles despite being written in assembly language and it remains to find larger tileable rectangle.

References

1. S. W. Golomb, *Polyominoes* (Scribner, 1965).
2. M. Gardner, *Mathematical Magic Show* (Allen & Unwin, 1977).

A "Y" Hexomino Tiling

David S. Mazel

Beauty in Functions

Figure 1 illustrates some of the variety of functions that may be created with fractal interpolation. In both diagrams, a family of curves was created, graphed and reflected about the x axis. To create each curve, a set of interpolation points is chosen, say, $\{(x_i, y_i) : i = 0, 1, \ldots, M\}$ and a subset of these points is chosen as address points, say, $\{(\tilde{x}_{j,k}, \tilde{y}_{j,k}) : j = 1, 2, \ldots, M; k = 1, 2\}$. For each consecutive pair of interpolation points, we map the address points to them with an affine map of the form:

$$\mathrm{w}_l \begin{pmatrix} x \\ y \end{pmatrix} = \begin{pmatrix} a_l & 0 \\ c_l & d_l \end{pmatrix} \begin{pmatrix} x \\ y \end{pmatrix} + \begin{pmatrix} e_l \\ f_l \end{pmatrix} : l = 1, 2, \ldots, M .$$

These maps are constrained so that

$$\mathrm{w}_l \begin{pmatrix} \tilde{x}_{l,1} \\ \tilde{y}_{l,1} \end{pmatrix} = \begin{pmatrix} x_{l-1} \\ y_{l-1} \end{pmatrix} \quad \text{and} \quad \mathrm{w}_l \begin{pmatrix} \tilde{x}_{l,2} \\ \tilde{y}_{l,2} \end{pmatrix} = \begin{pmatrix} x_l \\ y_l \end{pmatrix} \tag{1}$$

and $|\tilde{x}_{l,2} - \tilde{x}_{l,1}| > |x_l - x_{l-1}|$ and $|d_l| < 1$ which is known as the contraction factor. The parameter d_l is chosen arbitrarily and Eq. (1) is then used to compute the other map parameters.

Once the maps and addresses are known, the graph is constructed with the deterministic algorithm. Let A_0 be a single-valued function with support over the interval $[x_0, x_M]$, and apply the maps to A_0 in sequential order with the restriction that each map acts only on function values within the address interval associated with the map given by the constraints in Eq. (1). Call this new function A_1, discard A_0, and repeat this process for A_1. The series of functions $\{A_0, A_1, A_2, \ldots\}$ collapses to the final function which is guaranteed to be single-valued.

The top diagram in Fig. 1 consists of the interpolation points $\{(0, 0), (500, 1), (1000, 0)\}$ and each map has the same address points, namely: $\{(0, 0), (1000, 0)\}$. When the end-points of the interpolation points are the address points, the resulting curve is said to be "self-affine". The contraction factors were the same for each map for each curve and were varied from 0 to 0.5 in twenty uniform increments.

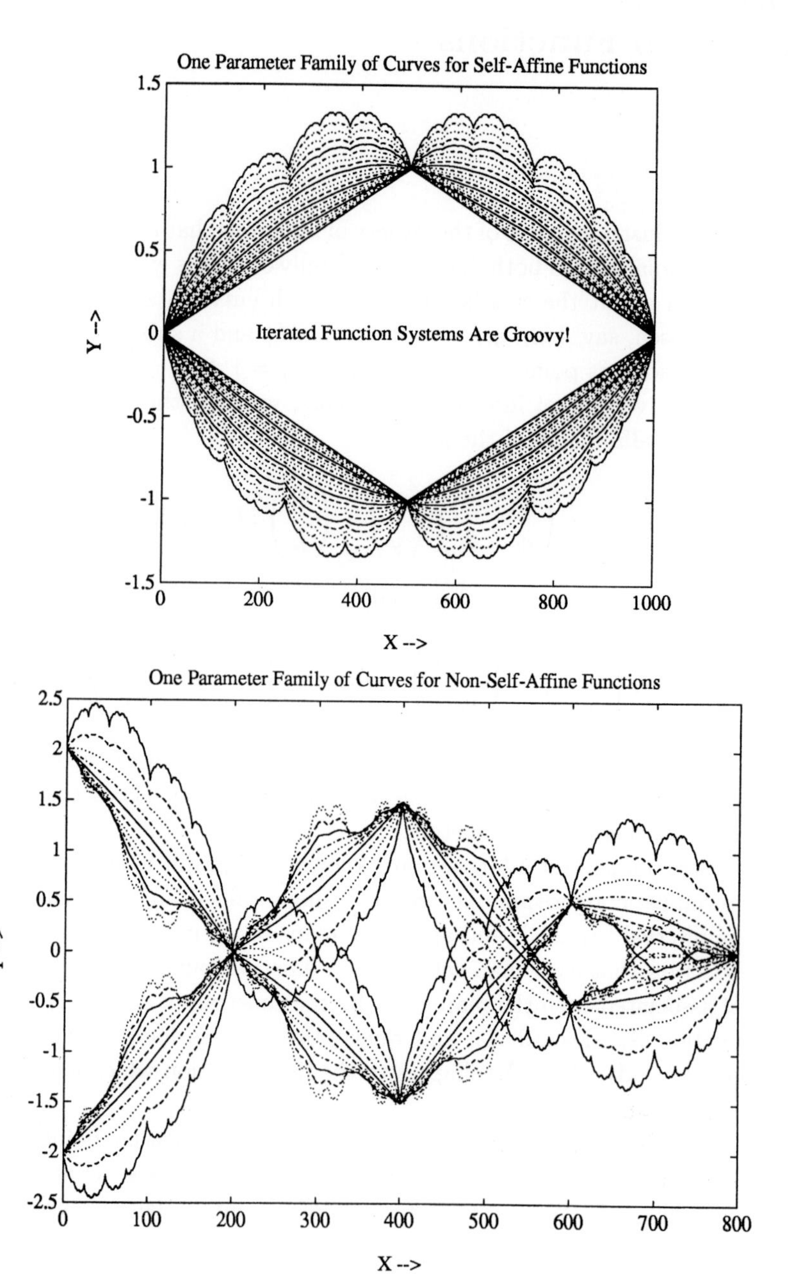

Figure 1. Families of single-valued functions.

The bottom diagram consists of the interpolation points {(0, 2), (200, 0), (400, −1.5), (600, 0.5), (800, 0)}, and the address intervals associated with each map were the same: {(400, −1.5), (800, 0)}. (Other address intervals could have been used just as well.) For each graph, the contraction factor was varied from −0.55 to 0.55 in ten uniform steps, and the value of each was negated for the odd-numbered maps for each curve, e.g., $d_1 = 0.55$, $d_2 = -0.55$, $d_3 = 0.55$, $d_4 = -0.55$ for the first curve.

Reference

1. M. F. Barnsley, *Fractals Everywhere* (Academic Press, 1988).

L. Kerry Mitchell

Logistic 3

Described here is a pattern showing the result of three iterations of the logistic difference equation, $x_n + 1 = r * x_n * (1 - x_n)$. The logistic difference equation is the canonical example of period-doubling and chaos in a discrete system, and is described widely in the technical and popular literature. See, for example, Peterson [1] or Reitman [2].

Iterating the equation maps the initial point x_0 into x_1, x_2, x_3, etc., with each x_0 leading to a different succession of x_i. Whereas previous work has concentrated on the bifurcation diagram, that is, showing the long-term behavior of x_i as a function of r, this figure shows how an interval is mapped into itself after three interations. In the present case, the value of the parameter r is set equal to -2. The horizontal line along the top of the figure represents 801 different initial conditions in the range $-2 < x_0 < 2$. At the bottom of the figure, each point on the lower line represents a value of x_3, the result after 3 iterations. A diagonal line joins each point x_3 with its corresponding value of x_0, and gives the viewer the impression of a curved line pattern. The beginnings and beauty of chaotic behavior are seen; small areas of near-constant x_0 are swept into regions encompassing the entire interval in only 3 iterations.

References

1. I. Peterson, *The Mathematical Tourist* (Freeman, 1988).
2. E. Rietman, *Exploring the Geometry of Nature* (Windcrest Books, 1989).

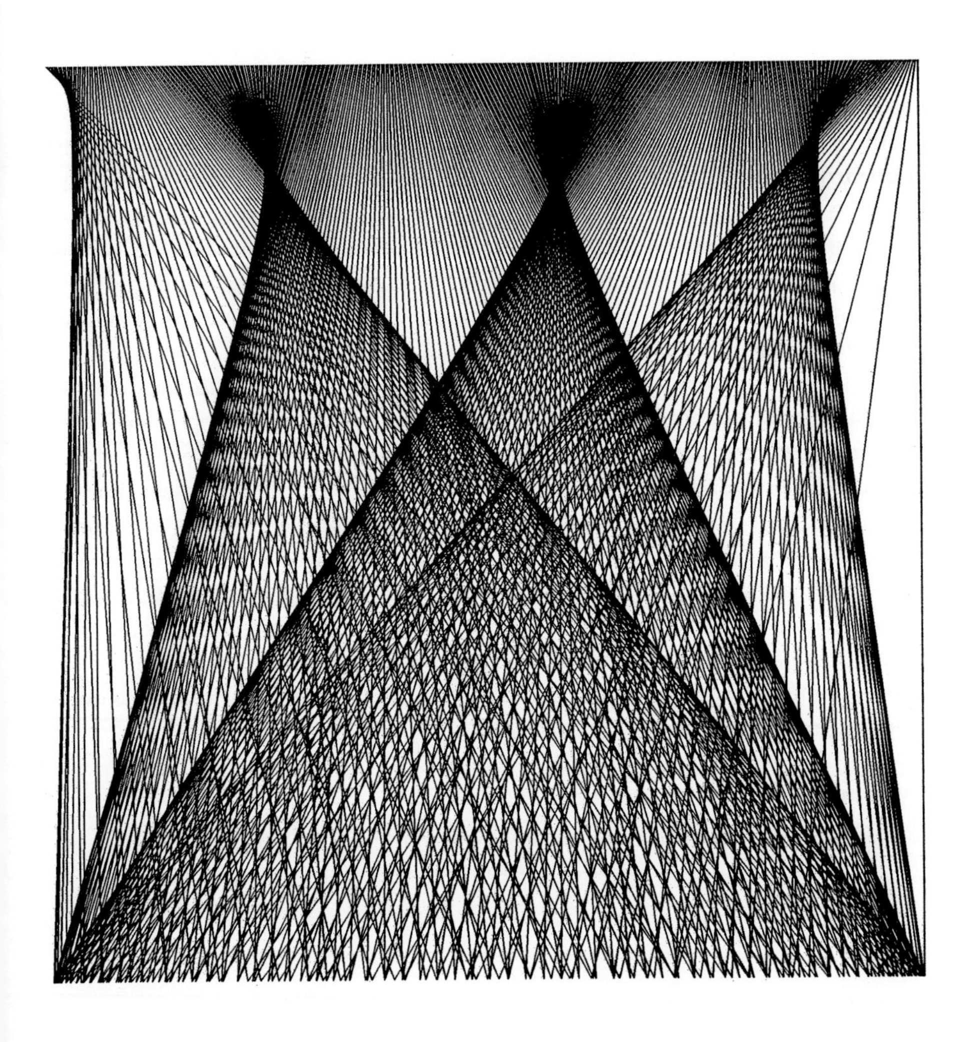

L. Kerry Mitchell

Inflation Rules 3

Described here is a pattern showing the result of six levels of recursion based on the "inflation rules" of Penrose tiles.

In 1977, British physicist and cosmologist Roger Penrose discovered a pair of quadrilaterals that, when joined according to certain rules, force a nonperiodic tiling of a plane. "Inflation rules" guarantee that the tiling will cover an infinite plane, because each of the tiles can be decomposed into smaller copies of the two. The two tiles have come to be known as "kites" and "darts", and have been discussed widely in the popular literature. See, for example, Gardner [1] and Peterson [2].

In the present work, only the decomposition aspect of the titles has been of interest. Beginning with a kite, it can be decomposed into two smaller kites, two half-darts, and an even smaller kite, as shown in Fig. 1. Each of these kites can be broken down into three more kites and two half-darts, continuing down as far as one is able. The result is a tiling composed of kites and darts of various sizes. Figure 2 shows the result after six levels of recursion. Areas where adjacent half-darts butt up against each other show where full darts would be used in the actual construction of this tiling. The fivefold symmetry of the tiles has been exploited by piecing five of the initial kites together to form a bounding dodecahedron, and a fascinating ribbon that weaves its way around the interior star.

References

1. M. Gardner, *Penrose Tiles to Trapdoor Ciphers* (Freeman, 1989).
2. I. Peterson, *The Mathematical Tourist* (Freeman, 1988).

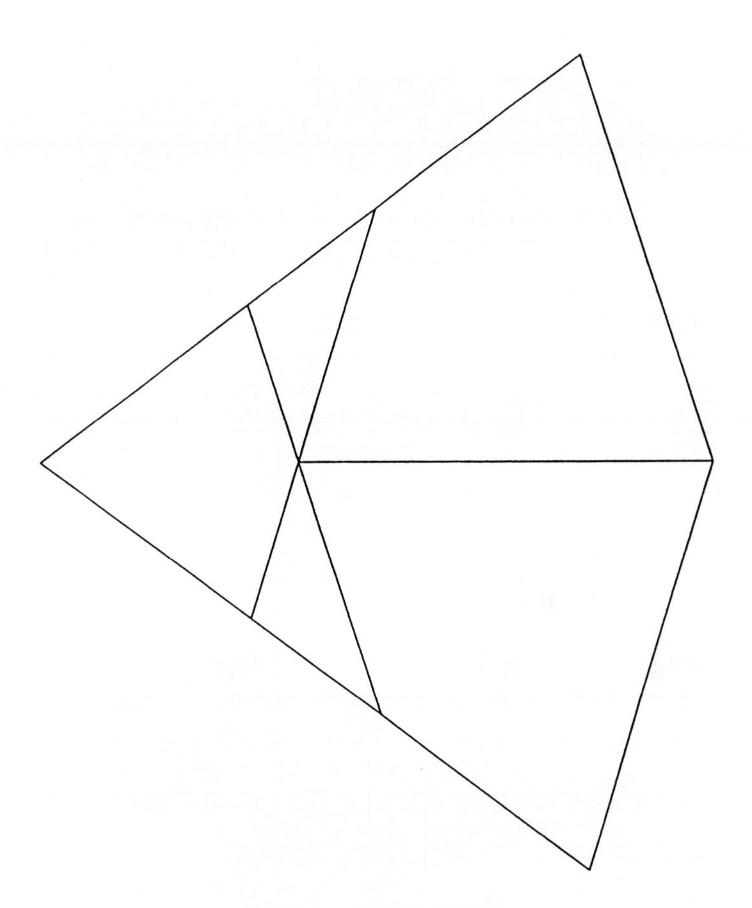

Figure 1. Inflation rules 3.

Figure 2. Inflation rules 5.

V. Molnar and F. Molnar

Hommage à Dürer

The starting point of the drawing shown herewith is a fourth-order magic square which Dürer judged worthy to present in his famous engraving "Die Melancholie", among other objects of geometrical significance. The magic square, which fascinated so many minds for so long, is a grid square filled in apparent disorder, with consecutive whole numbers. A more thorough examination reveals the repetition of the same total in the main directions of the gird. (Horizontals, verticals, and diagonals.) The square chosen by Dürer has another particularity as well: the two middle consecutive numbers of the fourth line 15 14, correspond to the year this masterpiece was created.

The simple idea emerging from the curiosity of people interested in images, consisted of transforming this grid, the matrix, to an oriented diagram, connecting the numbers in order from 1 to 16. This graph obtained in 1948, is a Hamiltonian graph. In fact, each summit is visited but only once. This much for the mathematicians. For the painter or for the aesthete, it is a beautiful drawing. (Left matrix.) In any case, to disappoint the theoreticians, it has to be mentioned that this drawing's beauty cannot be attributed to the magic square, since other drawings originating from other magic squares of the same order are undeniably less beautiful. (Right matrix.)

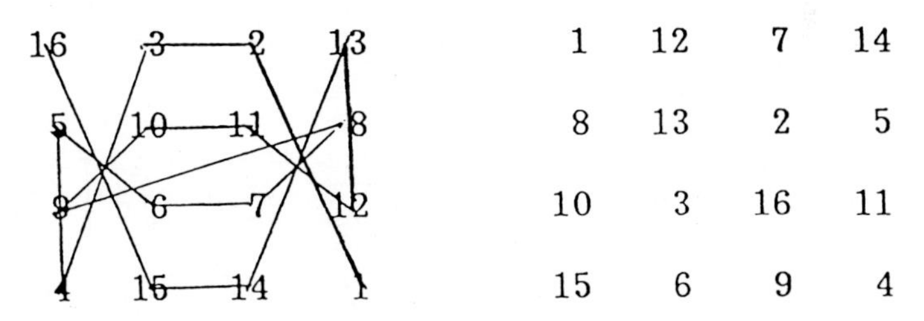

Since the beauty of the drawing is independent of the magic square, we tried to connect the 16 apexes of the matrix at random. Among the patterns obtained, there are some which are comparable to the beauty of the original pattern. The pattern obtained in this way maintains a certain regularity and a certain order. According to Birkhoff's well-known formula (which as a matter of fact should be verified), the measure M of beauty is expressed as follows:

$$M = O/C$$

where O signifies order and C complexity. It seemed necessary to us to increase the complexity of the pattern. Therefore, we have displaced the apex of the graph on the off chance. Thus, we added 5% noise to the drawing.

With this pattern we constructed a table of 16×16 variations. These 256 patterns constitute a sample obtained by chance among the more than 10^{13} possible permutations. The image reproduced is a detail of this work.

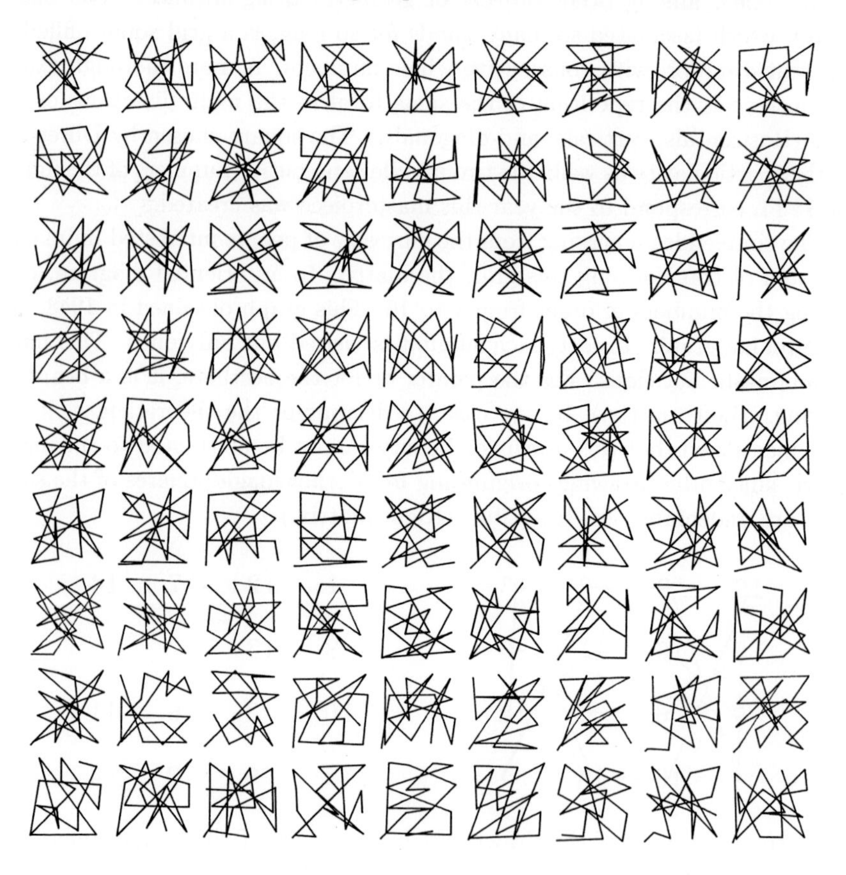

Douglas H. Moore

The Center of Dewdney's Radiolarian

Described here is a computer-generated image inspired by an article by A. K. Dewdney. The article appeared in [1]. It included a number of images called biomorphs. One in particular, Dewdney called a radiolarian. I was interested in obtaining a detailed view of the center of the radiolarian. Figure 1 is the result, using the program below with an EGA-EPSON screen dump. Dewdney provided a program to generate biomorphs which "follows Pickover's basic algorithm". The program is suggestive and not in a ready-to-run form. The program below runs in QUICKBASIC where screen 9 may be used for better resolution. The purpose of the aspect ratio, 5/6, at the end of line 1000 is to make the image come out square on the printer. The program will also run in BASIC using screen 2, using 100 in place of 175 in line 1000, and using the appropriate aspect ratio.

The parameter, A, determines the size of the image on the monitor screen. B determines the size of the square scanned in the complex plane. RC and IC are the real and imaginary parts of a given complex number, c. For Fig. 1, $A = 408$, $B = 1.6$, $RC = .5$, and $IC = 0$. Use $B = 10$ to obtain the full radiolarian in [1]. The program computes $z^3 + c$ in lines 500 and 600. Other functions there will produce other biomorphs.

```
50 SCREEN 9 : CLS
75 INPUT "A,B", A,B
100 INPUT "RC,IC", RC, IC
200 FOR J = 0 TO A : FOR K = 0 TO A
300 X=−B+2*B/A*J : y=−B+2*B/A*K
400 N=1
500 X1=X*X*X−3*X*Y*Y : Y1 = 3*X*X*Y−Y*Y*Y
600 X1=X1+RC : Y1 = Y1+IC
700 X2=ABS(X1) : Y2 = ABS(Y1)
800 Y3=X1*X1+Y1*Y1 : N=N+1 : IF N>10 THEN GOTO 1000
850 X=X1 : Y = Y1
900 IF Y3>100 THEN GOTO 1000 ELSE GOTO 500
```

```
1000 IF X2<10 OR Y2<10 THEN PSET (320−A/2+J,175−5*(A/2−K)/6)
1100 NEXT K : NEXT J
```

Reference

1. A. K. Dewdney, "Computer recreations", *Scientific American* **261**, 1 (1989) 110–113.

A,B 408,1.6
RC,IC .5,0

Figure 1. The center of Dewdney's radiolarian.

J. Peinke, J. Parisi, M. Klein and O. E. Rössler

2D Feigenbaum

Described here is a pattern showing smooth frontiers invading on a fractal support into a basin of attraction. The formerly connected basin is split into a Feigenbaum sequence of 2, 4, 8, 16,... parts. To obtain this pattern, we started with an iterative mapping very similar to the complex logistic map:

$$x_{n+1} = x_n^2 - y_n^2 + ax_n + b; \quad y_{n+1} = 2x_n y_n + c. \tag{1}$$

x_n and y_n denote the variables (with index $n = 0, 1, 2, \ldots$), while a, b, and c represent constants. Due to the term ax_n, Eq. (1) is not analytic any more. On the basis of this map, basins of attraction (filled-in Julia-like sets) and Mandelbrot-like sets can be calculated, reflecting self-similarity and fractal structures [1, 2].

The pattern of the figure shown here is obtained from a series of Julia-like sets. At first, the initial conditions were chosen as $-1.5 \leq x_0 \leq 1.5$, $-1.5 \leq y_0 \leq 1.5$ and divided up into 600×1200 points. Each point in the plane of the figure stands for a pair of initial conditions (x_0, y_0) and was iterated 500 times with double precision (FORTRAN). Only such points that fulfilled the condition $x_n^2 + y_n^2 \leq 10$ were marked by dots. This procedure was performed for $a = 0.3$, $b = 0.1225390$, $c = 0$. Next, the same procedure was done once more with $b = 0.1225392$ (a and c remained constant). But this time, nondivergent points marked before were now erased. That procedure was then repeated three times, each time increasing the last digit of the value of b by 2 and inverting the color of the previous step. With the help of this technique, it is possible to visualize how a Julia(-like) set vanishes under the variation of a control parameter (here, b) corresponding to a transition over a borderline of an appropriately chosen Mandelbrot set.

References

1. C. Kahlert and O. E. Rössler, "Instability of Mandelbrot set", *Z. Naturforsch.* **42a** (1987) 324.
2. J. Peinke *et al.*, "Smooth decomposition of generalized Fatou set explains smooth structure in generalized Mandelbrot set", *Z. Naturforsch.* **42a** (1987) 263; *Z. Naturforsch.* **43a** (1988) 14; *Z. Naturforsch.* **43a** (1988) 287.

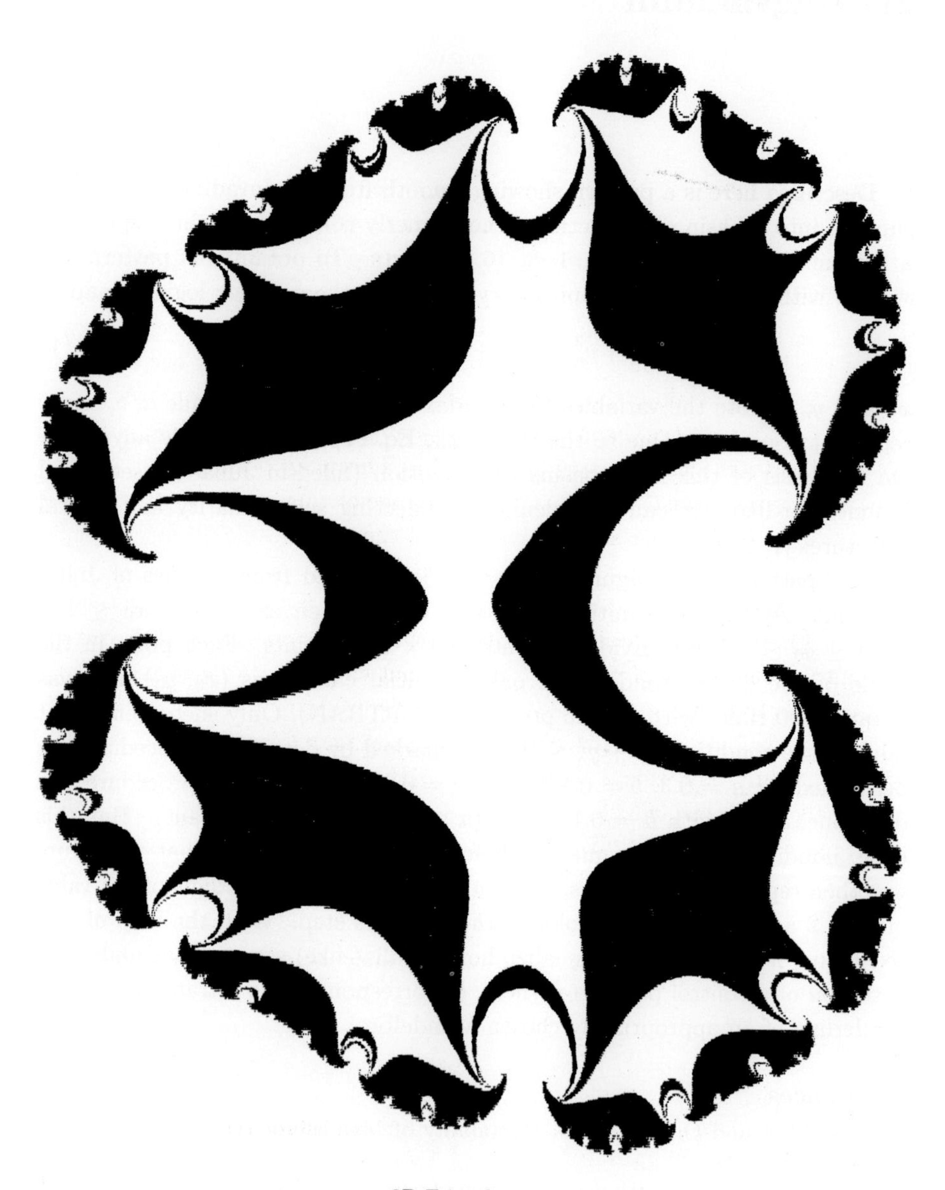

2D Feigenbaum

David Stuedell

Biomorphic Mitosis

Described here is a pattern showing graphical entities, which undergo mitosis (cell division), as a parameter in a mathematical feedback loop is varied. The term "Biomorph" is a name given to a computer graphic representation which resembles a biological morphology, hence, the name "Biomorph" [1]. As the term implies, biomorphs look like simple life forms, and the natural extension of the biomorph would obviously be their reproduction and cell division.

It is amusing to note that as one varies the constant c associated to the complex function $z^n + c$, the interior region goes through stages which resemble biological mitosis (cell division) (see Fig. 1). This is especially interesting in $z^2 + c$, as c varies from 0 to 2, because the change closely resembles the natural cell division. If one continues to increase c, the original biomorph will split into n new baby biomorphs, where n is the power to which z is raised.

The mitosis of the biomorph may lead computer graphics investigators to create a mathematical "tissue". In this process, Biomorphs in each cell go through various stages of mitosis. The constant could then represent the time it takes cells to divide and die.

The pictures were created using the same algorithm described in *Scientific American* (July 89) issue 2 with a loop for the constant c. These forms are related to Julia sets, which have been described in [3]. The graphics were generated on a Macintosh SE and an Imagewriter printer.

Acknowledgements

I thank J. Harper for the useful ideas and encouragement.

References

1. C. Pickover, "Biomorphs: Computer displays of biological forms generated from mathematical feedback loops", *Computer Graphics Forum* **5**, 4 (1987) 313–316.
2. A. Dewdney, "Computer recreations", *Scientific American* (July 1989) 110–113.
3. B. Mandelbrot, *The Fractal Geometry of Nature* (Freeman, 1982).

Figure 1. Biomorphic mitosis, produced by the iteration of $z = z^3 + c$ in the complex plane as c varies from 0.7 to 1.0.

Jim Nugent

Triangular Numbers and the Distribution of Primes

The pattern in Fig. 1 demonstrates the nonrandom distribution of primes in the first 100,000 numbers when they are plotted as a triangular array. This is similar to Fig. 2, the square spiral display of primes in the first 65,000 numbers that Ulam [1] published in 1964.

The image in Fig. 1 was generated on an IBM 50Z microcomputer with VGA graphics which have a resolution of 640 pixels across and 400 pixels high. Starting with a black screen, plot the first 100,000 numbers so that prime numbers are white while composites are left black. One number is plotted in the first or bottom row, two numbers in the second row, three in the next, etc. Each row contains one more number than the last. Each row is up one unit and to the left one-half unit creating a symmetric triangle.

Vertical columns with a heavy distribution of primes stand out in the triangular array. The strongest column is offset to the right of center by 29. $+12, +18, -17$, and -38 are offsets for other prominent columns. Since the number of dots in each horizontal row in the triangle is one more than the row before, the value for any vertical column of dots can be computed by finding the sum of the numbers up to the row in question, adding half to get to the center of that row, and adding or subtracting the offset.

Fainter lines are also apparent running parallel to and perpendicular to the left edge. The values of the dots forming the bright line parallel to the left edge are equal to a triangular number plus a constant. The rows that are perpendicular to the left edge terminate at the right edge of the triangle and therefore are finite. These lines of primes are the sum of an nth triangular number plus a regularly increasing amount.

Reference

1. M. L. Stein, S. M. Ulam and M. B. Wells, "A visual display of some properties of the distribution of primes", *Math. Monthly* **71**, 5 (1964) 516–520.

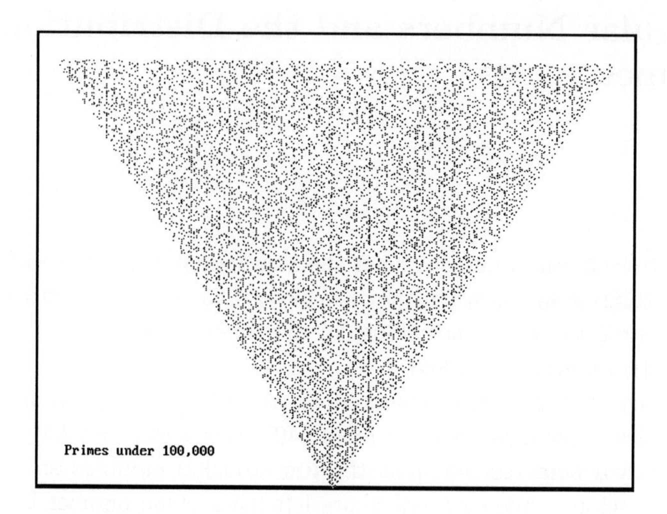

Figure 1.

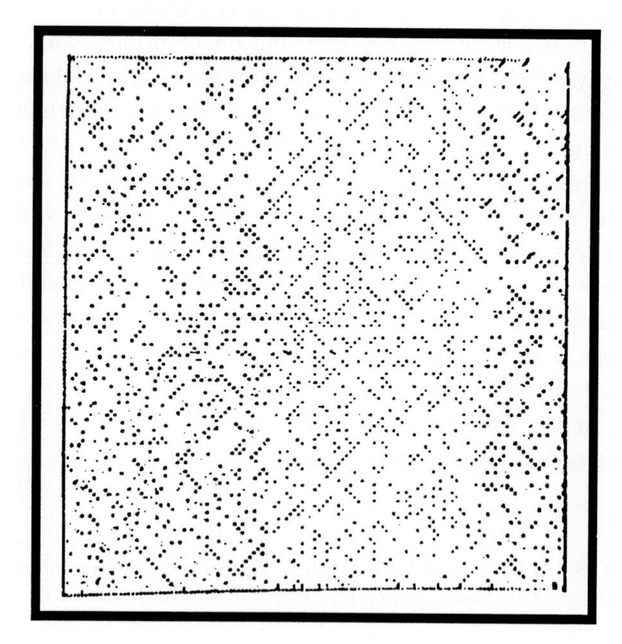

Figure 2.

Thomas V. Papathomas

Bistable Tiling Patterns with Converging/Diverging Arrows

Described here is a class of patterns showing interesting tessellations of the plane. The constituent micropatterns whose repetitions in space produce the tiling patterns were introduced in [1]. These micropatterns are shown separately at the extreme right in the figure. The three basic regular polygons that tessellate the plane are used: triangle (top), square (middle), and hexagon (bottom). These micropatterns all share the following common property, which will be illustrated for the isolated triangular shape, shown at the top right of the figure. This image can be perceived in two stable states: either as a solid black three-pronged object with three diverging arrowheads, seen against a white background, or as three white arrowheads converging towards the center of a solid black triangle that lies beneath. Similar bistable percepts exist for the single square and hexagonal images on the right of the figure.

In producing the tiling patterns of the figure, the key observation was that, when the micropatterns are repeated in two-dimensional space, an interesting image is generated. The converging white arrows of one micropattern are extended and joined with the white arrows of adjoining micropatterns to form white bars with arrowheads on both ends. These double-ended arrowheads are emphasized by drawing the black thin edges joining the adjacent micropatterns on the left side of the figure. The visual effect of these edges is impressive. Unfortunately, lack of space does not allow us to show a tiling pattern *without* these edges; the interested reader is invited to "white them out" on a copy of the figure and compare the resulting image with the original one to appreciate their role. What is remarkable about the tessellated patterns is that they also exhibit the bistability characterizing the micropatterns which generated them. The viewer can perceive either the black multipronged solid objects against a white background or the white arrows converging over black polygons.

For more information on multistable percepts and figure-gound reversals of simple or tiling patterns, the reader is referred to [2].

References

1. T. V. Papathomas, S. C. Kitsopoulos and J. I. Helfman, "Arrows-anchors: Figure-ground reversals", *Perception* **18**, 5 (1989) 689.
2. F. Attneave, "Multistability in perception", *Scientific American* **225**, 6 (1971) 63–71.

Tessellations of the plane with bistable ambiguous micropatterns: triangular (top), square (middle) and hexagonal (bottom).

Thomas V. Papathomas

Vivid Depth Percepts from Simple Gray-Level Line Patterns

Described here are two patterns, formed by white and black lines of either horizontal or vertical orientation against a gray background. It is remarkable that these images unexpectedly give rise to vivid three-dimensional (3D) percepts, as if they are renderings of anaglyph terrains. Both images are members of a class of stimuli conceived by the authors [1], for studying the interaction of visual attributes (color, luminance, orientation, spatial frequency, etc.) in motion and texture perception. The two attributes employed in this figure are *orientation* (horizontal and vertical lines) and luminance polarity, or simpler, *polarity* (black and white lines) on a gray background). These patterns were generated on a computer by the algorithm described in [1]; the algorithm can produce much more complex patterns, depending on the goal of the visual experiment, but its details need not concern us here. The top figure shows a pattern in which only two types of lines exist: white horizontal (WH) and black vertical (BV). These lines are joined in such a way so as to alternate and to form staircases. Unexpectedly, however, they give rise to a strong 3D percept; some observers describe the image as thin cardboard slabs, cut along staircase contours, stacked one behind the other; others perceive the scene as rows of seats, much like in an ancient stadium. In the bottom figure, which also appears in [3] in a different context, we have all four possible combinations of lines: WH, BV (as before), plus black horizontal (BH), and white vertical (WV). These lines form a regular orthogonal grid pattern which, quite unexpectedly, produces a strong depth percept, in which two types of squares can be seen, in a checkerboard format, i.e., half of the squares seem to pop out (in depth) in front of the rest.

The best explanation that we can give for these percepts involves the inherent assumption of the human visual system of a single light source shining from above (see [2] for a recent reference). Thus, white lines are naturally interpreted as edges that are lit directly by the source; black lines are interpreted as edges that are occluded by something interposed between themselves and the source. This explanation accounts for the vivid 3D percept obtained

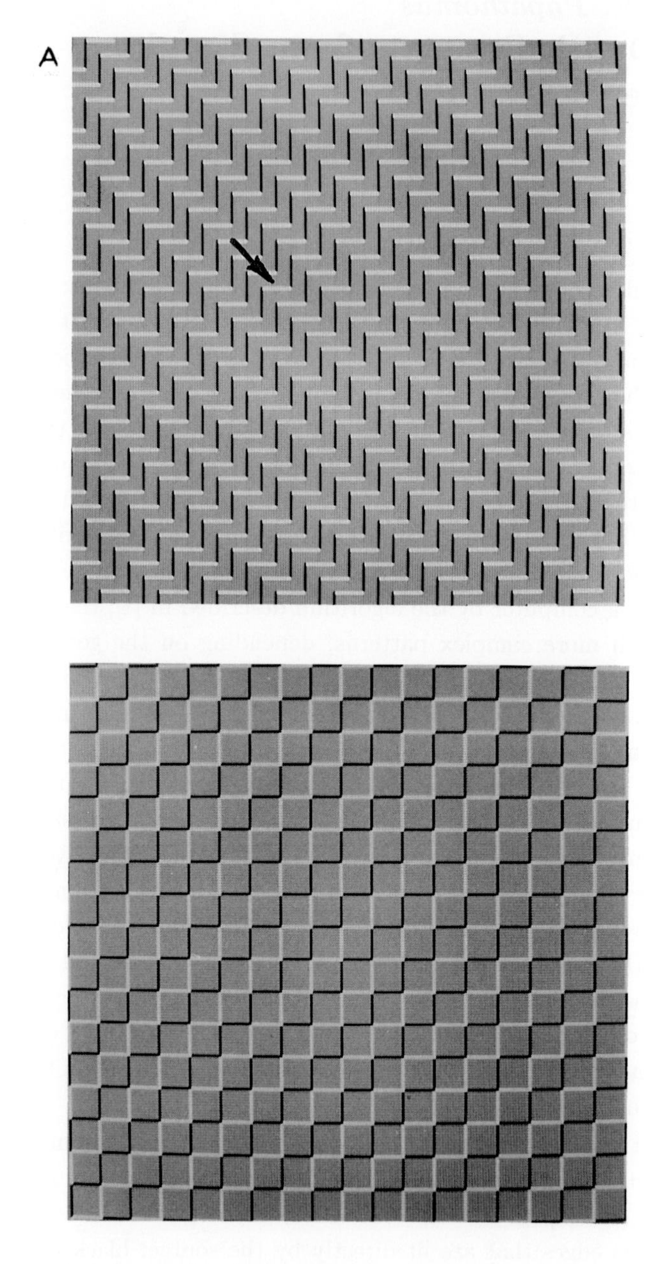

Simple patterns of black and white lines against a gray background that produce vivid depth percepts.

with these patterns. In addition, it explains the following two phenomena: 1) An interesting property of the top figure is that, when it is turned 90° counterclockwise, it produces two mutually exclusive stable percepts, i.e., as one moves his/her gaze from the corner labeled A toward the opposite corner, as shown by the black arrow, one may perceive the thin stacked slabs as either approaching toward or as receding away from him/herself. This is because the rotated image has white vertical edges, meaning that the source must be located either to the right or to the left, resulting in the bistable ambiguous percept. 2) In the bottom figure, it is the squares that are bounded above by a WH edge that "pops out" in front of the others, in agreement with the overhead source assumption. The reader is invited to rotate the bottom figure by 90° and observe that this property still holds.

References

1. T. V. Papathomas and A. Gorea, "Simultaneous motion perception along multiple attributes: A new class of stimuli", *Behav. Research Methods, Instruments and Computers* **20** (1988) 528–536.
2. V. S. Ramachandran, "Perception of shape from shading", *Nature* **331** (1988) 163–166.
3. D. Marr, *Vision* (W. H. Freeman, 1982) p. 77.

Kenelm W. Philip

Mandelbrot Set Spiral Tile Mosaics

Described here are patterns showing mosaics constructed from higher-order spiral patterns found at high magnifications within the Mandelbrot Set. They were conceived as possible selections for the endpapers of a book about the M-Set.

A common structure in the Mandelbrot Set is an object which can be referred to as a "double spiral", or "DS", which forms a link between the successive elements of so-called "sea horse tails" and other similar spiral features. High magnification of the arms of a double spiral can reveal higher-order spiral features. The upper left subpanel in Fig. 1 depicts a double/double spiral (DDS) from Sea Horse Valley, which was found by A. G. D. Philip (personal communication) at a magnification[a] of 3.75×10^{15} diameters. Subsequent investigation by both of us has turned up any number of these structures, and higher order spirals (DDDS, DDDDS) have been found at magnifications up to 10^{20} diameters. The quadruple spirals at the center of double spirals can also show higher-order structure.

Examination of these intricate structures gave me the idea that patterns derived from them might have artistic merit. The procedure adopted was as follows (see Fig. 1): a black and white image is made of the feature (using the program MandelZot, by Dave Platt, on a Macintosh II. MandelZot uses the standard Mandelbrot Set algorithm based on iterating the equation $Z \to Z^2 + C$ and testing for the distance of Z from the origin). Then the dwell-band width adjustment feature of MandelZot, which can combine a number of iteration-number contours into a single contour, is used to highlight the spiral structure. The image is captured from the screen to a paint program and trimmed as needed, and then three copies are made and rotated as shown (flipped horizontally, vertically, and both ways at once) so they may be combined into the unit tile for the mosaic. The tile is then moved to a drawing program, and replicated to cover the page.

Six such mosaics were generated, using higher-order spirals from Sea Horse Valley and East Valley in the Mandelbrot Set. Figure 1 illustrates the genesis

[a] "Magnification" is with reference to a standard view defined as 1X, running from -2 to $+0.5$ along the real axis.

of each mosaic — the exact composition of the basic tile is not always obvious from an inspection of the resulting mosaic. The final images were printed on an Apple Laserwriter. The mosaics have an almost organic quality which is not always found in computer-generated art, so further experimentation with this method of generating pattern tiles might be warranted.

The six mosaics (Figs. 2–7) were generated from the following M-set structures:

Figure 2. "Lozenge" — The first DDS, from the east side of Sea Horse Valley. Magnification: 3.75×10^{15} diameters.

Figure 3. "Ripple" — A DDS from East Valley, near the real axis at the right end of the M-set. Magnification: 4.5×10^{13} diameters.

Figure 4. "Hexagon" — A quad spiral with four attached double spirals, east Sea Horse Valley. Magnification: 3.22×10^{16} diameters.

Figure 5. "Bracket" — A DDS from the west side of Sea Horse Valley. Magnification: 1.47×10^{8} diameters.

Figure 6. "Eyes" — Another DDS from east Sea Horse Valley. Magnification: 2.17×10^{8} diameters.

Figure 7. "Trellis" — The first DDDS, from east Sea Horse Valley. Magnification: 8.42×10^{12} diameters.

Figure 1. Origin of mosaic patterns.

Figure 2. Lozenge.

Figure 3. Ripple.

Figure 4. Hexagon.

Figure 5. Bracket.

Figure 6. Eyes.

Figure 7. Trellis.

A. G. Davis Philip

Evolution of Spirals in Mandelbrot and Julia Sets

These pictures were made with Michael Freeman's (Vancouver, BC, Canada) Mandelbrot program, V63MBROT, which was written specifically for an IBM PC system with an ATI VGA Wonder video card. Four of the picture sections (1a, 1c, 1d and 3a) are pictures in the Mandelbrot Set; the remaining sections are Julia Sets. The parameters for each section are given in Table 1. The figures are calculated at a resolution of 800×600 pixels, with dwell values set in the range of 10000 to 44444. There are color editing routines in Freeman's program so one can edit the pictures to enhance the contrast when the pictures are printed in black and white on the HP II laser printer. My usual approach is to make the spiral arms black and the background, between the spiral arms, a very light color.

The names given to different parts of the Mandelbrot Set, referred to in this article, are described more fully in [1]. Additional introductory information concerning the Mandelbrot set can be found in [2]. Spirals found in the Mandelbrot and Julia Sets are described in [3, 4]. Readers can look at these articles for additional information.

Table 1. Parameters for pictures in Figs. 1–4.

Name	Dwell mag.	Julia point Real Imaginary	Region of picture Real (min.) Real (max.)	Imaginary (min.) Imaginary (max.)
WMDSL3	20000		+0.4052403616447634	+0.1467904399724176
	7.75E6		+0.4052407943423710	+0.1467907626532763
JSEREN7c	11111	−0.7485808956448470	−0.0053790642558517	−0.0038921296205975
	3.32E2	+0.0630646917147260	+0.0046459826973955	+0.0036329330185813
SERENDIP	20000		−0.7485808956487646	+0.0630646917776399
	4.17E15		−0.7485808956487637	+0.0630646917776405
DSH1735	10000		−0.7623822811227581	+0.0955630841818570
	2.08E13		−0.7623822811225580	+0.0955630841820070

Table 1. (*Continued*)

Name	Dwell mag.	Julia point Real Imaginary	Region of picture Real (min.) Real (max.)	Imaginary (min.) Imaginary (max.)
JSERE883	11111	−0.7485808977432403	−0.0003201355358612	+0.0010368809264518
	4.22E9	+0.0630647010452520	−0.0003201346506445	+0.0010368815183226
JSERE704	11111	−0.7485808845098242	−0.0002539236802797	−0.0001735677822053
	6.63E3	+0.0630646922498100	+0.0002479560170042	+0.0002033134851993
JSEREN7d	11111	−0.7485808956448470	−0.0002724383359874	−0.0001484423643783
	6.63E3	+0.0630646917147260	+0.0002294413612966	+0.0002284389030263
JSER7c25	44444	−0.7485808915423700	−0.0000010926324134	+0.0000009880033582
	3.59E9	+0.0630646994419824	−0.0000010915919383	+0.0000009887001405
SERENE14	22222		−0.7485808956487741	+0.0630646917776328
	1.67E15		−0.7485808956487541	+0.0630646917776478
JSERE474	11111	−0.7485808905342438	−0.0002344680001397	−0.0001899132623583
	6.63E3	+0.0630646895409476	+0.0002674116971442	+0.0001869680050463
JSERE375	11111	−0.7485808850175000	−0.0000021233783777	−0.0000015139790402
	8.10E5	+0.0630646979016339	+0.0000019646077406	+0.0000015741698952
JLSEREN7	20000	−0.7485832384672945	−0.0000001364451088	−0.0000000922844701
	1.35E6	+0.0630769377174541	+0.0000001236243267	+0.0000000928929387
JDSH17a	30000	−0.7623822811227579	−0.0263674783922023	−0.0183397690507255
	6.83E1	+0.0955630841819817	+0.0263675425337977	+0.0183398774872745
JDSH17c	30000	−0.7623822811227579	−0.0002636430338323	−0.0001833440144155
	6.83E3	+0.0955630841819817	+0.0002637071754277	+0.0001834524509645
JDSH17d	30000	−0.7623822811227579	−0.0000263354396653	−0.0000182856049945
	6.82E4	+0.0955630841819817	+0.0000263995812607	+0.0000183940415435
JDSH17e	30000	−0.7623822811227579	−0.0000026046802486	−0.0000017797640524
	6.82E5	+0.0955630841819817	+0.0000026688218440	+0.0000018882006014

The Patterns

Figure 1. Examples of spiral types.

Figure 1a. A Single Spiral found in the Sea Horse Valley of the main "warped" midget in a tendril radiating from Radical 7. An interesting feature to note is the series of double spirals (DS) found on the lower edge of the main spiral. As one goes from left to right, the DS wind up clockwise producing DS with many turns in the spiral arms. On the upper and lower edges of the spiral there are series of single spirals. Near the middle of the spiral arm there is another series of smaller DS. If the spiral arm is magnified, spirals of various types will be found in great profusion.

Figure 1b. A Double Spiral (DS) in Sea Horse Valley. In the center of the DS, there is a smaller spiral, below the limits of the resolution of this picture. At the end of each arm can be seen a single spiral.

Figure 1c. A Double/Double Spiral (DDS) in Sea Horse Valley. The end of each spiral arm contains a double spiral.

Figure 1d. An Octuple Spiral in Sea Horse Valley. At least 13 spiral arms can be counted in the outer arms. At the very center of the figure a small midget can be seen.

Figure 2. Spirals with holes in their centers. Each of the four sections in Fig. 2 is a Julia Set calculated from a region in Sea Horse Valley.

Figure 2a. A double spiral, somewhat like Fig. 1b, but there is no connection between the two halves.

Figure 2b. A variation of the DDS with an empty region at the very center of the pattern.

Figure 2c. Another variation of the DDS. The double spirals in the arms are very small relative to the arms. Again, there is an empty region at the center of the pattern.

Figure 2d. Another variation of the DDS. The double spirals at the ends of each arm are larger relative to the spiral arms. But this time the spiral arm pattern does not go through the center of the pattern.

Figure 3. Spirals of different types. Each of the four sections of Fig. 3 is from Sea Horse Valley.

Figure 3a. A double spiral with a smaller, well-resolved spiral at its center.

Figure 3b. A DDS with a bar at the center.

Figure 3c. A DDS with a smaller DDS at its center.

Figure 3d. A very complex spiral object. At the center is a Quad spiral with single spirals in one pair of arms and double spirals in the other pair of arms.

Figure 4. A Zoom series. This series is taken from Sea Horse Valley.

Figure 4a. A single spiral with a figure eight-like feature in its center.

Figure 4b. The very center of Fig. 1a magnified 100 times. There are four sets of arms, inside of which is a single spiral inside of which is a Quad Spiral.

Figure 4c. The Quad Spiral at the center of Fig. 4b, magnified 10 times. Each of the four arms of the Quad end in single spirals. The center of Fig. 4c is shown in Fig. 4d.

Figure 4d. A spiral with an empty center. There are two Quad spirals on each side of the center; each Quad contains a replica of the whole picture. This picture is a tenfold magnification of Fig. 4c.

References

1. A. G. D. Philip and K. W. Philip, "The taming of the shrew", *Amygdala* **19** (1990) 1–6.
2. A. G. D. Philip, "An introduction to the Mandelbrot Set", in *CCDs in Astronomy*, Vol. II, eds. A. G. D. Philip, D. S. Hayes, and S. J. Adelman (L. Davis Press, 1990) pp. 275–286.
3. C. A. Pickover, *Computers, Pattern, Chaos, and Beauty — Graphics From an Unseen World* (St. Martin's Press, 1990).
4. A. G. D. Philip, "The evolution of a three-armed spiral in the Julia Set and higher order spirals", in *Spiral Symmetry*, eds. I. Hargittai and C. A. Pickover (World Scientific, 1992) pp. 135–163.

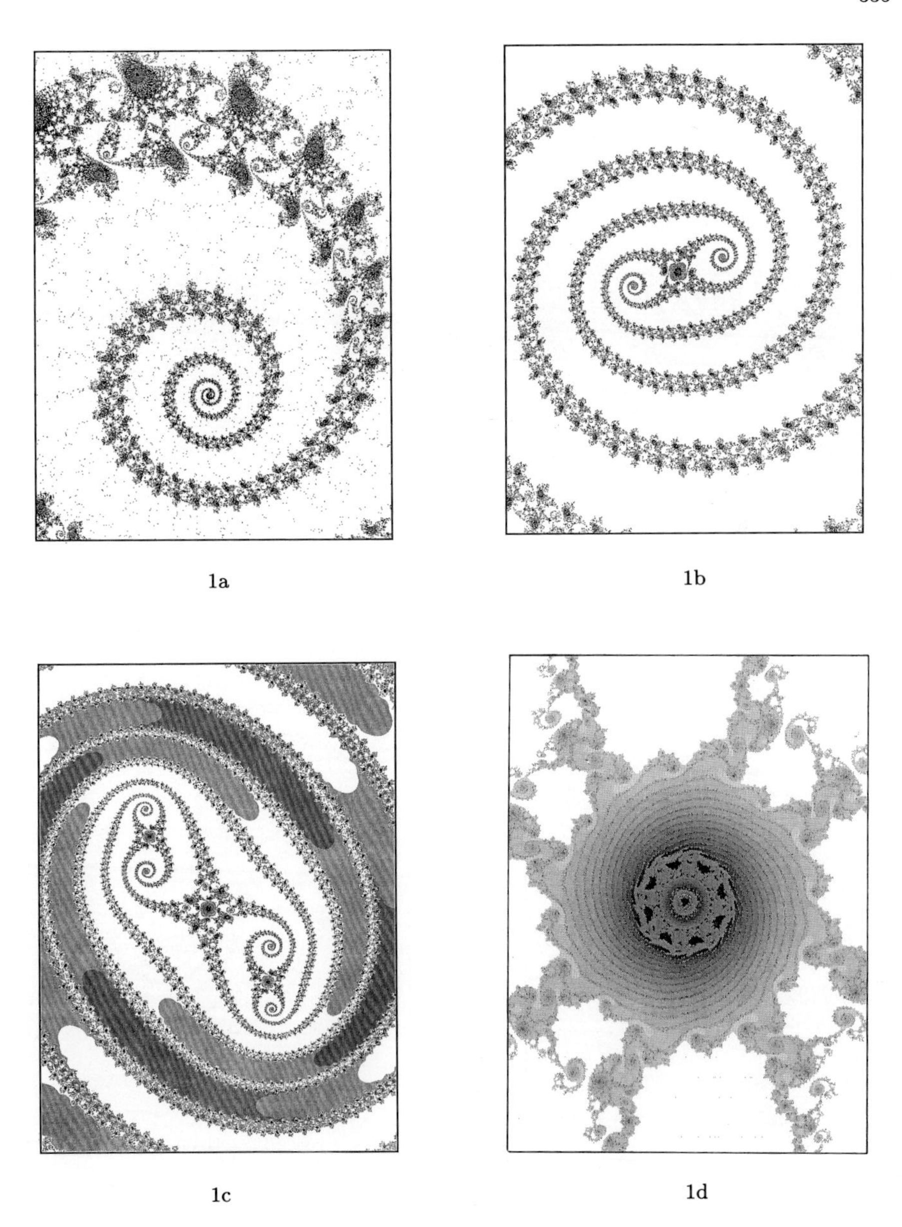

Figure 1. Examples of spiral types.

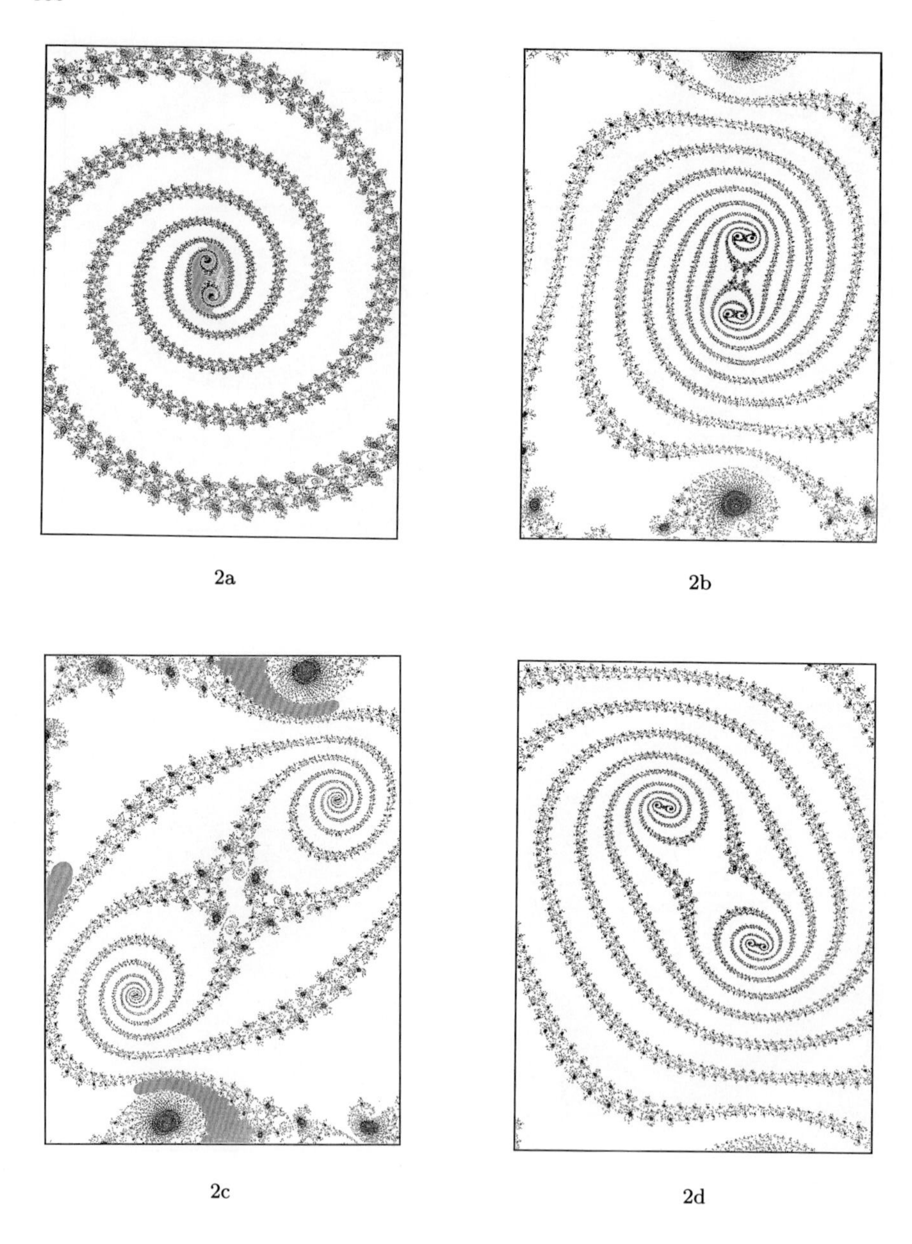

Figure 2. Spirals with holes in their centers.

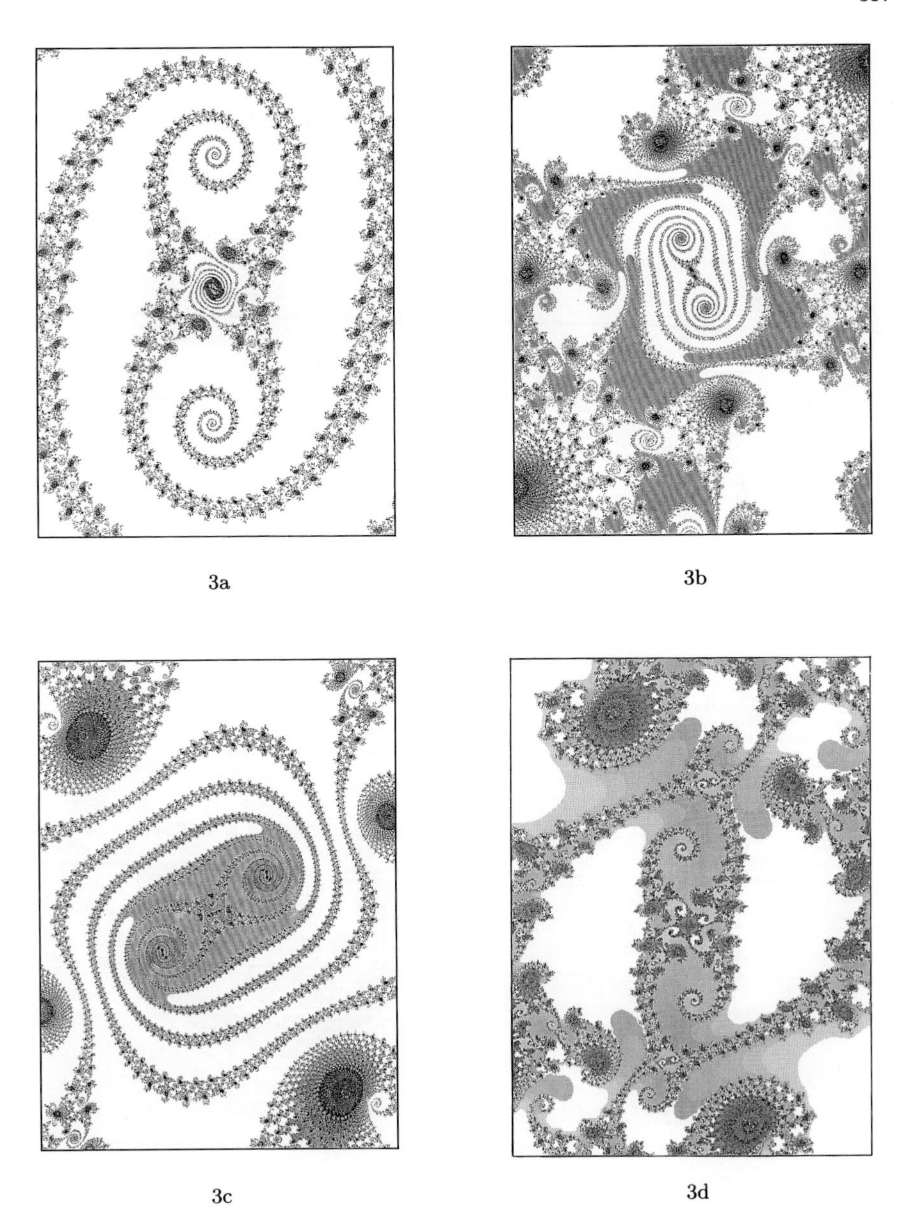

Figure 3. Spirals of different types.

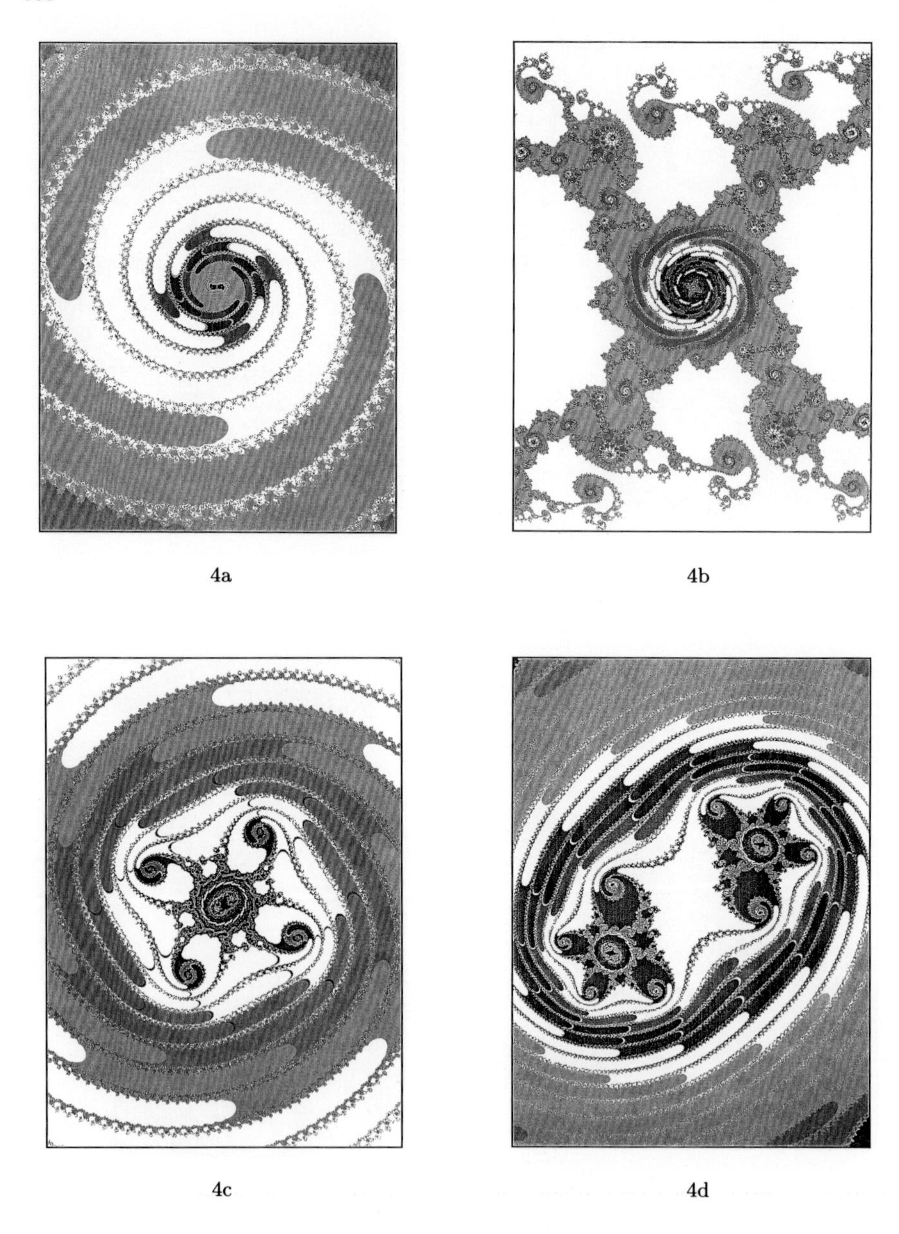

Figure 4. A Zoom series.

Uwe Quasthoff

Hyperbolic Tilings

Described here are patterns showing tilings of the circle as a model of the hyperbolic plane. Some of M. C. Escher's famous wood cuts use similar tilings but beautiful shapes instead of our chessboard pattern. Figure 1 first appeared in 1897 in [1].

From the mathematical point of view, these patterns are symmetric with respect to some reflections. There are both reflections with respect to straight lines and circles. Our figures can be obtained using these reflections in an iterated function system with condensation.

All patterns presented here are generated by three reflections. For a given starting point we calculate its images under these reflections and use these points as new starting points. A point is drawn if we applied an even number of transformations counting from the starting point. This procedure is repeated for all original starting points of a given so-called fundamental region. This region has the property that its images under iterated transformation will just fill the whole circle.

The Algorithm

The following tree parsing algorithm computes orbits as far as it meets an already-drawn point. For all points of the fundamental region, do the following:

Step 1. (Initialize). Set even-flag=TRUE and put the point of the fundamental region and even-flag on the stack.

Step 2. (End?). If the stack is empty, the processing of the considered starting point is finished.

Step 3. (Next). Remove a point and its even-flag from the stack. If even-flag=TRUE then draw the point. Compute all of its images under the considered reflections.

Step 4. (Test). For every image point check whether it is already drawn. If not, put in and NOT (even-flag) on the stack.

Step 5. (Loop). Go to Step 2.

Last we present the data used to generate the figures.

Figure 1.

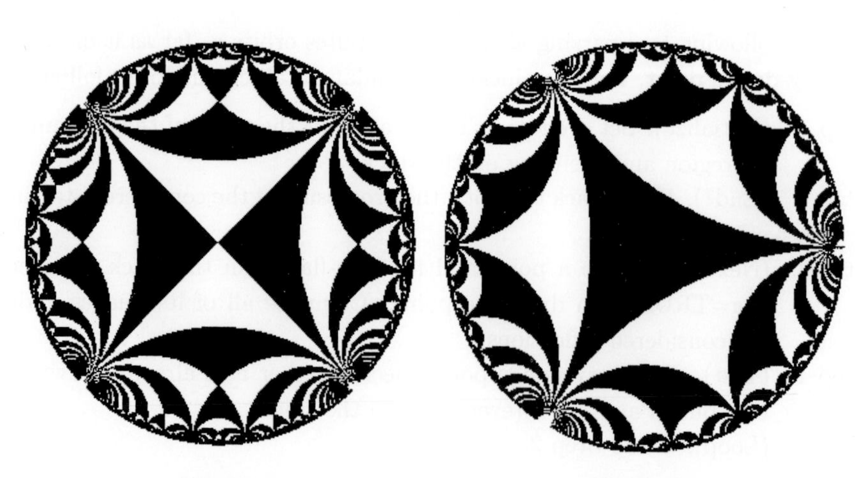

Figure 2. **Figure 3.**

Figure 1. Reflection w.r.t. the straight line through (0, 0) and (1, 0); reflection w.r.t. the straight line through (0, 0) and ($\cos(\pi/7)$, $\sin(\pi/7)$); reflection w.r.t. the circle of radius $2 * \sin(\pi/7)$ around (1, 0). As a fundamental region, we take a large black triangle near the center of the picture. It has the angles ($\pi/2$, $\pi/3$, $\pi/7$).

Figure 2. Reflection w.r.t. the straight line through (0, 0) and (1, 1); reflection w.r.t. the straight line through (0, 0) and (1, −1); reflection w.r.t. the circle of radius 1/SQR(2) around (1, 0). As a fundamental region, we take a large black triangle near the center of the picture. It has the angles ($\pi/2$, 0, 0).

Figure 3. Reflection w.r.t. the circle of radius SQR(3)/2 around (1,0); reflection w.r.t. the circle of radius SQR(3)/2 around (−0.5, SQR(3)/2); reflection w.r.t. the circle of radius SQR(3)/2 around (−0.5, −SQR(3)/2). As a fundamental region, we take the central triangle. It has the angles (0, 0, 0).

Reference

1. R. Fricke and F. Klein, *Vorlesungen über die Theorie der Automorphen Funktionen I* (Teubner, 1897).

Clifford A. Reiter

Products of Three Harmonics

Here are patterns produced by color contouring the height of functions that are the products of three harmonics; one harmonic in the x, y, and radial directions each. In particular, Fig. 1 is colored by the height of the function $f(x, y) = \cos(x)\cos(y)\cos\sqrt{(x^2+y^2)}$ in the domain: $-32 \leq x \leq 32$, $-24 \leq y \leq 24$. The colors are chosen to vary through blue, light blue, light cyan, green, yellow, and white as the function height varies from -1 through 1. Figure 2 is a color contour plot for the function $f(x, y) = \sin(x)\sin(y)\sin\sqrt{x^2+y^2}$ on the same domain with the same coloration scheme. Notice the vertical, horizontal and circular 0-level sets in both figures.

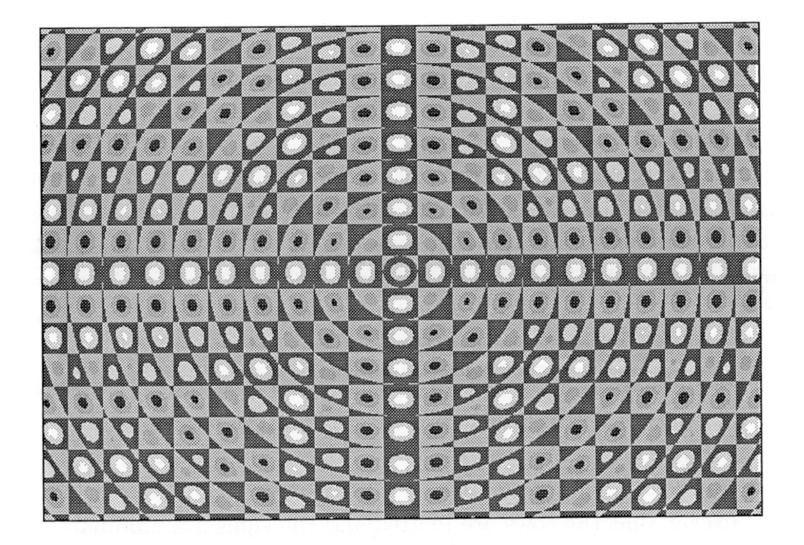

Figure 1. Products of three cosines.

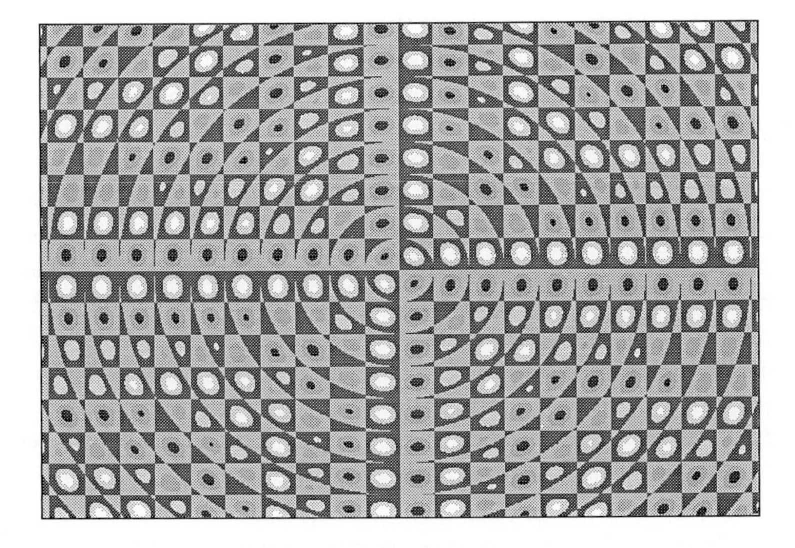

Figure 2. Products of three sines.

Donovan E. Smith

Pascalian Cellular Automata

Described here are patterns that evolve from a single "defect" in bounded, one-dimensional, cellular automata in which the number of cells in the row is equal to the number of possible states, n.

Each cell's neighborhood consists of two adjacent cells, but the cell at each end of the row lacks one neighbor. The state of each cell in generation g is coded by an integer from 0 to $(n-1)$.

The state numbers are cyclic $(0, 1, 2, \ldots (n-1), 0, 1, 2, \ldots)$ because the state number of each cell in generation $(g+1)$ is the sum, modulo n, of the state numbers in that cell's neighborhood in generation g.

Each automaton thus generates successive rows of integers as in Pascal's triangle, except that the rows are limited and the sums are modular.

Each cell is displayed on the computer's screen as a pixel with color number = state number, modulo 16. The state numbers are printed with zeros shown as blanks to enhance the Pascalian pattern.

Figures 1 and 2 show examples of the repeating truncated-triangle patterns of the state numbers that evolve if and only if (at least in hundreds of tests) the number of cells and states, n, is prime. If the prime is of the form $(4x-1)$, then generation 0 recurs every $2(n-1)$ generations, as in Fig. 1. If the prime is of the form $(4x+1)$, then the period is still $2(n-1)$ generations, but generation 0 never recurs, as in Fig. 2. With any prime value of n, the same pattern with a period of $2(n-1)$ generations evolves from any value of the initial defect.

Figure 3 shows the early stages of the an example of the deterioration into patternless "debris" that occurs whenever the number of cells and states is composite, and the defect in generation 0 is state 1. The periods, p, of such automata always exceed $2(n-1)$ and have no consistent relationship to n. For example, if $n = 6$, then $p = 182$; if $n = 10$, $p = 48,422$; if $n = 12$, $p = 1,638$; and if $n = 20$, $p = 7,812$. With certain values of the defect greater than 1, the periods are not only shorter, but even predictable if the composite n is a power of a prime.

Two-dimensional analogs of these cellular automata behave similarly. If the bounded grid is square, with n cells on each side, n states, and the initial defect in the center, then if and only if n is prime:

$p = 2(n-1)$ if each cell has four orthogonal neighbors,

$p = n - 1$ if each cell has four diagonal neighbors, and

$p = n^2 - 1$ if each cell has eight adjacent neighbors.

(Edge and corner cells lack some neighbors, of course.)

None of these automata is a practical means of identifying prime numbers. They present, however, the intellectual challenge of explaining why these automata are so orderly when n is prime, but so disorderly when n is composite.

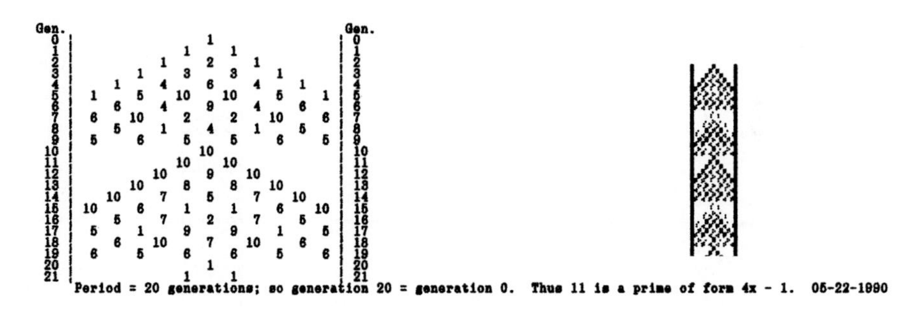
Gen.
Period = 20 generations; so generation 20 = generation 0. Thus 11 is a prime of form 4x - 1. 05-22-1990

Figure 1. Pascalian cellular automaton with 11 cells, 11 states, and defect = 1.

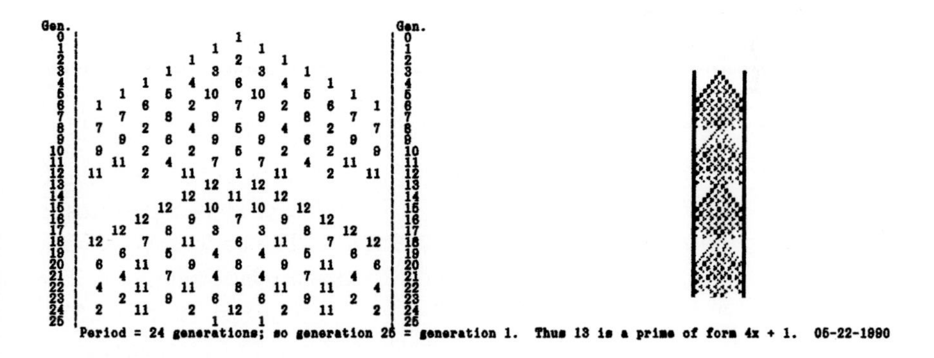
Gen.
Period = 24 generations; so generation 25 = generation 1. Thus 13 is a prime of form 4x + 1. 05-22-1990

Figure 2. Pascalian cellular automaton with 13 cells, 13 states, and defect = 1.

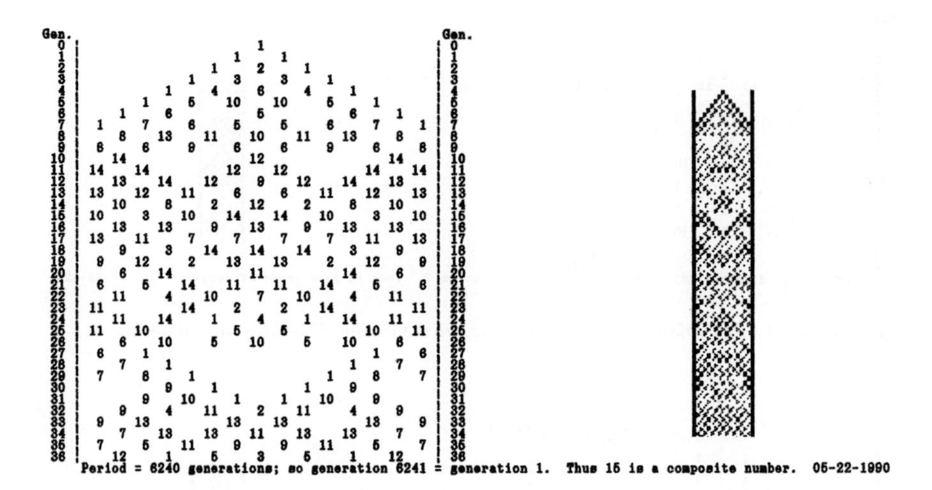
Gen.
Gen.
Period = 6240 generations; so generation 6241 = generation 1. Thus 15 is a composite number. 05-22-1990

Figure 3. Pascalian cellular automaton with 15 cells, 155 states, and defect = 1.

Franz G. Szabo

Inverse Mandelbrot Art Prints

Described here are art prints which are created by using the representation of the Mandelbrot Set in the $1/c$-plane. Using different drawing programmes written for ATARI-ST computers, the prints are various combinations of the original set.

The algorithm used is the same as computing the Mandelbrot Set with only one exception: Instead of the complex number z and c, one uses the inverse complex numbers $z' = 1/z$ and $c' = 1/c$. In the computation one has to take precautions for the case that z or c are equal to 0 [1].

An example of the algorithm in BASIC runs as follows:

```
INPUT "Left border of complex plane=";Cx_Min
INPUT "Right border of complex plane=";Cx_Max
INPUT "Lower border of complex plane=";Cy_Min
INPUT "Upper border of complex plane=";Cy_Max
INPUT "Maximum iteration per pixel=";Maxiter

X_Step=(Cx_Max-Cx_Min)/Pixel_Horizontal
Y_Step=(Cy_Max-Cy_Min)/Pixel_Vertical

FOR I=0 TO Pixel_Horizontal
 Cx=Cx_Min+I*X_Step
 FOR J=0 TO Pixel_Vertical
  Cy=Cy_Min+J*Y_Step
  Denominator=Cx*Cx+Cy*Cy
  IFDenominator=0 THEN
    Cx_New=1000000
    Cy_New=Cx_New
  ELSE
    Cx_New=Cx/Denominator
    Cy_New=-Cy/Denominator
  ENDIF
```

```
  X=0
  Y=0
  N=0
  REPEAT
   X_New=X*X-Y*Y+Cx_New
   Y_New=2*X*Y+Cy_New
   X=X_New
   Y=Y_New
   N=N+1
  UNTIL (N=Maxiter) OR ((X*X+Y*Y)>4)
  IF (N=Maxiter)OR (N MOD 2<>0) THEN DRAW (I,Pixel_Vertical-J)
 NEXT J
NEXT I
```

Reference

1. K.-H. Becker and M. Dörfler, *Dynamische Systeme und Fraktale* (Friedr. Vieweg & Sohn Braunschweig, 1988).

Figure 1. Inverse Mandel art 1.

Figure 2. Inverse Mandel art 2.

Figure 3. Inverse Mandel art 3.

Franz G. Szabo

Bio Art Prints

Described here are art prints which are created by using the representation of Barry Martin's sets of iteration of real numbers.

The algorithm in pseudocode lists as follows:

```
INPUT num
INPUT a,b,c

x=0
y=0
FOR i=1 TO num
  LOT(x,y)
  xx=y-SIGN(x)*(ABS(b*x-c))^(1/2)
  yy=a-x
  x=xx
  y=yy
ENDFOR
```

Using different drawing programmes written for ATARI-ST computers, the prints are various combinations of the original sets and show completely new patterns.

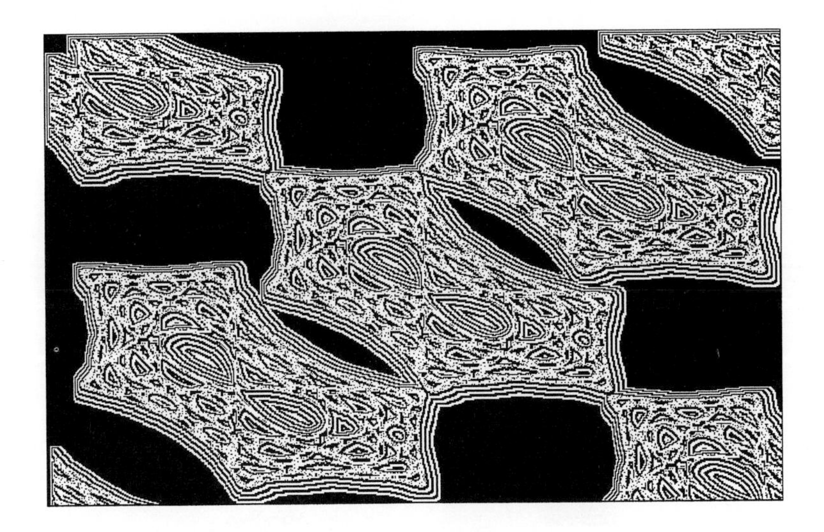

Figure 1. Bio art 1.

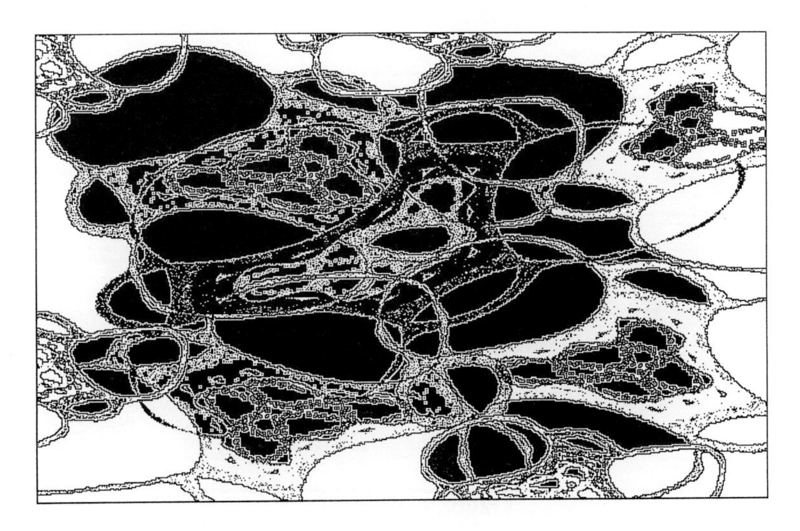

Figure 2. Bio art 2.

Figure 3. Bio art 3.

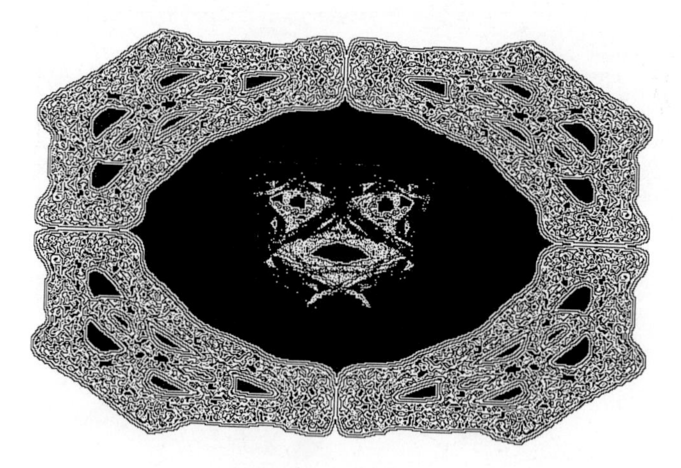

Figure 4. Bio art 4.

Franz G. Szabo
Rotation

Described here is an ornament which was created by playing around with lines in a vector-oriented drawing programme. After various rotations, the vector-object was turned into a pixel-object, shaded differently and arranged to produce the final pattern.

Mieczyslaw Szyszkowicz

Wallpaper from GCONTOUR Procedure

Procedure GCONTOUR [1] produces contour plots, in which values of three variables x, y, z are represented in two dimensions. One of the variables is a contour variable. Presented here is a pattern showing the values

$$z = \text{mod}\ (A * A + B * B,\ 11.35)$$

at the point (A, B). z is the contour variable. The result of the function mod(argument1, argument2) is the remainder when the quotient of argument1 divided by argument2 is calculated. For example,

mod(6, 3)	returns the value 0
mod(10, 3)	returns the value 1
mod(11, 3.5)	returns the value .5
mod(10, −3)	returns the vlaue 1

Such a function is realized in the SAS package [1]. Figure 1 shows the values of z for $0 < A,\ B < 40$, with 0.77 increment in x's direction and 0.79 increment in y's direction. The test of $A * A + B * B$ by modulo arithmetic in the plane was proposed by John Connett. This method is described by Dewdney in his book [2] in the chapter entitled *Wallpaper for the Mind.*

References

1. SAS Institute Inc., *SAS/GRAPH Guide for Personal Computers*, Version 6 Edition (SAS Institute Inc., 1987) p. 534.
2. A. K. Dewdney, *The Armchair Universe* (Macmillan, 1988).

Figure 1. Wallpaper from SAS subroutine.

Mieczyslaw Szyszkowicz

Pattern Generated by XOR

Presented here is a pattern showing the values

$$z = A * A \text{ XOR } B * B\,.$$

For the given initial point $(A,\ B)$, z is calculated from the above formula and its value is displayed in this point. XOR means exclusive alternative. Here, XOR is used to test bits of binary representation of A and B. This logical operator is defined as follows:

arguments: X	Y	values: X XOR Y
1	1	0
1	0	1
0	1	1
0	0	1

Other logical operators may be used to create interesting patterns. Figure 1 shows z for $-6 < A,\ B < 6$. Figure 2 is a variant of Fig. 1 in gray scale.

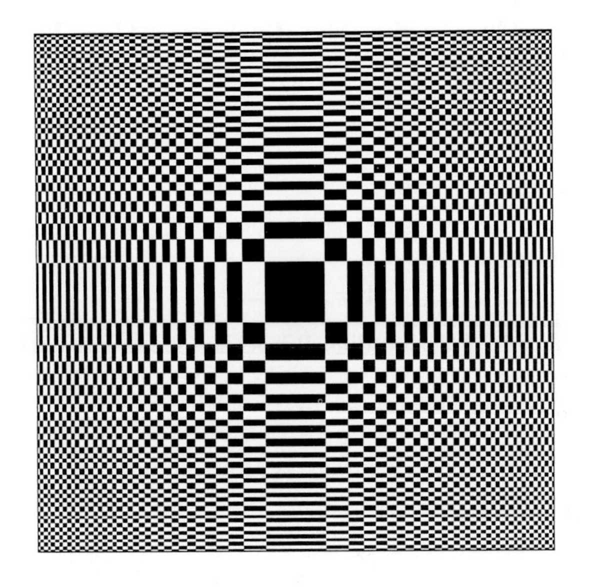

Figure 1. XOR in the plane — z modulo 2 is displayed.

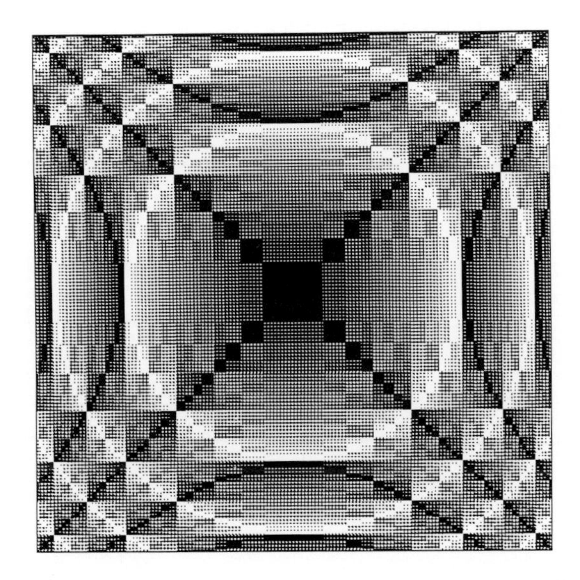

Figure 2. XOR in the plane — z modulo 16 is displayed.

Mieczyslaw Szyszkowicz

Representation of a Surface on the Plane

A surface can be described by a single relationship satisfied by the Cartesian coordinates of its points

$$\Phi(x, y, z) = 0 .$$

For example, the sphere with radius R has the following equation:

$$\Phi = x^2 + y^2 + z^2 - R^2 = 0 .$$

An alternative description of a surface is Monge's form

$$z = f(x, y) .$$

In this form our sphere has the description

$$z = \sqrt{R^2 - x^2 - y^2} .$$

Computer graphics technique can be used to represent $z = f(x, y)$ in the xy-plane. If we have only black and white, then we may represent the value z modulo 2 (See Fig. 1), where z may be scaled if it is necessary. More colors may be applied to distinguish the positive and negative values of z and more accurately show the values of the function $f(x, y)$.

Figure 1. A surface $z = \sin(x + y)\cos(x^2 + y^2)$ in the xy-plane.

Mieczyslaw Szyszkowicz

Composition of Color Pieces

Presented here is a pattern representing the partition of the rectangle by using curves. The partition is done by a simple algorithm realized by a computer (see [1]). The algorithm starts from a rectangular area limited by a frame composed of four lines. From the point (x, y) which lies on the line or curve, a curve is drawn until it meets another curve or line. The points (x, y) are chosen randomly. Figure 1 illustrates the rectangle with "colored" pieces.

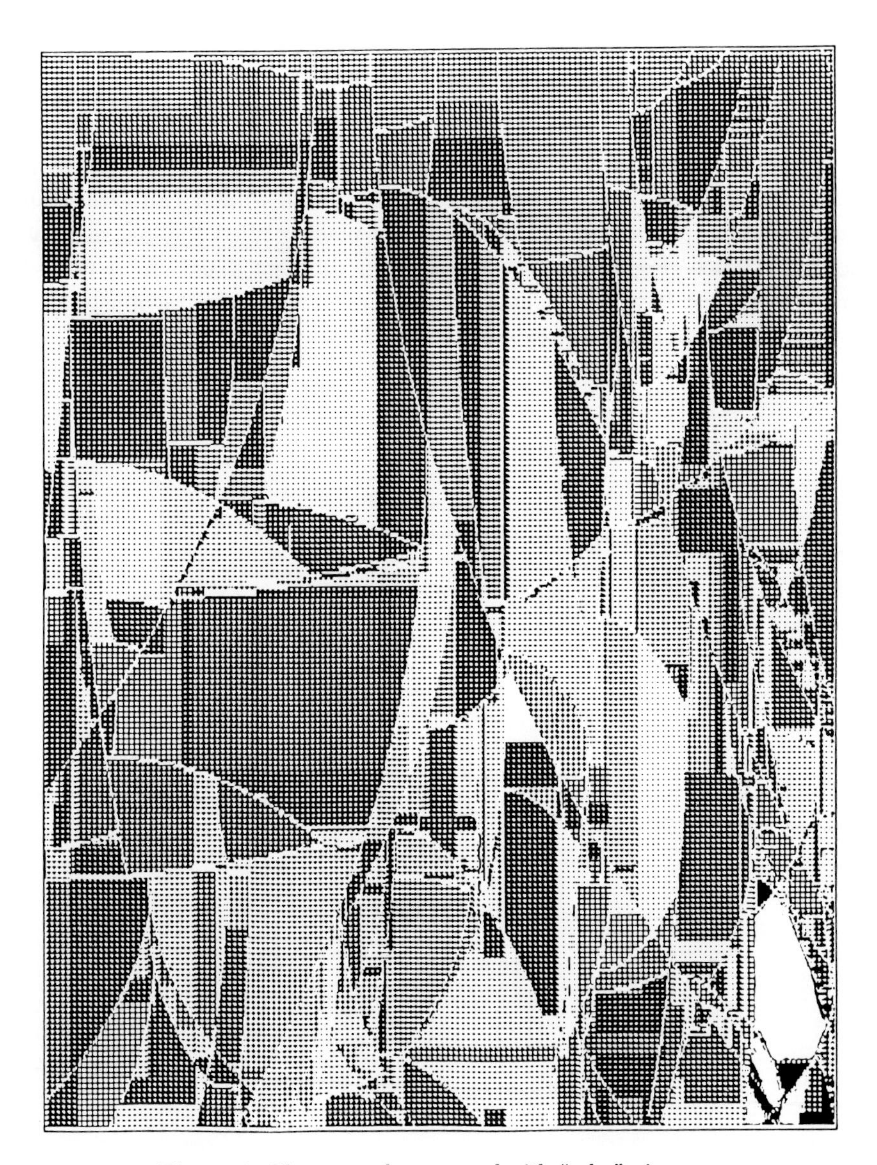

Figure 1. The rectangle composed with "color" pieces.

Mieczyslaw Szyszkowicz

Patterns Composed with Squares

Presented here are two patterns generated by using squares as a basic structural theme. The pattern (Fig. 1) illustrates what I call "constrained randomness". It was created by randomly placing filled squares so they do not quite touch. Figure 2 shows a drastically different situation. A new placing square must have one common point with at least one previously located square. For more information, see [1].

Figure 1. Squares which do not kiss.

Figure 2. Squares which kiss.

Daniel T. H. Tan, G. N. Toh, L. C. Liong and S. K. Yek

Eucledian Geometry using IFS

Described here are patterns showing some new and interesting manifestations of chaotic behavior arising from parametric equations dynamics of the form

$$x = a_0x^2 + a_1y^2 + a_2xy + a_3x + a_4y + a_5 \tag{1}$$

$$y = b_0x^2 + b_1y^2 + a_2xy + b_3x + b_4y + b_5 \tag{2}$$

where a_n, b_n for $n = 0, 1, 2, \ldots, 5$ are real constants. Generally, such patterns are irregular and disorderly. However, it can be shown that Eqs. (1) and (2) are capable of generating a wide range of highly regular patterns. Consider Eqs. (3) and (4):

$$x_n = x_{n-1}^2 - y_{n-1}^2 + a_5 \tag{3}$$

$$y_n = 2x_{n-1}y_{n-1} + b_5 \tag{4}$$

they represent the seed equations for iterations to generate the well-known Mandelbrot Set [1] of the complex dynamical function $f : z = z^2 + c$, where z and c are complex variables.

In this paper, a new family of objects (*called rectifs*) will be described that uses the Eqs. (5) and (6):

$$x_n = x_{n-1}^2 + y_{n-1}^2 + a_5 \tag{5}$$

$$y_n = 2x_{n-1}y_{n-1} + b_5 \tag{6}$$

The escape time algorithm [2] is used to generate these patterns. The area in which the patterns are found in a square area is $-2 < x < 2$, $-2 < y < 2$. This area is divided into a grid of 1000×1000 points.

Iteration begins with the $x_0 = -2$, $y_0 = -2$, $a_5 = -2$, $b_5 = -2$. The values x_1 and y_1 are evaluated using Eqs. (5) and (6). The process is repeated with each subsequent time using the output (x_n, y_n) as the input for the next iteration to compute (x_{n+1}, y_{n+1}). The values (a_5, b_5) are kept constant. The

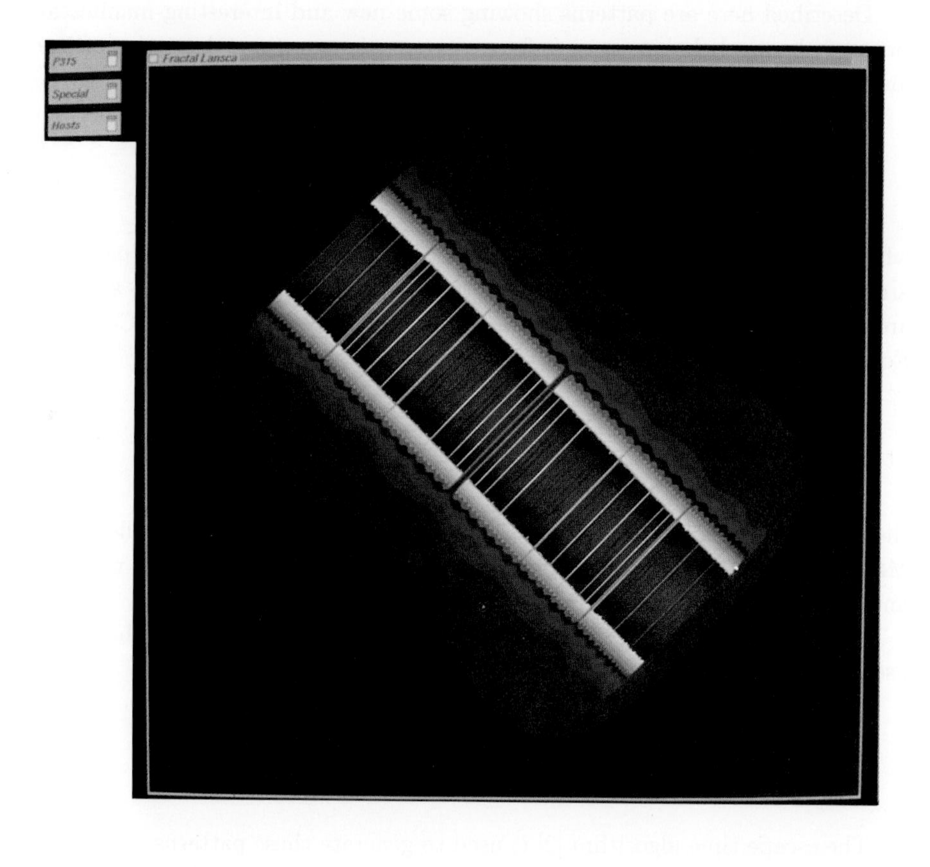
P315
Special
Hosts
Fractal Lansca

divergence test used is $x_n^2 + y_n^2 > 4$. Once the divergence test is satisfied or the number of iterations has reached 255, the process stops.

The number of iterations required to satisfy the divergent test is known as the *escape time*. The point on the monitor corresponding to $(-2, -2)$ is colored according to its escape time.

A new iteration cycle begins with the next point $(-2, -1.996)$, that is, $x_0 = -2$, $y_0 = -1.996$, $a_5 = -2$, $b_5 = -1.996$. The iteration process is repeated to obtain the escape time and hence, the color of the point. The iteration cycle is repeated till all the points in the area of interest are used.

Figure 1 shows the patterns obtained using this method. They are remarkably simple and exhibits high regularity and Eucledian properties. The entire real plane is divided into two regions. The central black region marks the stable region. The points in this region will never flee to infinity. The unstable region consists of points that escape to infinity with different escape times. The boundary of the two regions appears as rectangles. It is surprising that such simple structures demonstrate the complicated behavior of a chaotic system. (Upon closer magnification of the boundary areas, one will see similar regular structures.)

References

1. B. B. Mandelbrot, *The Fractal Geometry of Nature* (W H Freeman and Company, 1982).
2. M. F. Barnsley, *Fractals Everywhere* (Academic Press, Inc., 1988).
3. C. Pickover, "Chaotic behavior of the transcendental mapping $z - \cosh(z) + u$)", *The Visual Computer* **4** (1988) pp. 243–246.

David Walter

The Galactic Virus Attack

Described here is a pattern which is interpreted as showing a galaxy launching a viral attack on an intruder.

The set is a 490747.2 times magnified portion of a Mandelbrot Set created using the cubic equation:

$$Z = 3 * Z^3 - 3 * A^2 * Z + B\,,$$

$$A = (C * C + 1)/3 * C\,,$$

$$B + 2 * A^2 + (C - 2)/3 * C\,.$$

It is located at

$$\begin{aligned} X\text{min} &= -1.230040437\text{E} - 0002 \\ X\text{max} &= -1.229497048\text{E} - 0002 \\ Y\text{min} &= 2.647697323\text{E} + 0000 \\ Y\text{max} &= 2.647701398\text{E} + 0000. \end{aligned}$$

The set was computed using software purchased from Sintar Software, P O Box 3746, Bellevue, WA 98009, USA. A 26Mz Beam Technology 286 computer with a color VGA screen was used to compute the set. The maximum iteration count was 2000.

The colored image was printed in black and white on a Hewlett Packard (HP) Deskjet Plus having a resolution of 300 dpi. This necessitated the use of a special GRAPHICS.COM and GRAPHICS.PRO supplied by HP in Singapore, and which replace files of the same names supplied with DOS 4. The command line: GRAPHICS DESKJET GRAPHICS.PRO/R enables SHIFT PRINT SCRN to print the Graphics screen on the Deskjet, which is a relatively inexpensive, quiet, high resolution black and white printer. The /R option enables reverse video, which of course causes white on the screen to print white.

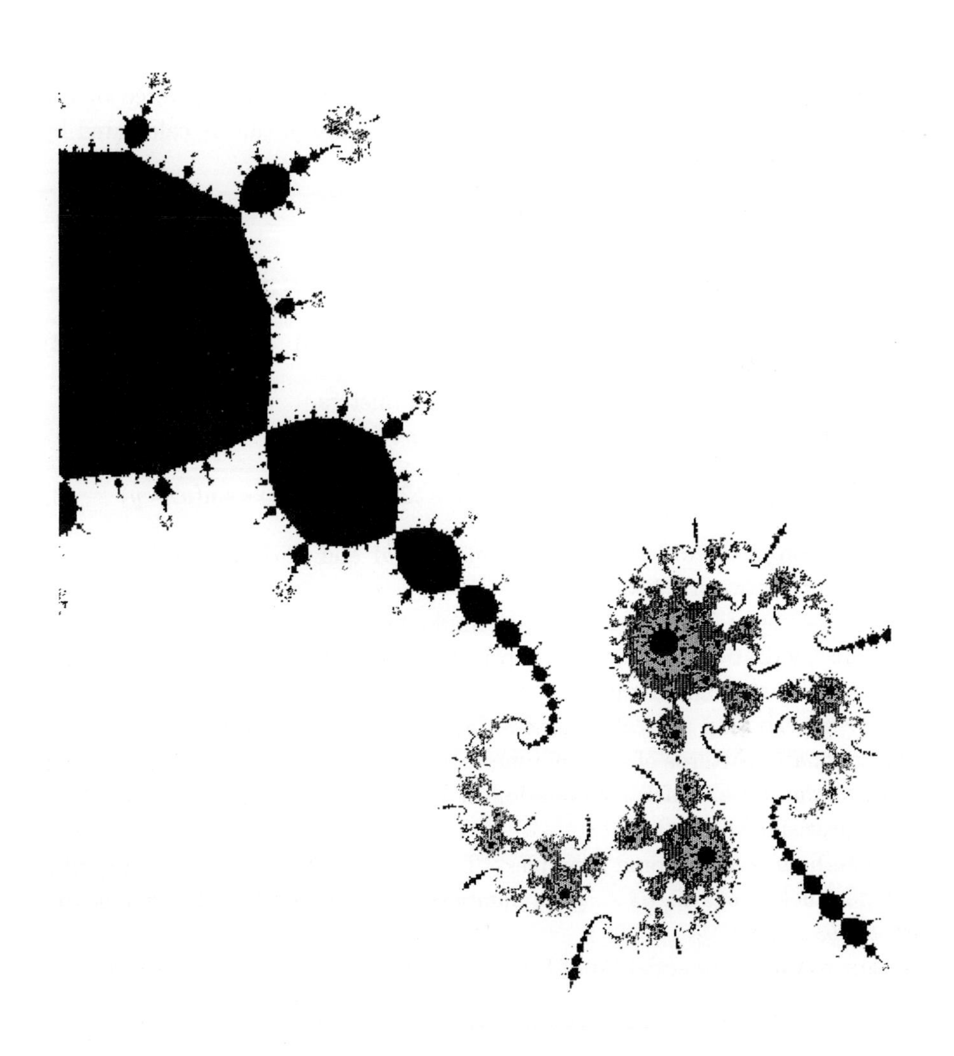

Wilfried Weber

Sound Ornaments

The patterns of an oscillating quadratic membrane, first discovered by the German physicist E. Chladni in the early 19th century, can be calculated by solving the partial differential equation

$$s'' + f * s = 0\,. \tag{1}$$

The solution for a membrane with fixed boundaries is given by the function

$$s(x,\, y) = \sin(a1 * x) * \sin(a2 * y) \tag{2}$$

where $a1$ and $a2$ denote arbitrary integer parameters.

As the parameters are interchangeable, the function

$$s(x,\, y) = b1 * \sin(a1 * x) * \sin(a2 * y) + b2 * \sin(a1 * x) * \sin(a2 * y) \tag{3}$$

is a solution of the differential equation, too. $b1$ and $b2$ denote arbitrary real parameters between -1 and $+1$ [1].

The zeroes of (3) computed, for example, by Newton's algorithm and plotted in a quadratic area ($x,\, y = 0 \ldots \text{pi}$) show the pattern of a membrane oscillating with one single frequency.

By superposing several equations (3) with different parameters, one gets the enigmatic patterns of a membrane oscillating with two or more frequencies at the same time. Each combination of values of the parameters leads to a new surprising tile pattern.

The figures show some examples. The numbers in the first line describe the parameters $a1$, $a2$, $a3 \ldots$, the numbers in the second line the parameters $b1$, $b2$, $b3 \ldots$.

For example the equation of Fig. 1 is

$$\sin(3 * x) * \sin(5 * y) + \sin(5 * x) * \sin(3 * y)$$
$$- \sin(9 * x) * \sin(13 * y) - \sin(13 * x) * \sin(9 * y) = 0$$

Figures 1, 2, and 3 show the superposing of two frequencies, Figs. 4 to 9 the superposing of three frequencies.

Reference

1. H. Courant, *Methoden der Mathematischen Physik* (Springer, 1968).

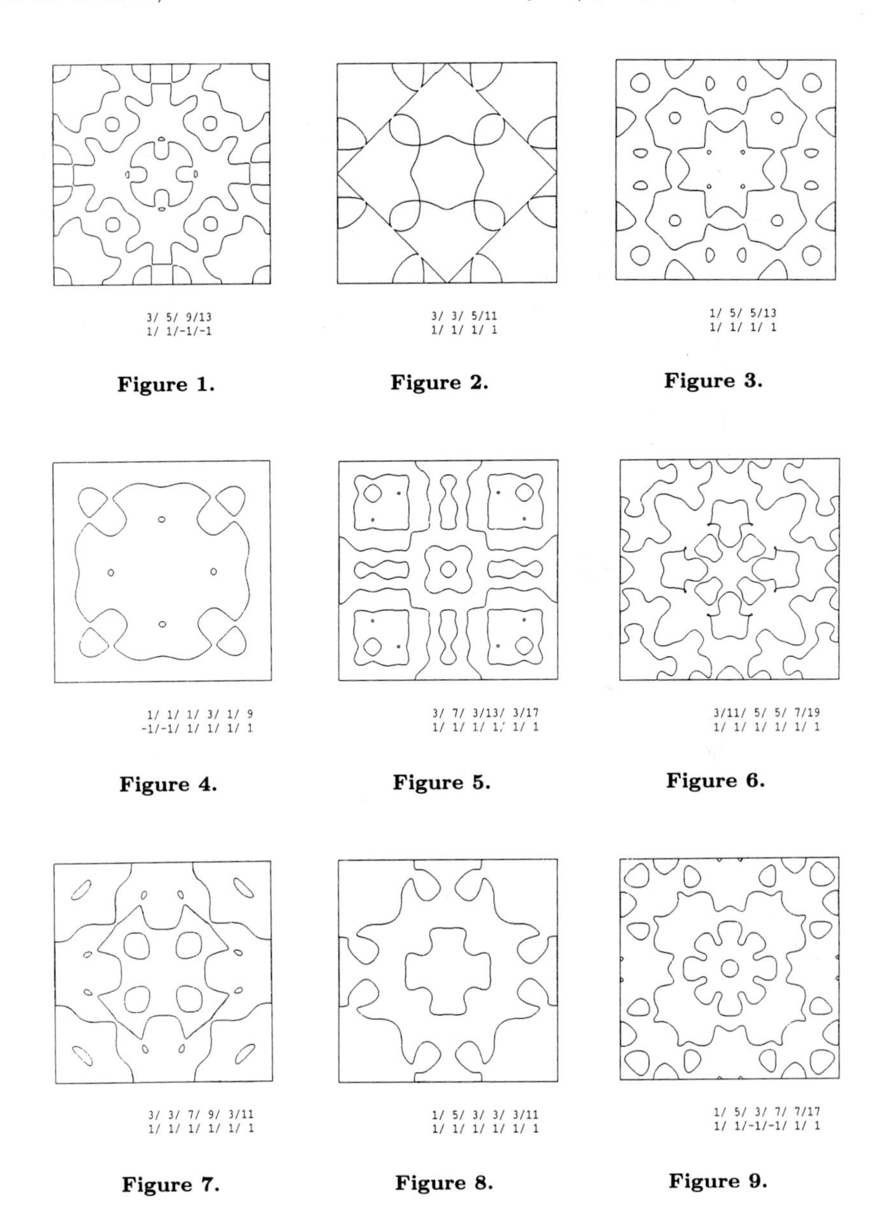

Figure 1. **Figure 2.** **Figure 3.**

Figure 4. **Figure 5.** **Figure 6.**

Figure 7. **Figure 8.** **Figure 9.**

PART III
HUMAN ART

A. K. Dewdney

An Informal Tesselation of Cats

A close friend once sighed, “How nice it would be, on cold winter evenings, to have a cat blanket.”

“What”, I asked, “is a cat blanket?”

“It’s a blanket of cats that all come up on your bed, hundreds of cats. They arrange themselves into a blanket, leaving no spaces between them. Then they all go to sleep, purring through the night”. She seemed to purr at the idea.

“What kind of cats?”

“All kinds. Black, white, calico, tabby, Abissynian, marmalade, patches, gray, all kinds.”

The idea was clearly unworkable but not if the cats were stuffed. That is to say, not if the cats were fiber-filled cloth replicas sewn into a blanket. Inspired, I set about to design the blanket.

As I worked, I thought of Escher and Bach. (Godel is irrelevant here.) A strict tesselation, like a strict fugue, involves a basic figure (graphical or musical) that has to obey so many constraints in matching up with itself over space or time, that there is almost no aesthetic freedom left. It is almost certain to be ugly. If we ignore the wonderful patterns that Escher and Bach have made and study the constituent figures instead, we will find a certain number of uncomfortable angularities. Isolated, they are not particularly pleasant to see or hear.

But the freedom that cats in different sizes and different positions gave me allowed for more natural elements. I am tempted to have the design stitched into a blanket, each cat a bas-relief in a fabric that reflects its fur. But perhaps it would be wiser to have a real artist redesign the blanket, someone with more than three hours to spend on the project.

Jean Larcher

Op Art 1

This abstract pattern is composed of repeated shapes and parts. Try coloring the pattern in different ways to create your own unique art.

Reference

1. J. Larcher, *Op Art Coloring Book* (Dover, 1975).

Paulus Gerdes

Extension(s) of a Reconstructed Tamil Ring-Pattern

Described here is a pattern that I found when analyzing the geometry of traditional Tamil designs [1].

Tamil women of South India used a mnemonic device for the memorization of their standardized pictograms. After cleaning and smoothing the ground, they first set out an orthogonal net of equidistant points. Each design is normally monolinear, i.e., composed of one smooth line that embraces all points of the reference frame. Figure 1 shows a *pavitram-* or ring-pattern reported by Layard [2] that does not satisfy the Tamil standard as it is made out of three superimposed closed paths. In [1], I suggested that the reported *pavitram* design could be a "degraded" version of an originally monolinear pattern. The reconstructed design shown in Fig. 2 is probably the original pattern. Figure 3 displays a reduced version of Fig. 2, i.e., without "border ornamentation". Applying the same *geometric algorithm* as used in Fig. 3, one always obtains monolinear patterns for dimensions $(4m + 1) \times (4n + 1)$ of the rectangular reference frame, where $4m + 1$ denotes the number of dots in the first row and $4n + 1$ the number of dots in the first column (m and n represent arbitrary natural numbers). Figure 4 illustrates this geometric alogrithm in the case 9×9 and Fig. 5 shows the 17×17 version with a rotational symmetry of 90°.

References

1. P. Gerdes, "Reconstruction and extension of lost symmetries: Examples from the Tamil of South India", *Computers and Mathematics with Applications* **17**, 4–6 (1989) 791–813.
2. J. Layard, "Labyrinth ritual in South India: Threshold and tattoo designs", *Folk-Lore* **48** (1937) 115–182.

Figure 1. Reported pavitram-design. **Figure 2.** Reconstructed pavitram-design.

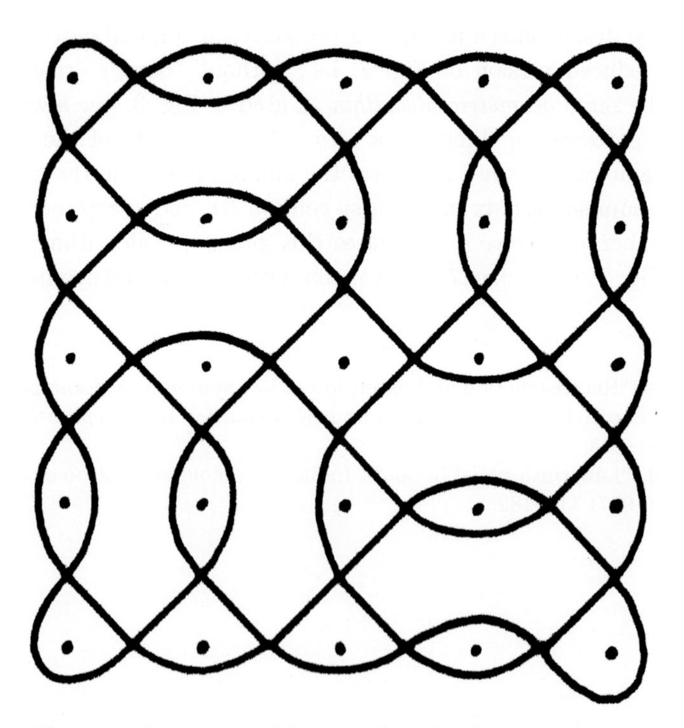

Figure 3. Reconstructed design without border ornamentation.

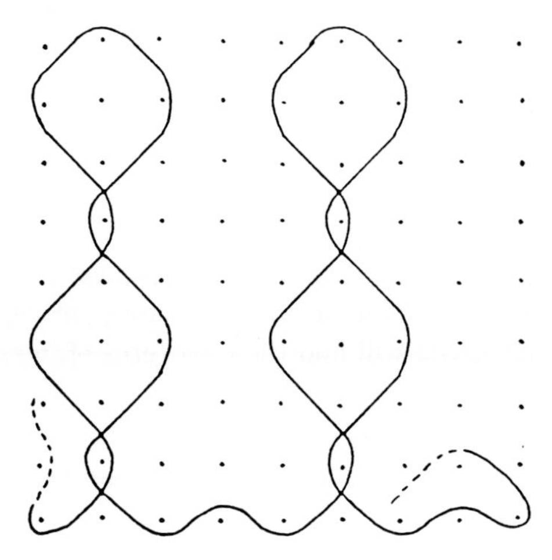

Figure 4. Geometrical algorithm illustrated in the case 9×9.

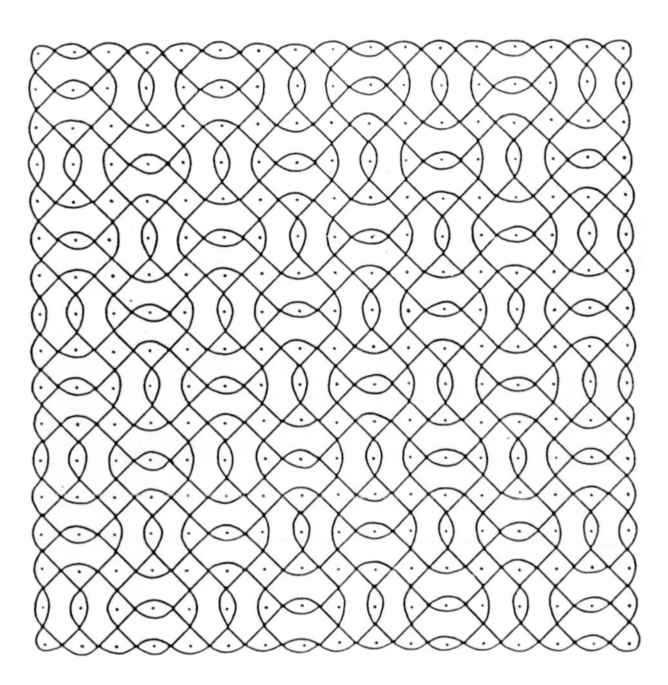

Figure 5. 17×17 version.

Hajime Ouchi

Japanese Optical and Geometrical Art 1

These designs were created for the Leading Part Company of Osaka. Designs in this series employ lines, circles, dots, squares, and ellipses. Some patterns seem to recede and shimmer in the manner of "op" art. Hopefully, designers and graphic artists will find these patterns of interest.

Reference

1. H. Ouchi, *Japanese Optical and Geometric Art* (Dover, 1977).

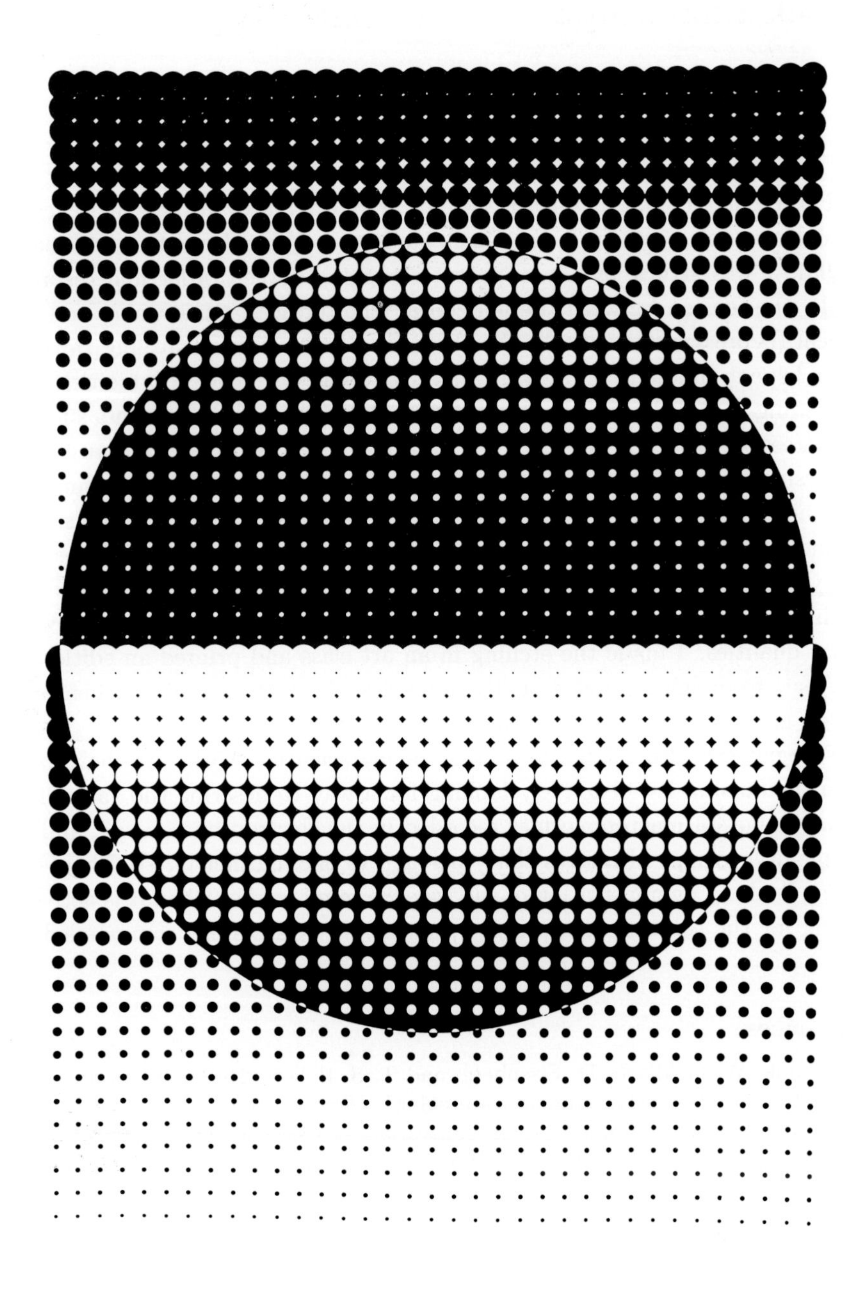

Joachim Frank

Greek Landscape — A Visual Diary

This etching (Fig. 1), which I made in 1978, originates in the experience of seeing large-scale patterns in the Greek landscape. When I was first in Greece, in 1967, I was struck by the apparent struggle between order and disorder and by the tension between the horizontal (the terraces, stone walls) and the vertical (cypresses and pines). For three months, as I was traveling through Crete, I spent all idle time covering a piece of paper with tiny new hieroglyphs, reinventing the landscape as a tapestry by seriation of basic elements such as suns, trees, birds, and eyes in many variations. Ten years later I transferred this design — somewhat modified — onto a wax-covered zinc plate, using the finest needle I could find. To convey something of the richness of a landscape, I found I had to create a texture with features too small to be resolved by the eye in the normal viewing distance. Only in this case, the experience of viewing the etching could invoke the experience of viewing a nature scene with fractal qualities. I made the etching in an art class and printed an edition of 10. Apart from using translational symmetries, the pattern explores themes of texture and the creative role of variations in a design.

This pattern also relates to a stage of my scientific work. At the time I made the etching I was actively looking for a way to describe the variability among images of macromolecules recorded in the electron microscope, and to average over groups of images that "belong" to one another [1, 2]. Thus, the etching is also a demonstration of the way the mind works: when it is actively looking for the solution of a problem, all of its resources are commanded at once.

References

1. J. Frank, W. Goldfarb, D. Eisenberg, and T. S. Baker, "Reconstruction of glutamine synthetase using computer averaging", *Ultramicroscopy* **3** (1978) 283–290.
2. M. Van Heel and J. Frank, "Use of multivariate statistics in analysing the images of biological macromolecules", *Ultramicroscopy* **6** (1981) 187–194.

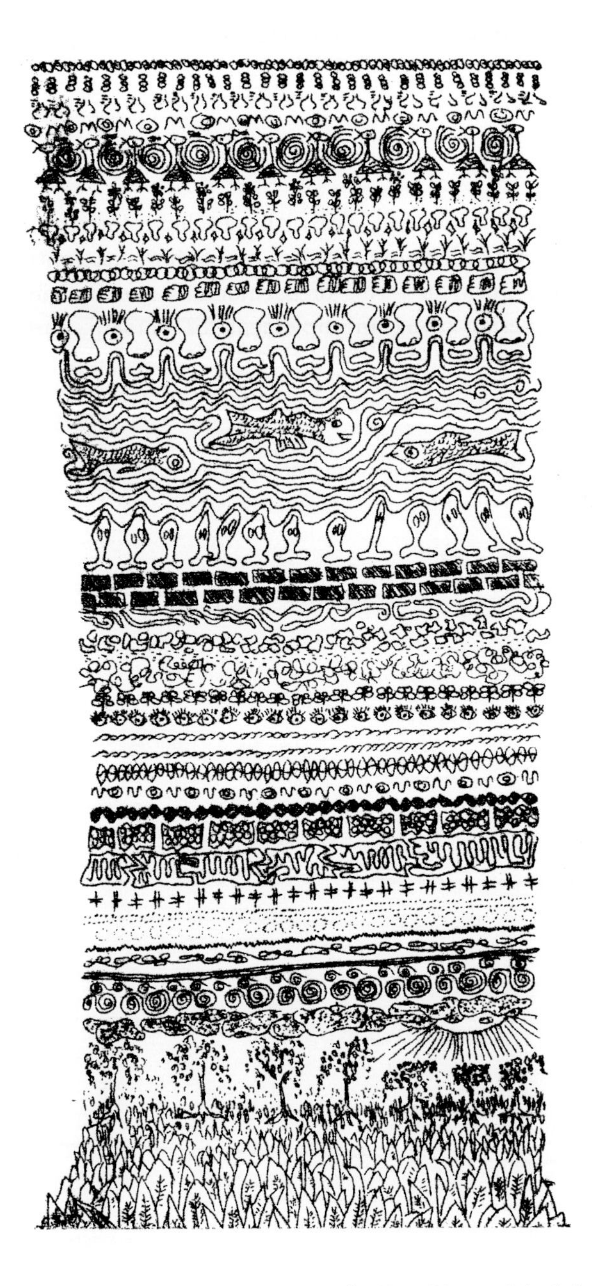

Figure 1. Greek landscape — a visual diary (1978 etching; original size 2" × 4").

Ali Dowlatshahi

Persian Designs and Motifs

Shown here are Persian wall tiles with metallic colors (from the 13th and 14th centuries). For a millennium, Iranian artists have attempted to state their feelings about the world in terms of harmonious and imaginative patterns, every sketch and design being an implement for worship, and a reflection of inner strength.

Reference

1. A. Dowlatshahi, *Persian Designs and Motifs* (Dover, 1979).

Wall tiles with metallic colors; 13th and 14th centuries.

Alan Mackay

Symmetric Celtic Sycophancy

Since the earliest times (of, for example, Aristophanes) the ultimate demand made by a man, dominant in a social group, for recognition by his sycophants (the word itself being an obvious euphemism), has been "Kiss my arse". Almost all European languages have equivalent expressions [1]. The Gaelic version, "Pogue Mahone", is widely known as it is used as the name of a British pop group.

Pictish/Celtic art is characterized by complicated interlaced designs deriving from the forms of animals and represents the extreme Western end of a band of common styles extending geographically, via the Scythians, to Siberia. The Picts themselves were tattooed with symmetric designs, rather than painted as their name (given by the Romans) suggests. Such tattooed skin has been recovered from tombs in the permafrost of the Altai. The characteristic Pictish talent for elaborate symmetric designs is seen here (in the figure) in a detail from one of the stones (Meigle 26) from Meigle, Perthshire, Scotland [2]. It dates from the 7–8th century AD. The symmetry is combined with sharp social comment on the Condorcet paradox of voting, where A can quite rationally be preferred to B and B to C, but also C to A. Niemi and Riker [4] have discussed this paradox in the context of Carter, Ford, and Reagan. On their tombstones, the Picts had an elaborate graphic symbolism for denoting social categories [3].

The human swastika motif (of the figure) occurs on several Irish crosses and in the Book of Kells [2], the latter being one of the masterpieces of Celtic symmetric design. It is evident that the rotational symmetries of orders 2,3,4, and 6 are by far the main ones used in Celtic and Pictish artifacts, although 10 and 7 are very occasionally to be seen [5] used rather unhappily. Since most patterns in relief represent interlaced motifs, there are few true mirror planes, but many designs have approximate mirror planes.

References

1. C. Kunitskaya-Peterson, *International Dictionary of Obscenities* (Scythian Books, 1981).
2. S. Cruden, *The Early Christian and Pictish Monuments of Scotland* (HMSO, 1962).

3. C. Thomas, "The interpretation of Pictish symbols", *Archaeological J.* **CXX** (1963).
4. R. G. Niemi and W. H. Riker, "The choice of voting systems", *Scientific American* **234**, 6 (1976) 21–27.
5. S. Young, ed., *The Work of Angels* (British Museum Publications, 1989).

Figure 1. Detail from a Pictish stone relief (Stone No. 26, Meigle, Perthshire). Reproduced by courtesy of HMSO.

Erhard Schoen
Horoscope

Interest in astrology and astronomy began thousands of years ago in Mesopotamia, Egypt, Central America, and the Far East. Today, we still use Assyrian-Babylonian and Egyptian astronomical symbols for sun, moon, and stars. Greek and Roman names signify the planets and constellations, while zodiac signs are Chaldean.

Reference

1. E. Lehner, *Symbols, Signs, and Signets* (Dover, 1969).

Carol B. Grafton

Bizarre and Ornamental Alphabets

In the past, decorative letters were used in books, serving as the initial letters of paragraphs, chapters, and headings. Modern printers and designers no longer make extensive use of decorative letters. Letters are now generally employed to create special effects in advertising or striking book cover designs. Shown here is a sample from an unusual alphabet integrating human body parts and architectural themes.

Reference

1. C. Grafton, *Bizarre and Ornamental Alphabets* (Dover, 1981).

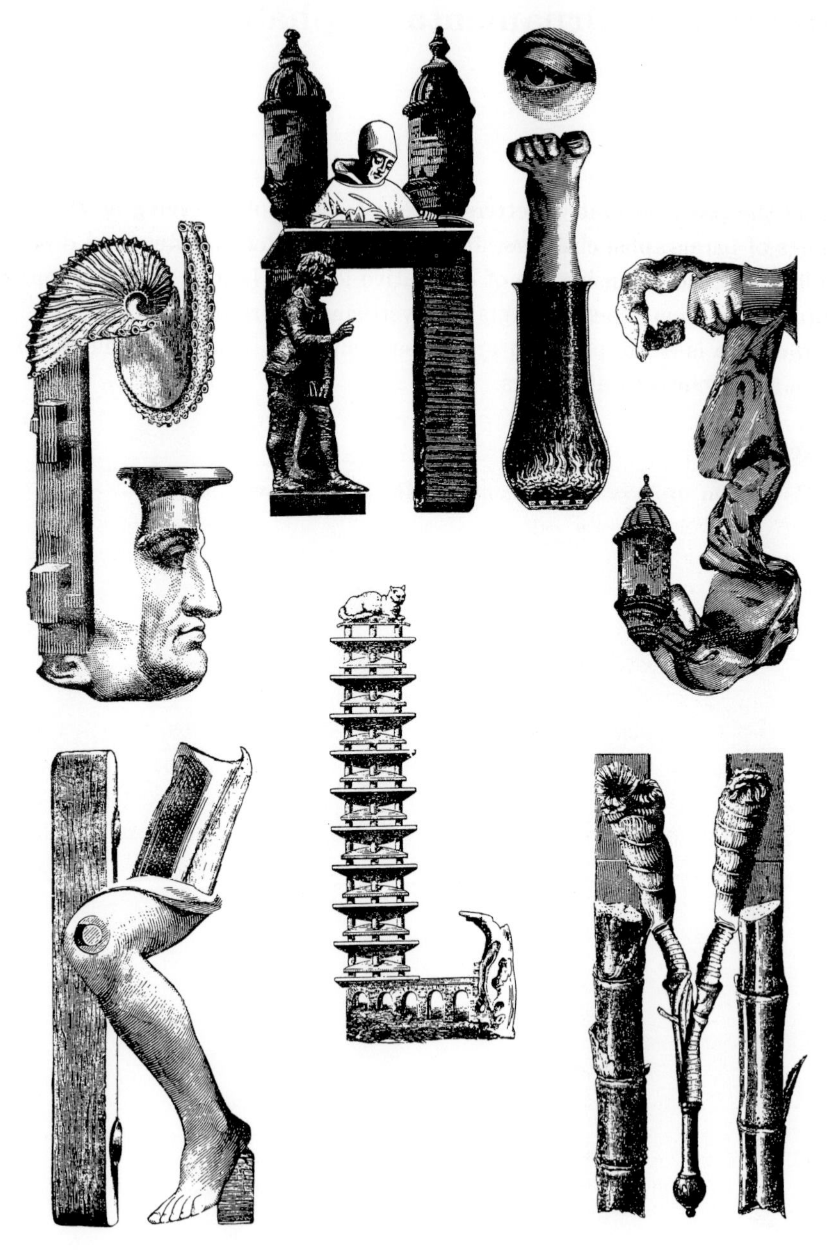

Bizarre and Ornamental Alphabets

Peter Hettich

Ambiguous Art

The patterns presented here are based on ambiguous, cubist shapes. Ambiguous features, visual illusions, and the depiction of impossible objects have fascinated artists and lay people through the centuries. Famous past artists who have experimented with these visually interesting art forms come easily to mind: Escher, Magritte, da Vinci, Picasso, Dali, Albers, Vasarely, and most recently, Shepard [1]. See [1–6] for more information. Past researchers have even used such geometrical art as a probe of the human perceptual-cognitive system [1].

Representing three-dimensional objects on a 2D surface naturally creates certain perceptual ambiguities. In the real world, viewers can walk around 3D objects and also use their binocular vision to help resolve certain spatial ambiguities. This is not possible with paintings. In particular, I enjoy working with *depth ambiguities*. The ambiguities arise because the object is portrayed from a viewing position that yields alternative interpretations about where objects are located and what is in front of what.

My art (see figures) often reduces complicated objects (for example, human portraits and bodies) to simple basic elements, such as the ambiguous cube and rectangle (Fig. 1). I think that artists of the future will often experiment with this kind of art using traditional media and computer graphics. Rapidly growing interest and understanding of the brain's visual and perceptual system makes figures such as these of particular interest in the 21st century.

References

1. R. Shepard, *Mind Sights* (Freeman, 1991).
2. D. Hofstadter, *Godel, Escher, Bach* (Klett, 1985).
3. E. Gombrich, *Kunst und Illusion* (Belzer, 1986).
4. K. Boff, L. Kaufman and J. Thomas, *Handbook of Perception and Human Performance*, Vols. 1 and 2 (Wiley, 1986).
5. G. Cagliotti, *Symmetriebrechnung und Wahrnehmung* (Vieweg, 1990).
6. P. Hettich, "The future of ambiguous art", in *Visions of the Future*, ed. C. A. Pickover (St. Martin's Press, 1992).

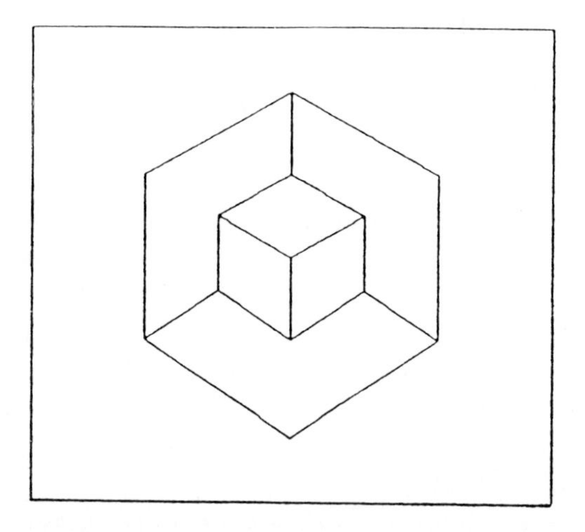

Figure 1. The ambiguous cube. Is the cube in front of, and inside of, the corner of a room? Is it in front of another cube? Is it touching the background "floor" or floating above it?

Figure 2. Gitte (1990; 49" × 55").

Figure 3. Hans (1991; 60" × 84").

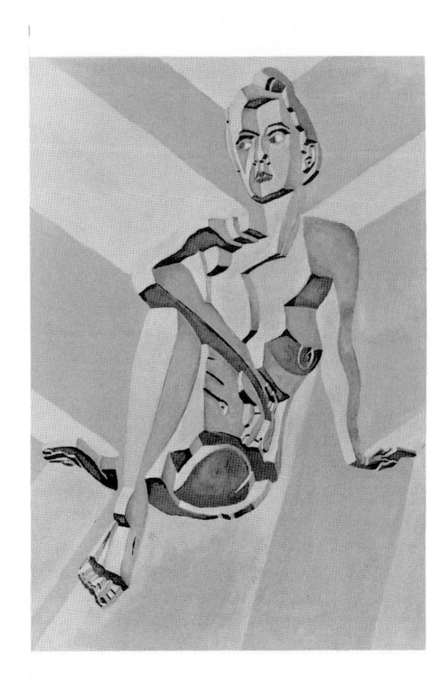

Figure 4. Yellow nude (1990; 75" × 110").

W. and G. Audsley

Japanese Diaper Ornaments

Diaper ornaments may be generally defined as designs in which features occur at regular intervals, and which are enclosed or connected by geometrical flowing lines. Shown here are Japanese diaper ornaments.

Reference

1. W. Audsley and G. Audsley, *Designs and Patterns from Historic Ornament* (Dover, 1968).

Harold J. McWhinnie

Satanic Flowers

Described here is a pattern showing the shapes of a flower as translated into the electronic environment. It differs from some of the other computer-based designs, in that a drawing of a flower was first made and then placed into the computer by the use of a mouse and the software program "MacPoint". The basic pattern has been then manipulated by the computer-based software to form a set of variations on a simple theme, in this case, the "Satanic Flower". The final results are printed out on a laser printer. I used the MacPoint part of Hypercard and employed a Mac SE computer to create these images.

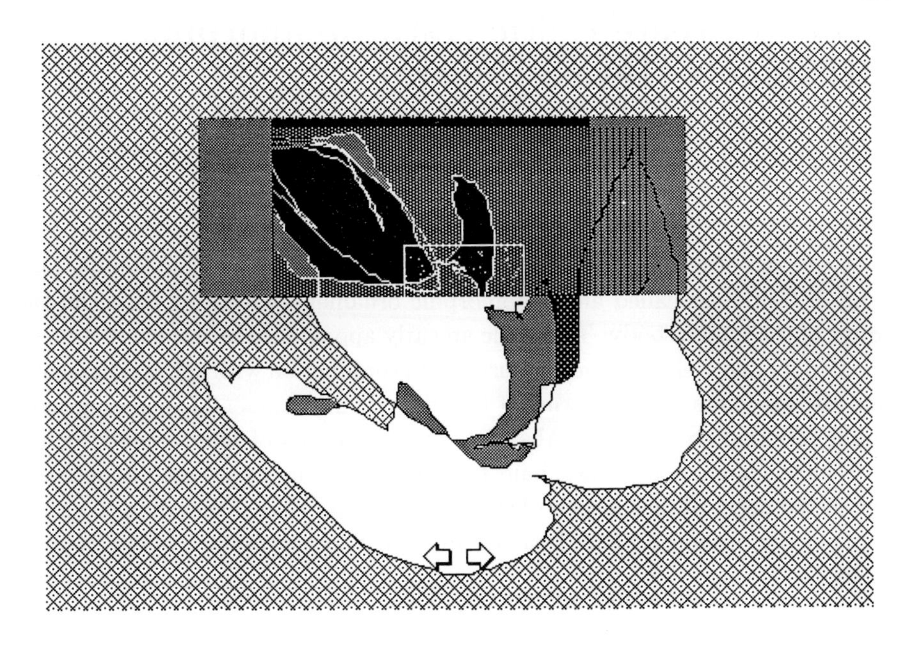

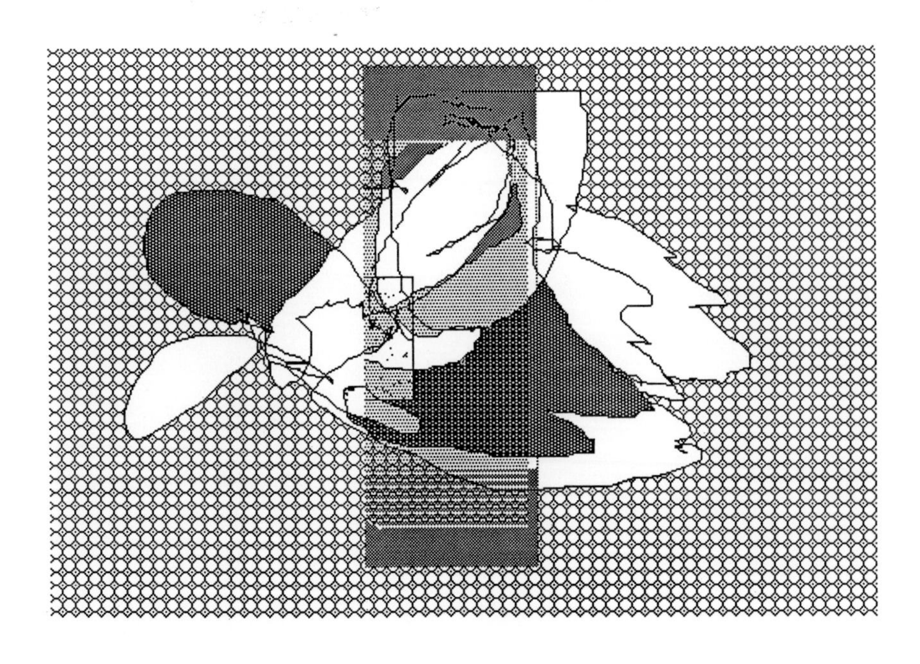

George Bain

How to Create Celtic Anthropomorphic Ornaments (Beard-Pullers)

Celtic art often consists of intricate knots, interlacements, and spirals. One form of Celtic art called anthropomorphic ornaments are those based on the forms of the human body. The made an early appearance in the art of Bronze-age Britain and Ireland. They are usually in conjunction with spiral ornaments, and the leg-joints and rib-forms of the animals in ornamental rendering have spiral terminal treatments. Abstract representations of human male figures with interlacing limbs, bodies, hair, top knots, and beards are used to decorate the sacred pages of the Gospels of the Book of Kells. Here, I show how to design some of the ornaments found on the stone at Clonmacnoise and the Book of Kells.

Reference

1. G. Bain, *Celtic Art: The Methods of Construction* (Dover, 1973).

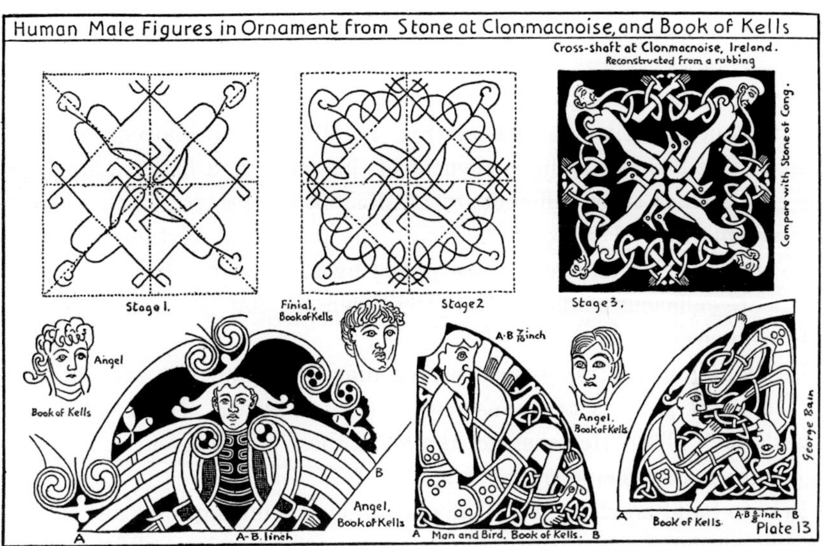
Human Male Figures in Ornament from Stone at Clonmacnoise, and Book of Kells
Cross-shaft at Clonmacnoise, Ireland.
Reconstructed from a rubbing
Compare with Stone of Cong.
Stage 1.
Finial, Book of Kells
Stage 2
Stage 3.
Angel
Book of Kells
A·B 7/16 inch
Angel. Book of Kells
George Bain
B
Angel, Book of Kells
A
A-B. 1 inch
A Man and Bird. Book of Kells. B
A
Book of Kells.
A·B 5/8 inch B
Plate 13

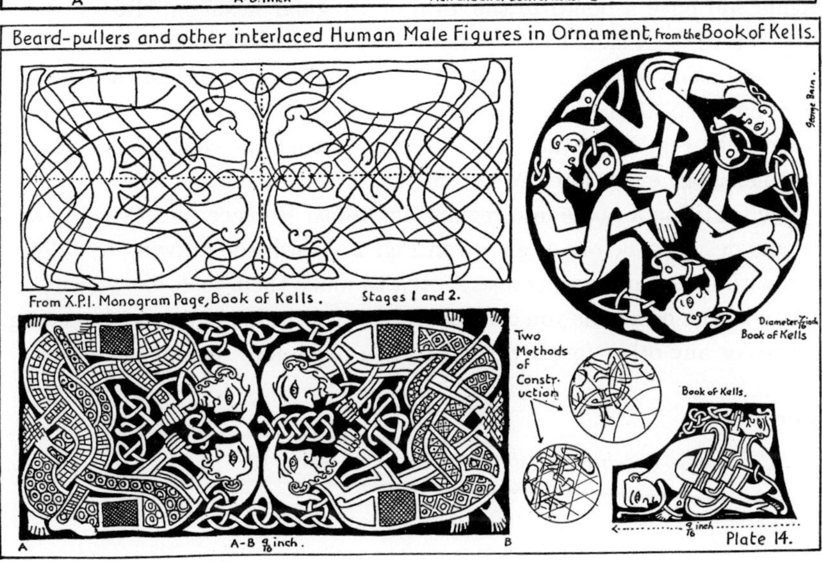
Beard-pullers and other interlaced Human Male Figures in Ornament, from the Book of Kells.
George Bain.
From X.P.I. Monogram Page, Book of Kells.
Stages 1 and 2.
Diameter 7/16 inch
Book of Kells
Two Methods of Construction
Book of Kells.
A
A-B 9/16 inch.
B
9/16 inch
Plate 14.

Stephan Kreuzer

Clowns

Described here is a pattern consisting of identically formed figures which cover a plane in a tile-like manner (Fig. 1). There is no blankspace between them. Figure 2 shows a single element of this pattern, a little smiling clown. This element has three different orientations in the pattern. You will notice some symmetries. For example, rotate the pattern at certain points (e.g., where the three hands meet at one point) by 120 degrees. In addition, three colours are sufficient to differentiate the clowns so that two neighbors are not of the same color.

Designing such a repetitious partitioning of a plane by natural-looking forms such as persons or animals was invented by the Dutch artist M. C. Escher (1898–1972). He constructed a lot of these patterns and he even wrote a book about this theme [1].

How to build an element for the pattern is explained graphically in Fig. 3. Imagine you have a board in the form of a regular hexagon. Take a saw and cut a section from side 1 out of the board (in our example it is the head of the clown). Now attach it on side 2. You will get a form like the one shown in Fig. 3a. Repeat the whole procedure with the other sides of the board. It will look like Fig. 3b at the end. One can see that the clown is the same size as the hexagon: every part we cut out we attach to another side.

Now partition a plane into regular hexagons. Replace each hexagon by the elements designed before as shown in Fig. 1. Finish the drawing by filling the forms with eyes, bow-ties, trousers, etc., to make them look like clowns.

Apart from hexagons you can construct patterns beginning with rectangles, triangles or any other forms which divide a plane into regular partitions.

Reference

1. M. C. Escher, *Regelmatige Vlakverdeling* (De Roos, 1958).

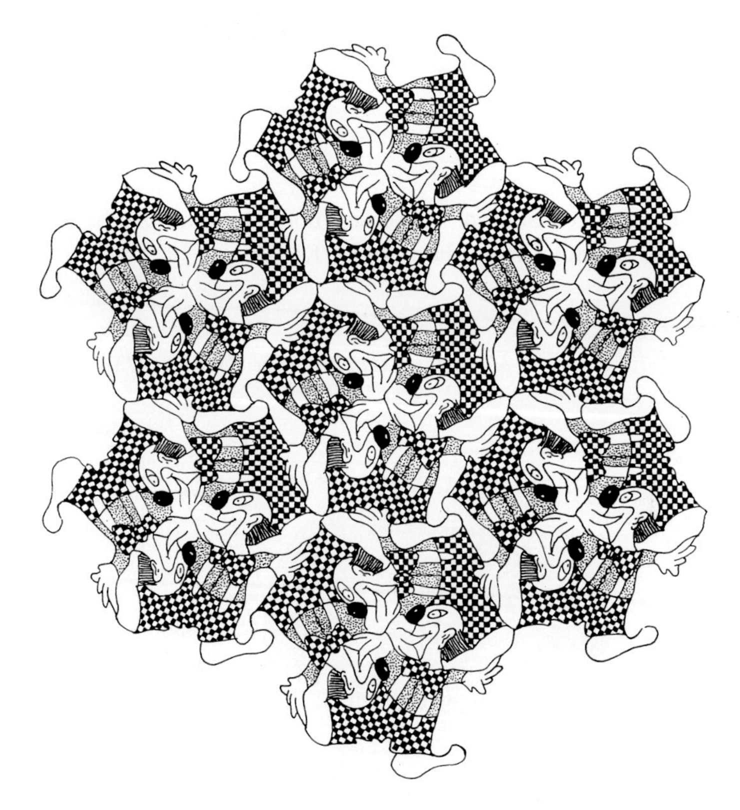

Figure 1. Clowns.

Figure 2. A single element of the pattern.

Figure 3a. The first step of the construction.

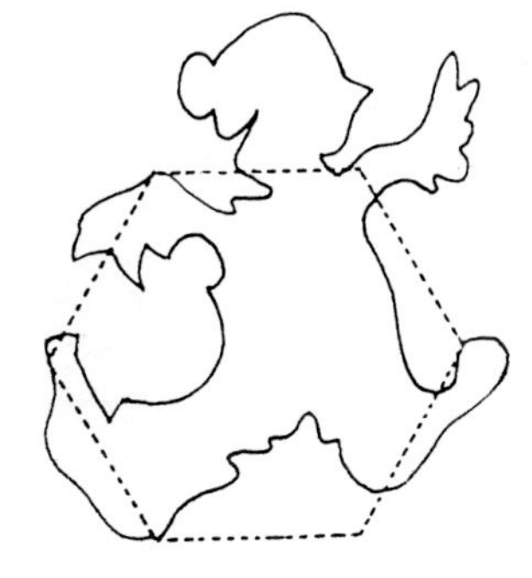

Figure 3b. The final form and the original hexagon.

George Bain
Celtic Plants Emerging from Pots

Celtic art often consists of intricate knots, interlacements, and spirals. Here, are some examples from the Book of Kells.

Reference

1. G. Bain, *Celtic Art: The Methods of Construction* (Dover, 1973).

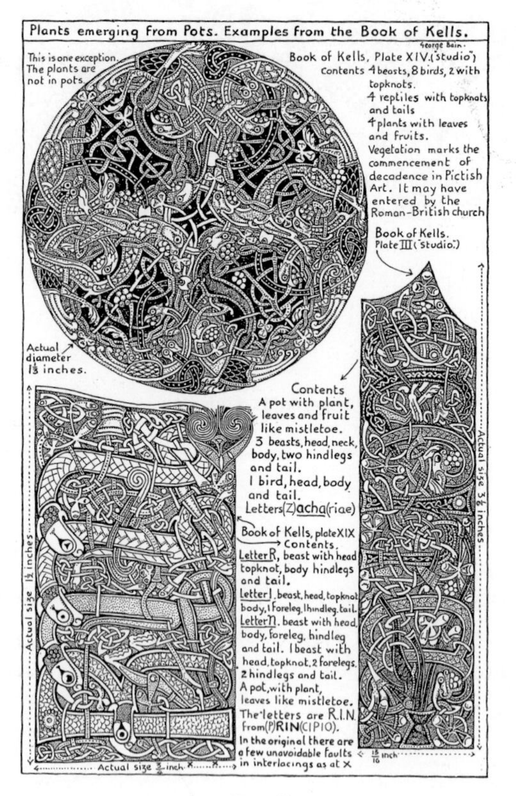

Plate W

Figure 1. Celtic art.

Yuzo Nakano

Interactive Patterns Between Light and Matter

It was a small discovery in my childhood that there were so many similar patterns existing in nature. This realization has since evolved into the concept of what I call the "law of similarity" in the natural world.

My long speculation on this topic eventually brought me to an idea — *evolution of light* — which became the title of my visual book completed in 1979. As illustrated in the volume, patterns of transformative motion are the basis of a theory of evolution based on the law of similarity. In this process, patterns oscillate and bind together to form matter which is actually only the temporary consequence of light transformations.

I did my research as an artist, not as a scientist. The result, as shown in my book, is that the reality of nature is the shape of patterns formed by the continual oscillation and transformation of energy. One problem in such research is determining what kind of energy a pattern needs to sustain itself. Even though the shape of matter observed in the physical world is merely a temporary, fictitious form, the energy that sustains the shape of the pattern formed by the transformation of light can actually be considered the remains of the original light in each pattern. This remaining energy should tend to follow or attract the original light. Perhaps the real power to sustain the shape of each pattern is in this attracted energy. "Matter makes an effort to become light" poets tell us.

If the attracted energy does not exist in the natural world, then the speed of light — 300,000 km/second — should be infinite, and it would not be possible for matter to retain its shape.

Through the atomic to the visual level, each pattern created in the process of the transformation of light has a distinct life span. Therefore, in the law of similarity, the force (energy) with which each pattern interactively oscillates must enable us to see the vital form of the world. If this is so, traces of the time lapse during the process of light transformation must be recorded in patterns in the interior structure or the surface of both organic and inorganic matter. We should then be able to see them.

An example of the pattern proposed here is one not duplicated from nature, but one created according to an abstract system. Each broken line (colored) is simple matter (pigment) and occupies a characteristic field and time.

In the abstract system I use, any part of the whole is always in relationship to the intervals between parts. This relationship is continuous and generates scattered similar figures.

The interval that exists between parts of the whole defines a kind of symmetry in which each interval permits the light and creates the motion of light, making a visible fluctuating light pattern which will vary depending on certain view points (distance, angle, motion). This traveling light through the interval can be called "neo time" compared with "real time", which is actually colored broken lines. The relationship between the spectroscopic reflection and absorption on each part through the spectrum of the light source and its artificial control (i.e., systems, illumination) causes the "whole" to optically produce kinetic effects in the interaction with light patterns.

I am a painter and an electronic musician. Right now I am very interested in brain nerve physiology in order to research my hypothesis that the mechanism of the interpretive or memory domain in the brain might in principle be formed by the law of similarity. For example, when one discovers some unknown fact (i.e., pattern), one synthesizes this information by associating it with similar past experiences or knowledge of another similar fact (i.e., pattern). This is a conscious discovery which will be the cause of a nerve impulse that produces an electro-chemical signal in order to operate the exchange for inner or outer information.

Marcia Loeb

Art Deco Design 1

Art Deco, with its bold color schemes and strong geometrical patterning, has recently been rediscovered with tremendous impact, and is of major importance to designers today. Art Deco is the name now generally applied to the most typical artistic production of the 1920s and 1930s. This name comes from the large exhibition held in Paris in 1925, called the *Exposition Internationale des Arts Decoratifs et Industriels Modernes.* The art of these years had many sources, including turn-of-the-century art nouveau and such archeological interests of the day as ancient Egyptian and Mayan Art.

Reference

1. M. Loeb, *Art Deco Designs and Motifs* (Dover, 1972).

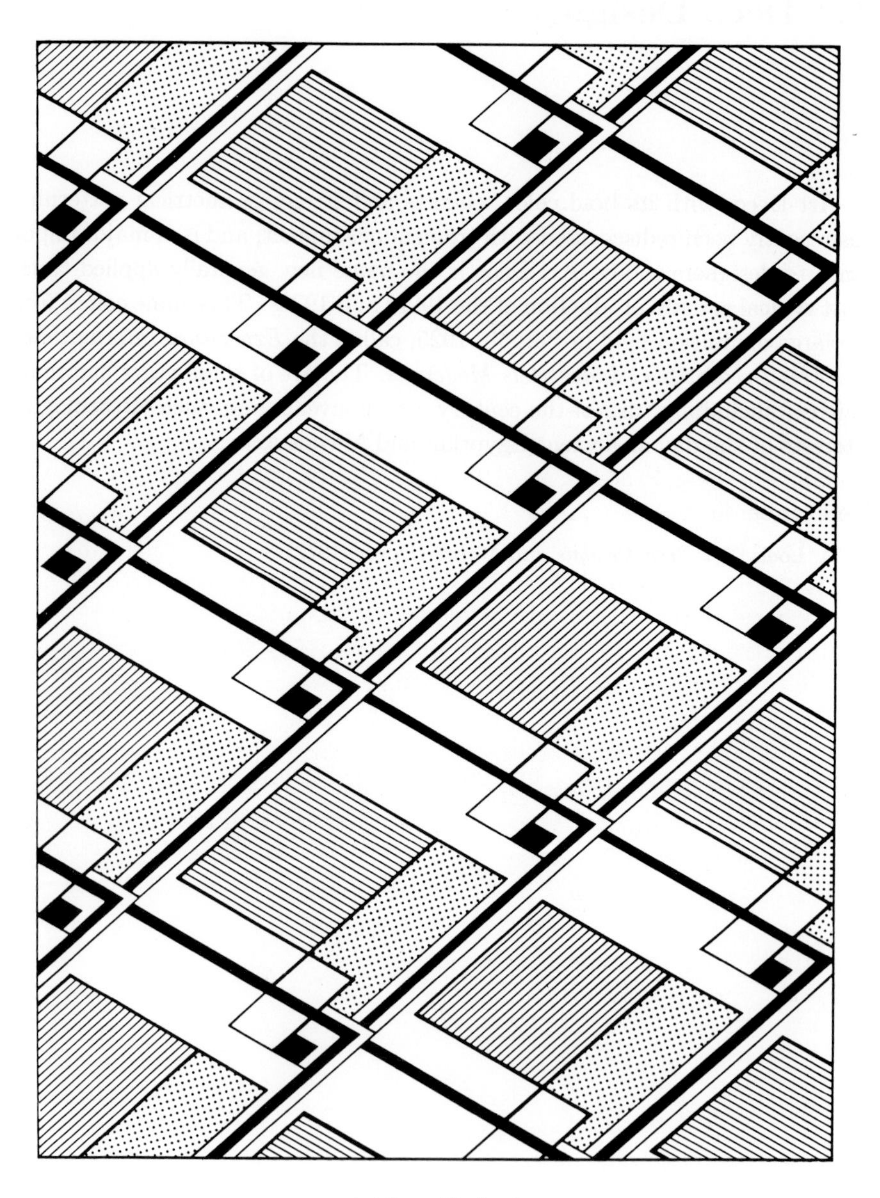

Art Deco Design 1

Federico Fernández

Periodic Pattern of Class 19

Shown here is a pattern belonging to an infinite series of periodic patterns. These are generally produced in the tables of f values of quadratic and higher degree binary forms when the multiples of a number are selected. The patterns result from the overlay of nets of points with integer coordinates.

The pattern in Fig. 1 corresponds to the binary form of 8th degree: $X^8 + 3X^6Y^2 + 8X^4Y^4 + 3X^2Y^6 + Y^8 = f$.

The values f are calculated, and when these numbers are multiples of 19, dots are marked in the corresponding cells of a table of grid in which the coordinate axes X, Y are drawn. The square cells with dots are then colored. The ornamental motif is completed inside a square of 19×19 units that repeats, producing the periodic pattern.

An alternative method to produce this pattern consists of the overlay of eight nets of points with integer coordinates. The points are changed to colored squares of the cell size as shown in Fig. 2. In the overlays, the four-cell colored squares at the vertices A, B, C, D of each net have to coincide. The overlay is made of two sets of four congruent nets each of the class 19.

The author has classified the nets of points with integer coordinates. There are twenty nets in the class 19. These twenty nets can be combined in groups of several nets at a time. The overlays produce enormous numbers of periodic patterns. Since there are as many classes of nets as natural numbers, the possible combinations are infinite in theory.

With selection, a great number of artistic patterns can be obtained in a short time. With computers, the economical mass production of ornamental geometric patterns is today a fact.

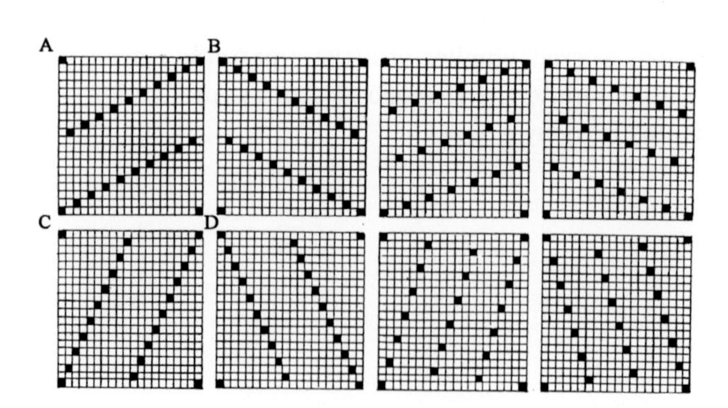

Periodic Pattern of Class 19

William Tait

Turtle Moon Artists' Logo and Chop Mark

Described here is a pattern showing multiple copies of my logo and printers' chop mark, assembled in a nonstatic pattern. This image was originally created by me as my artists' logo and is composed of the following significant elements.

The turtle as a totem animal. That is, I personally relate to the slow but deliberate pace of the turtle. In Chinese mythology, for one thousand years the turtle held an elephant on its back while the elephant held the universe, so the ability to undertake protracted, arduous tasks enters in. The moon signifies the cyclic or patterned movement of life through time and space, being a new moon relates to constant renewal. For me art is a pattern of creative tasks undertaken in the spirit of renewal, which sometimes needs the deliberate pace of the turtle coupled with the constant renewal of cycles.

At one point in my artistic journey, I totally immersed myself in printmaking (etching, lithography) for several years and continue now to practice various graphic disciplines. One detail in the creation of hand made prints is blind embossing a mark or *chop mark* on the sheet of paper the image is printed on. I had a blind embossing stamp made and embossed the turtle moon image on my handmade prints. As the nature of printmaking is the creation of multiples of an image (pattern) I feel it is an appropriate symbol for this.

Turtle Moon

Glossary

Abacus
Instrument for performing calculations by sliding beads along rods or grooves.

Abscissa
The horizontal coordinate of a point in a plane rectangular coordinate system (the x axis).

Acoustics
The study of sound.

Affine transformation
Loosely speaking, an affine transformation acts by shrinking, enlarging, shifting, rotating, or skewing an original pattern, set of points, or object.

Algebraic function
A function containing or using only algebraic symbols and operations such as $2x + x^2 + \sqrt{2}$.

Algebraic operations
Operations of addition, subtraction, multiplication, division, extraction of roots, and raising to integral or fractional powers.

Amplitude (of a wave)
The absolute value of the maximum displacement from zero value during one period of an oscillation. The "height" of the wave.

Amino acid
Basic building block of proteins.

Analog-to-digital converter
Electronic device that transforms continuous signals into signals with discrete values.

Analytic function
An analytic function is differentiable throughout a neighborhood of each point.

It can be shown that an analytic function has continuous derivatives of all orders and can be expanded as a Taylor series. Functions with a power series expansion are analytic.

Angstrom
A unit of measure corresponding to one ten-billionth of a meter.

Articulation
Movements of the vocal tract to produce speech sounds.

Attractor
Predictable attractors correspond to the behavior to which a system settles down or is "attracted" (for example, a point or a looping closed cycle). The structure of these attractors is simple and well understood. A *strange attractor* is represented by an unpredictable trajectory, where a minute difference in starting positions of two initially adjacent points leads to totally uncorrelated positions later in time or in the mathematical iteration. The structure of these attractors is very complicated and often not well understood.

Autocorrelation
For the acoustic applications in this book, the autocorrelation function for data describes the general dependance of the values of the data at one time on the values at another time.

Autonomous
The behavior of an autonomous dynamical system is expressed by an equation which is independent of time. If a time-dependent term is added, this represents an "external influence" which drives the system away from this equilibrium, for example, by adding or subtracting energy. Systems with a time-dependent term are nonautonomous (an unsteady fluid flow is such a system).

Bifurcation
Any value of a parameter at which the number and/or stability of steady states and cycles changes is called a bifurcation point, and the system is said to undergo a bifurcation.

Bilateral symmetry
The property of having two similar sides. Each side is a "mirror image" of the other.

Binomial coefficients
The coefficients in the expansion of $(x+y)^n$. For example, $(x+y)^2 = x^2 + 2xy + y^2$ so that the binomial coefficients of order 2 are 1, 2, and 1.

Catenary
The curve assumed by a uniform, flexible chain hanging freely from its ends. Its equation is $f(z) = \cosh(z) = (e^z + e^{-z})/2$.

Cellular automata
A class of simple mathematical systems that are becoming important as models for a variety of physical processes. Although the rules governing the creation of cellular automata are simple, the patterns they produce are complicated and sometimes seem almost random, like a turbulent fluid flow or the output of a cryptographic system. Cellular automata are characterized by the fact that they act on a discrete space or grid as opposed to a continuous medium.

Center
See *limit* and *fixed* point.

Chaos
Irregular behavior displaying sensitive dependence on initial conditions. Chaos has been referred to by some physicists as the seemingly paradoxical combination of randomness and structure in certain nonperiodic solutions of dynamical systems. Chaotic behavior can sometimes be defined by a simple formula. Some researchers believe that chaos theory offers a mathematical framework for understanding much of the noise and turbulence that is seen in experimental science.

Chaotic trajectory
A chaotic trajectory exhibits three features:
(1) the trajectory, or motion, stays within a bounded region — it does not get larger and larger without limit, (2) the trajectory never settles into a periodic pattern, (3) the motion exhibits a sensitivity to initial conditions. See also *Chaos.*

Cilia
Minute hair-like projections.

Complex number
A number containing a real and imaginary part, and of the form $a + bi$ where $i = \sqrt{-1}$.

Conservative dynamical systems
In mechanics, conservative dynamical systems, also known as Hamiltonian dynamical systems, are frictionless. These systems do not entail a continual decrease of energy. See also *dissipative dynamical systems.*

Converge
To draw near to. A variable is sometimes said to converge to its limit.

Cycle
The cycle describes predictable periodic motions, like circular orbits. In phase plane portraits, the behavior often appears as smooth, closed curves.

Damp
To diminish progressively in amplitude of oscillation.

Differential equations
Equations often of the form $dx_i/dt = f_i(x)$ where $x_i(t)$ represents the ith variable and the function $f_i(x)$ gives the time, or spatial, evolution of $x_i(t)$. Mathematical models in the physical and biological sciences are often formulated as differential equations.

Dissipative dynamical systems
These are systems typical of the macroscopic engineering world in which some resisting source causes energy loss. In dissipative dynamical systems the volume of phase space occupied by an ensemble of starting points decreases with time. See also *conservative dynamical systems.*

Disulfide bond
A strong sulfur-to-sulfur chemical bond that may cross-link two sections of a folded polymer or protein.

Dynamical systems
Models containing the rules describing the way a given quantity undergoes a change through time or iteration steps. For example, the motion of planets about the sun can be modeled as a dynamical system in which the planets move according to Newton's laws. A discrete dynamical system can be represented mathematically as $x_{t+1} = f(x_t)$. A continuous dynamical system can be expressed as $dx/dt = f(x, t)$.

Feedback
The return to the input of a part of the output of a system.

Fibonacci sequence
The sequence 1, 1, 2, 3, 5, 8, 13, ... , $(F_n = F_{n-2} + F_{n-1})$, which governs many patterns in the plant world. Each term is the sum of the last two.

Finite difference equations
Equations often of the form $x_i(t+1) = f_i(x(t))$ where $x_i(t)$ represents the value of the ith component at a time, or other coordinate, t.

Fixed point
A point which is invariant under a mapping (i.e., $x_t = x_{t+1}$ for discrete systems, or $x = f(x)$ for continuous systems). A particular kind of fixed point is a *center*. For a center, nearby trajectories neither approach nor diverge from the fixed point. In contrast to the center, for a *hyperbolic fixed point*, some nearby trajectories approach and some diverge from the fixed point. A *saddle point* is an example of a hyperbolic fixed point. An *unstable fixed point* (or repulsive fixed point or repelling fixed point) x of a function occurs when $f'(x) > 0$. A *stable fixed point* (or attractive fixed point) x of a function occurs when $f'(x) < 0$. For cases where $f'(x) = 0$ higher derivatives need to be considered.

Focus (of a conic section)
A conic section is a set of points for which the distances of each from a fixed point called the *focus* and from a fixed line called the *directrix* are in constant ratio.

Folium of Descartes
A plane curve represented in rectangular coordinates by $x^3 + y^3 = 3axy$.

Fourier analysis

The separation of a complex wave into its sinusoidal components.

Fractals

Objects (or sets of points, or curves, or patterns) which exhibit increasing detail ("bumpiness") with increasing magnification. Many interesting fractals are self-similar. B. Mandelbrot informally defines fractals as "shapes that are equally complex in their details as in their overall form. That is, if a piece of a fractal is suitably magnified to become of the same size as the whole, it should look like the whole, either exactly, or perhaps only after slight limited deformation."

Gasket

A piece of material from which sections have been removed. *Mathematical gaskets*, such as Sierpiñski gaskets, can be generated by removing sections of a region according to some rule. Usually the process of removal leaves pieces which are similar to the initial region, thus the gasket may be defined recursively.

Gaussian white noise

White noise which is subsequently altered so that it has a bell-shaped distribution of values. In this book, Gaussian noise is often approximated by summing random numbers. The reader may also wish to use the following formula which will generate a Gaussian distribution: $r = \sqrt{-\log(r_1)} \times \cos(2\pi r_2)\sigma + \mu$, where r_1 and r_2 are two random numbers from a (0, 1] uniform distribution, and σ and μ are the desired standard deviation and mean of the Gaussian distribution.

Invariant curve

Generalization of a *fixed point* to a line (in this case, a *curve* is invariant under the map or flow).

Helix

A space curve lying on a cylinder (or sphere, or cone) which maintains a constant distance from a central line (i.e. a "spiral extended in space").

Henon map

The Henon map defines the point (x_{n+1}, y_{n+1}) by the equations $x_{n+1} = 1.4 +$

$0.3y_n - x_n^2$, $y_{n+1} = x_n$. Note that there are various expressions for the Henon map, including $x_{n+1} = 1 + y_n - \alpha x_n^2$, $y_{n+1} = \beta x_n$.

Homeomorphism
This is best explained by an example. Consider a circle inside a square. Draw a line from the circle's center out through the circle and square. The line intersects the circle at point P and the square at P'. The mapping $g : P \to P'$ assigns to each point of the circle a point of the square and vice versa. In addition, two adjacent points on one shape are mapped to two adjacent points of the other. The mapping g is a *homeomorphism.* The circle and triangle are *homeomorphic.*

Iteration
Repetition of an operation or set of operations. In mathematics, composing a function with itself, such as in $f(f(x))$, can represent an iteration. The computational process of determining x_{i+1} given x_i is called an iteration.

Julia set
Set of all points which do not converge to a fixed point or finite attracting orbit under repeated applications of the map. Most Julia sets are fractals, displaying an endless cascade of repeated detail. An alternate definition: repeated applications of a function f determine a trajectory of successive locations x, $f(x)$, $f(f(x))$, $f(f(f(x)))$, ..., visited by a starting point x in the complex plane. Depending on the starting point, this results in two types of trajectories, those which go to infinity and those which remain bounded by a fixed radius. The Julia set of the function f is the boundary curve which separates these regions.

Limit
In general, the ultimate value towards which a variable tends.

Linear transformation
A relation where the output is directly proportional to the input. A function satisfying two conditions: (1) $F(\vec{p}+\vec{q}) = F(\vec{p}) + F(\vec{q})$ and (2) $F(r\vec{p}) = rF(\vec{p})$.

Logistic equation
The nonlinear equation $x_{n+1} = kx_n(1-x_n)$ is called the logistic equation, and it has been used in ecology as a model for predicting population growth.

Lotka-Volterra equations

The Hamiltonian system defined by $(dx/dt = ax - bxy,\ dy/dt = -cy + dxy)$, which was one of the first predator-prey equations which predicted cylic variations in population.

Lorenz attractor

$\dot{x} = -10x + 10y,\ \dot{y} = 40x - y - xz,\ \dot{z} = -8z/3 + xy.$

Initially the system starts anywhere in the three-dimensional phase space, but as transients die away, the system is attracted onto a two-lobed surface. For a more general formulation, the Lorenz equations are sometimes written with variable coefficients.

Lyapunov exponent

A quantity, sometimes represented by the Greek letter, Λ, used to characterize the divergence of trajectories in a chaotic flow. For a one-dimensional formula, such as the logistic equation, $\Lambda = \lim_{N\to\infty} 1/N \sum_{n=1}^{N} \ln |dx_{n+1}/dx_n|$.

Mandelbrot set

For each complex number μ let $f_\mu(x)$ denote the polynomial $x^2 + \mu$. The Mandelbrot set is defined as the set of values of μ for which successive iterates of 0 under f_μ do not converge to infinity. An alternate definition: the set of complex numbers μ for which the *Julia set* of the iterated mapping $z \to z^2 + \mu$ separates disjoint regions of attraction. When μ lies outside this set, the corresponding Julia set is fragmented. The term "Mandelbrot Set" is originally associated with this quadratic formula, although the same construction gives rise to a (generalized) Mandelbrot Set for any iterated function with a complex parameter.

Markov process

A stochastic process in which the "future" is determined by the "present".

Manifold

Curve or surface. The classical *attractors* are manifolds (they are smooth). Strange *attractors* are not manifolds (they are rough and fractal).

Newton's method

A method of approximating roots of equations. Suppose the equation is $f(x) =$

0, and a_1 is an approximation to the roots. The next approximation, a_2, is found by $a_2 = a_1 - f(a_1)/f'(a_1)$, where f' is the derivative of f.

Nonlinear equation
Equations where the output is not directly proportional to the input. Equations which describe the behavior of most real-world problems. The response of a nonlinear system can depend crucially upon initial conditions.

Perfect numbers
An integer which is the sum of all its divisors excluding itself. For example, 6 is a perfect number since $6 = 1 + 2 + 3$.

Period
The time taken for one cycle of vibration of a wave.

Periodic
Recurring at equal intervals of time.

Phase portrait
The overall picture formed by all possible initial conditions in the $(x, \dot{x})$ plane is referred to as the phase portrait. Consider a pendulum's motion which comes to rest due to air resistance. In the abstract two-dimensional *phase space* (with coordinates x, the displacement, and $\dot{x}$ the velocity) motions appear as noncrossing spirals converging asymptotically towards the resting, fixed state. This focus is called a *point attractor* which attracts all local transient motions.

Plosive
A type of consonant sound made by sudden release of air impounded behind an occlusion of the vocal tract.

Poincare map
A Poincare map is established by cutting trajectories in a region of phase space with a surface one dimension less than the dimension of the phase space.

Polynomial
An algebraic expression of the form $a_0x^n + a_1x^{n-1} + \ldots a_{n-1}x + a_n$ where n is the degree of the expression and $a_0 \neq 0$.

Quasiperiodicity
Informally defined as a phenomenon with multiple periodicity. One example is the astronomical position of a point on the surface of the earth, since it results from the rotation of the earth about its axis and the rotation of the earth around the sun.

Quadratic mapping
Also known as the *logistic map*, this famous discrete dynamical system is defined by $x_{t+1} = cx_t(1 - x_t)$.

Quaternion
A four-dimensional, "hyper" complex number of the form $Q = a_0 + a_1 i + a_2 j + a_3 k$.

Rational function
A function which can be expressed as the quotient of two polynomials.

Recursive
An object is said to be recursive if it partially consists of or is defined in terms of itself. A *recursive operation* invokes itself as an intermediate operation.

Sierpiñski gasket
See *gasket*.

Steady state
Also called equilibrium point or *fixed point*. A set of values of the variables of a system for which the system does not change as time proceeds.

Strange attractor
See attractor.

Tesselation
A division of a plane into polygons, regular or irregular.

Trajectory
A sequence of points in which each point produces its successor according to some mathematical function.

Transfinite number

An infinite cardinal or ordinal number. The smallest transfinite number is called "aleph-nought" (written as $\aleph_0$) which counts the number of integers.

Tractrix

A plane curve represented in parametric equations by $x = a(\ln \cot\phi/2 - \cos\phi)$, $y = a\sin\phi$.

Transcendental functions

Functions which are not algebraic, for example, circular, exponential, and logarithmic functions.

Transformation

The operation of changing (as by rotation or mapping) one configuration or expression into another in accordance with a mathematical rule.

Witch of Agnesi

A plane curve represented in rectangular coordinates by $y = 8a^3/(x^2 + 4a^2)$.

Index